Metal Complexes Containing Bioactive Ligands: Structure and Biological Evaluation

Metal Complexes Containing Bioactive Ligands: Structure and Biological Evaluation

Guest Editor

Dušan Dimić

MDPI

Basel • Beijing • Wuhan • Barcelona • Belgrade • Novi Sad • Cluj • Manchester

Guest Editor
Dušan Dimić
Faculty of Physical Chemistry
University of Belgrade
Belgrade
Serbia

Editorial Office
MDPI AG
Grosspeteranlage 5
4052 Basel, Switzerland

This is a reprint of the Special Issue, published open access by the journal *Inorganics* (ISSN 2304-6740), freely accessible at: https://www.mdpi.com/journal/inorganics/special_issues/8BPL1IE6US.

For citation purposes, cite each article independently as indicated on the article page online and as indicated below:

Lastname, A.A.; Lastname, B.B. Article Title. *Journal Name* **Year**, *Volume Number*, Page Range.

ISBN 978-3-7258-7671-6 (Hbk)
ISBN 978-3-7258-7672-3 (PDF)
https://doi.org/10.3390/books978-3-7258-7672-3

Contents

About the Editor

Dušan Dimić

Dušan Dimić is an Associate Professor at the University of Belgrade, Faculty of Physical Chemistry, Belgrade, Serbia, where he teaches courses in forensic physicochemical analysis. His research focuses on the structural and reactivity analysis of bioactive organic molecules and transition metal complexes using advanced spectroscopic (IR, Raman, UV–Vis, NMR, EPR) and theoretical methods (DFT, NBO, QTAIM). He is particularly engaged in experimental and theoretical work in forensic physical chemistry. He has authored over 90 publications in international peer-reviewed journals, with more than 1500 citations and an h-index of 25 (Scopus). He has also reviewed over 150 manuscripts for leading scientific journals and served as a Guest Editor for several special issues in international journals. He was included in the Stanford "World's Top 2% Scientists" list in the fields of Inorganic Chemistry and Chemical Physics for 2023 and 2024. Dr. Dimić has been actively involved in mentoring undergraduate, master's, and doctoral students and has participated in multiple national and international research projects, including collaborations with institutions in Germany, Sweden, Argentina, Israel, and the United States.

Preface

This Reprint focuses on the design, synthesis, characterization, and biological evaluation of bioactive ligands and their transition metal complexes. Particular emphasis is placed on compounds that exhibit activity against free radicals, cancer cell lines, and microbial systems, as well as their potential applications in medicinal and environmental chemistry. The scope of the issue includes modern synthetic strategies, detailed mechanistic insights, and comprehensive physicochemical characterization of newly obtained compounds. A central aim is to highlight the integration of experimental and computational approaches, especially Density Functional Theory, Natural Bond Orbital, and Quantum Theory of Atoms in Molecules analyses, in understanding structure, stability, and reactivity. In addition, contributions employing Molecular Docking, Molecular Dynamics, and Quantitative Structure–Activity Relationship studies are included to elucidate interactions with biomolecules and to predict biological activity and toxicity. The motivation for this collection arises from the growing need to rationalize the design of functional bioactive systems through combined theoretical and experimental methodologies. The Reprint brings together nine research articles that reflect current trends and advances in this interdisciplinary field. It is intended for researchers in chemistry, biochemistry, materials science, and related disciplines who are interested in bioactive compounds and computational modeling.

Dušan Dimić
Guest Editor

inorganics

Editorial

Metal Complexes Containing Bioactive Ligands: Structure and Biological Evaluation

Dušan Dimić

Faculty of Physical Chemistry, University of Belgrade, Studentski trg 12-16, 11000 Belgrade, Serbia; ddimic@ffh.bg.ac.rs

Currently, the development of medicinal chemistry depends on naturally occurring scaffolds, as nearly 50% of FDA-approved compounds are related to compounds already present in nature [1]. The metal chelation of bioactive molecules provides an important means of increasing the biological activity of related complexes. In this context, the design of metal-based therapeutics represents a promising strategy to overcome the limitations associated with purely organic drugs, such as poor bioavailability, limited selectivity, and resistance development [2–7]. The incorporation of transition metal ions can significantly modulate redox properties, coordination geometry, and lipophilicity, thereby influencing pharmacokinetics and mechanisms of action [8–12]. Consequently, the rational combination of bioactive ligands with metal centers continues to attract considerable attention, incorporating inorganic chemistry, medicinal chemistry, and computational modeling.

This Special Issue, titled "Metal Complexes Containing Bioactive Ligands: Structure and Biological Evaluation", was curated to showcase modern synthetic procedures for the chemical modification of bioactive compounds and their transition metal complexes. Research topics also include theoretical methods for elucidating the modification mechanism, compound stability, complexation modes, and interactions, using Density Functional Theory (DFT), Natural Bond Orbital Theory (NBO), and Quantum Theory of Atoms in Molecules (QTAIM). In the following parts, short outlines of the published papers are presented.

A review paper by Mihajlović, Trif, and Živković presents a comprehensive collection of papers regarding the metal complexes with hydroxyflavones and their anticancer and antimicrobial activity [13]. Hydroxyflavones represent a hydroxylated subgroup of flavones, compounds composed of three six-membered rings. These compounds are known for their antioxidant and antimicrobial activity due to their involvement in plant defense mechanisms [14,15]. Furthermore, the literature suggests that cytotoxic effects and antiviral, antidepressant, antiparasitic, and antiallergenic activity are shown by hydroxyflavones [16–21]. The complexation mode of these compounds strongly depends on the number and relative position of hydroxyl groups [22,23]. The reviewed literature suggests that a broad range of biological activities is shown by metal complexes containing hydroxyflavones, such as anticancer, antimicrobial, anti-inflammatory, and antimutagenic action [24]. The authors of the review also propose directions for the future development of these metal-based agents, as solubility in water remains a major challenge.

Yang, Li, and Luo examined the complexation abilities of two-sulfur-containing compounds, namely, diphenylacetyl disulfide and 3,3′-diaminodiphenyl sulfone [25]. Three mononuclear complexes were obtained in one-pot conditions ($[C_4H_{18}CuO_{12}S_2]$(I), $[C_{12}H_{18}N_4NiO_{11}S]$(II) and $[C_{24}H_{24}Cl_2N_4O_4S_2Zn]$(III)). The structures of the compounds were elucidated using X-ray analysis, FTIR, UV-Vis, NMR, and ESI-MS, while the thermal stability was determined by thermogravimetry. The cytotoxicity measurements were performed against the human cancer cell lines lung cancer A549, liver cancer SMMC-7721, breast cancer MDA-MB-231, and colon cancer SW480. The inhibition rates of complexes II and III were relatively high, around 20% in MDA-MB-231 and SW480.

A research team led by Prof. Jevtovic [26] presented structural, antioxidant, and protein/DNA-binding properties of a sulfate-coordinated Ni(II) complex with a pyridoxal-semicarbazone (PLSC) ligand. In this contribution, the structure of $[Ni(PLSC)(SO_4)(H_2O)_2]$ was solved via X-ray analysis, and the stabilization interactions within the crystallographic structure were described by Hirshfeld surface analysis [27,28]. The structure was later optimized by the B3LYP functional [29] in conjunction with the 6-311++G(d,p) basis set [30] for non-metallic atoms and two pseudopotentials (LanL2DZ and def2-TZVP [31,32]) for Ni in the Gaussian 09 program package [33]. The applicability of the selected level of theory was demonstrated by comparing experimental and theoretical bond lengths and angles. The QTAIM method [34] was used to assess the strength and type of interactions between central metal ions and ligand molecules. The antioxidant activity of the complex towards hydroxyl and ascorbyl radical was examined via electron paramagnetic resonance (EPR) spectroscopy, and the detected activity towards hydroxyl radical was higher than that of ascorbic acid. Interactions with Human Serum Protein (HSA) and DNA were examined using spectrofluorimetric titrations and molecular docking.

The interactions of four ruthenium(II) complexes of the type [(η^6-*p*-cymene)RuCl(dpm)], where dpm are hexa-(L3–L5) and meso-substituted (L6) dipyrromethene ligands, were prepared in the contribution by Murillo and coworkers [35]. The structures of these compounds were characterized via MS, FTIR, NMR, UV-VIS, and X-ray crystallography. For the optimization of structure, the authors have employed the B3LYP functional together with DGDZVP basis set for ruthenium atoms [36] and 6-31+G(d,p) basis set for the remaining atoms. The DNA-binding affinity was determined via UV-VIS titration, followed by computational calculations. The highest experimental binding constant was obtained for Ru-3 ($6.5 \pm 0.5 \times 10^{11}$ M^{-1}), with the binding energy of -61.8 kJ mol^{-1}. This result was further supported by docking studies and molecular-level interaction analysis.

De Matos and coworkers [37] obtained two Ru(II) polypyridyl complexes, *cis*-$[Ru(dmbpy)_2Cl(bpy)](PF_6)$ (Rubpy) and *cis*-$[Ru(dmbpy)_2Cl(bpe)](PF_6)$ (Rubpe) (dmbpy = 4,4′-Dimethyl-2,2′-dipyridyl, bpy = 4,4′-dipyridyl and bpe = 1,2-bis(4-pyridyl)ethane) that were spectroelectrochemically characterized. This type of complex has been extensively examined due to its lower toxicity, slower ligand exchange and dissociation, strong interactions with proteins and DNA, and cytotoxicity that includes different cancer models [38–40]. In the previous contributions, it was proven that the aromaticity of the terpyridine substituent in the complexes of the type $[Ru(R\text{-}tpy)(LL)Cl]_n^+$ (R = chloro or 4-chlorophenyl; LL = bidentate ligand) enhances the biological activity [41,42]. The complexes reported by de Matos and coworkers showed strong interactions with the transport proteins, as evidenced by titration experiments. The dominant mechanisms of interaction with DNA included partial intercalation and groove binding. The cellular uptake of the complexes was relatively high in HeLa and MDA-MB-231 cells.

In the paper by Akın-Polat and coworkers [43], six chloro[*N*-alkyl-*N*-cinnamyl-benzimidazol-2-yliden]silver(I) complexes were successfully synthesized, in which allyl (3a), methoxymethyl (3b), benzyl (3c), 3-fluorobenzyl (3d), 4-fluorobenzyl (3e), and 4-methyl-benzyl (3f) substituents were present on the benzimidazole ring. The structures were confirmed via MS, FTIR, and NMR, while X-ray crystallography was used to determine the structures of the 3c and 3d complexes. The activity of compounds against trophozoites and cysts of the pathogenic *Acanthamoeba castellanii* strain was investigated, with the following order of activity: 3d > 3c > 3f > 3a > 3b > 3e. High concentrations of compounds

(1000 μM) led to the near-total eradication of protozoa. This is the first contribution in the literature to use silver-*N*-heterocyclic carbene complexes against *Acanthamoeba castellanii*, one of the *Acanthamoeba* spp., free-living amoebae commonly found in soil, freshwater, and air [44,45].

Halevas et al. examined the antioxidant activity of four newly synthesized nickel and cobalt complexes with chrysin in reference [46]. Each of these compounds contained one of the bidentate chelating agents (2,2′-bipyridine and 1,10-phenanthroline), and the structures were verified using physicochemical and crystallographic methods. Chrysin is a flavonoid with two hydroxyl groups at the 5 and 7 carbons [47] and it can be found in plant extracts, propolis, and honey extracts [48,49]. Before these results, only a few other complexes with chrysin had been characterized via X-ray crystallography, such as in [50,51]. The activity of the compounds presented in the Special Issue was tested against the DPPH radical, and dose-dependent antioxidant potential was shown. All four compounds were more active than chrysin, and in general, the nickel complexes had higher antioxidant activity than cobalt complexes. The obtained activity was of the same order of magnitude as for the cobalt complexes containing non-steroidal anti-inflammatory drugs, as another class of bioactive ligands [52].

Mohammed and coworkers [53] published results regarding the synthesis and structural characterization of two ligands containing escitalopram and eugenol/curcumin and their complexes with Mn(II), Co(II), Ni(II), and Cu(II). Eugenol is a phenolic compound that acts as a bidentate ligand with potential antibacterial activity, which is reflected in the damage to the membranes of Gram-negative and Gram-positive bacteria [54]. Escitalopram belongs to the class of selective serotonin reuptake inhibitors (SSRIs), and it has previously been linked to form multifunctional ligands [55]. Curcumin is a naturally occurring compound, obtained from turmeric powder of *Curcuma longa*, with a strong chelating effect towards central transition metal ions [56]. Chemical modifications of bioactive compounds and their conjugation with other active components are expected to enhance therapeutic efficacy [57]. The complexes described in the papers in this Special Issue were characterized via FTIR, UV-Vis, 1H NMR, elemental analysis, molar conductivity, and magnetic susceptibility. The geometries were optimized using DFT methods. The diffusion method was used to assess the antimicrobial activity of complexes towards *Escherichia coli*, *Staphylococcus aureus*, and *Candida albicans*. Cobalt-containing complexes were the most active due to the increase in co-lipophilicity and improved diffusion through the cell membrane.

The Guest Editor hopes that this Special Issue will serve as a useful resource for researchers in bioinorganic and medicinal inorganic chemistry and encourage further collaborative work towards the development of new metal-based therapeutic and diagnostic approaches.

Data Availability Statement: This article is an Editorial and does not report original research data. No new data were created or analyzed in this study. Data sharing is not applicable.

Acknowledgments: The Guest Editor sincerely thanks all authors for their valuable contributions, the reviewers for their careful and constructive evaluations, and the editorial team of *Inorganics* for their professional support throughout the preparation of this Special Issue.

Conflicts of Interest: The author declares no conflicts of interest.

References

1. Newman, D.J.; Cragg, G.M. Natural Products as Sources of New Drugs over the Nearly Four Decades from 01/1981 to 09/2019. *J. Nat. Prod.* **2020**, *83*, 770–803. [CrossRef]
2. Marker, S.C.; King, A.P.; Granja, S.; Vaughn, B.; Woods, J.J.; Boros, E.; Wilson, J.J. Exploring the In Vivo and In Vitro Anticancer Activity of Rhenium Isonitrile Complexes. *Inorg. Chem.* **2020**, *59*, 10285–10303. [CrossRef] [PubMed]

https://doi.org/10.3390/inorganics14030075

3. Johnstone, T.C.; Suntharalingam, K.; Lippard, S.J. The Next Generation of Platinum Drugs: Targeted Pt(II) Agents, Nanoparticle Delivery, and Pt(IV) Prodrugs. *Chem. Rev.* **2016**, *116*, 3436–3486. [CrossRef] [PubMed]
4. King, A.P.; Gellineau, H.A.; Ahn, J.-E.; MacMillan, S.N.; Wilson, J.J. Bis(Thiosemicarbazone) Complexes of Cobalt(III). Synthesis, Characterization, and Anticancer Potential. *Inorg. Chem.* **2017**, *56*, 6609–6623. [CrossRef]
5. Palanimuthu, D.; Shinde, S.V.; Somasundaram, K.; Samuelson, A.G. In Vitro and in Vivo Anticancer Activity of Copper Bis(Thiosemicarbazone) Complexes. *J. Med. Chem.* **2013**, *56*, 722–734. [CrossRef]
6. Paprocka, R.; Wiese-Szadkowska, M.; Janciauskiene, S.; Kosmalski, T.; Kulik, M.; Helmin-Basa, A. Latest Developments in Metal Complexes as Anticancer Agents. *Coord. Chem. Rev.* **2022**, *452*, 214307. [CrossRef]
7. Casini, A.; Pöthig, A. Metals in Cancer Research: Beyond Platinum Metallodrugs. *ACS Cent. Sci.* **2024**, *10*, 242–250. [CrossRef] [PubMed]
8. Jevtovic, V.; Alshamari, A.K.; Milenković, D.; Marković, J.D.; Marković, Z.; Dimić, D. The Effect of Metal Ions (Fe, Co, Ni, and Cu) on the Molecular-Structural, Protein Binding, and Cytotoxic Properties of Metal Pyridoxal-Thiosemicarbazone Complexes. *Int. J. Mol. Sci.* **2023**, *24*, 11910. [CrossRef]
9. Jevtovic, V.; Alhar, M.S.O.; Milenković, D.; Marković, Z.; Marković, J.D.; Dimić, D. Synthesis, Structural Characterization, Cytotoxicity, and Protein/DNA Binding Properties of Pyridoxylidene-Aminoguanidine-Metal (Fe, Co, Zn, Cu) Complexes. *Int. J. Mol. Sci.* **2023**, *24*, 14745. [CrossRef]
10. Adhikari, S.; Nath, P.; Das, A.; Datta, A.; Baildya, N.; Duttaroy, A.K.; Pathak, S. A Review on Metal Complexes and Its Anti-Cancer Activities: Recent Updates from in Vivo Studies. *Biomed. Pharmacother.* **2024**, *171*, 116211. [CrossRef]
11. Soroceanu, A.; Bargan, A. Advanced and Biomedical Applications of Schiff-Base Ligands and Their Metal Complexes: A Review. *Crystals* **2022**, *12*, 1436. [CrossRef]
12. Abdolmaleki, S.; Aliabadi, A.; Khaksar, S. Riding the Metal Wave: A Review of the Latest Developments in Metal-Based Anticancer Agents. *Coord. Chem. Rev.* **2024**, *501*, 215579. [CrossRef]
13. Mihajlović, L.E.; Trif, M.; Živković, M.B. Metal Complexes with Hydroxyflavones: A Study of Anticancer and Antimicrobial Activities. *Inorganics* **2025**, *13*, 250. [CrossRef]
14. Treutter, D. Significance of Flavonoids in Plant Resistance and Enhancement of Their Biosynthesis. *Plant Biol.* **2005**, *7*, 581–591. [CrossRef]
15. Hernández, I.; Alegre, L.; Van Breusegem, F.; Munné-Bosch, S. How Relevant Are Flavonoids as Antioxidants in Plants? *Trends Plant Sci.* **2009**, *14*, 125–132. [CrossRef]
16. Cotelle, N.; Bernier, J.-L.; Catteau, J.-P.; Pommery, J.; Wallet, J.-C.; Gaydou, E.M. Antioxidant Properties of Hydroxy-Flavones. *Free Radic. Biol. Med.* **1996**, *20*, 35–43. [CrossRef]
17. Lemańska, K.; Szymusiak, H.; Tyrakowska, B.; Zieliński, R.; Soffers, A.E.M.F.; Rietjens, I.M.C.M. The Influence of pH on Antioxidant Properties and the Mechanism of Antioxidant Action of Hydroxyflavones. *Free Radic. Biol. Med.* **2001**, *31*, 869–881. [CrossRef] [PubMed]
18. Lewandowski, W.; Lewandowska, H.; Golonko, A.; Świderski, G.; Świsłocka, R.; Kalinowska, M. Correlations between Molecular Structure and Biological Activity in "Logical Series" of Dietary Chromone Derivatives. *PLoS ONE* **2020**, *15*, e0229477. [CrossRef]
19. Sanderson, J.T. Induction and Inhibition of Aromatase (CYP19) Activity by Natural and Synthetic Flavonoid Compounds in H295R Human Adrenocortical Carcinoma Cells. *Toxicol. Sci.* **2004**, *82*, 70–79. [CrossRef]
20. Loa, J.; Chow, P.; Zhang, K. Studies of Structure–Activity Relationship on Plant Polyphenol-Induced Suppression of Human Liver Cancer Cells. *Cancer Chemother. Pharmacol.* **2009**, *63*, 1007–1016. [CrossRef]
21. Torres-Piedra, M.; Ortiz-Andrade, R.; Villalobos-Molina, R.; Singh, N.; Medina-Franco, J.L.; Webster, S.P.; Binnie, M.; Navarrete-Vázquez, G.; Estrada-Soto, S. A Comparative Study of Flavonoid Analogues on Streptozotocin–Nicotinamide Induced Diabetic Rats: Quercetin as a Potential Antidiabetic Agent Acting via 11β-Hydroxysteroid Dehydrogenase Type 1 Inhibition. *Eur. J. Med. Chem.* **2010**, *45*, 2606–2612. [CrossRef]
22. Dell'Anna, M.M.; Censi, V.; Carrozzini, B.; Caliandro, R.; Denora, N.; Franco, M.; Veclani, D.; Melchior, A.; Tolazzi, M.; Mastrorilli, P. Triphenylphosphane Pt(II) Complexes Containing Biologically Active Natural Polyphenols: Synthesis, Crystal Structure, Molecular Modeling and Cytotoxic Studies. *J. Inorg. Biochem.* **2016**, *163*, 346–361. [CrossRef]
23. Raza, A.; Xu, X.; Xia, L.; Xia, C.; Tang, J.; Ouyang, Z. Quercetin-Iron Complex: Synthesis, Characterization, Antioxidant, DNA Binding, DNA Cleavage, and Antibacterial Activity Studies. *J. Fluoresc.* **2016**, *26*, 2023–2031. [CrossRef]
24. Samsonowicz, M.; Regulska, E.; Kalinowska, M. Hydroxyflavone Metal Complexes—Molecular Structure, Antioxidant Activity and Biological Effects. *Chem. Biol. Interact.* **2017**, *273*, 245–256. [CrossRef]
25. Yang, Y.; Li, D.; Luo, M. Synthesis, Characterization, and Cytotoxicity Research of Sulfur-Containing Metal Complexes. *Inorganics* **2025**, *13*, 26. [CrossRef]
26. Jevtovic, V.; Golubović, L.; Alshammari, O.A.O.; Alhar, M.S.; Alanazi, T.Y.A.; Radulović, A.; Nakarada, Đ.; Marković, J.D.; Rakić, A.; Dimić, D. Structural, Antioxidant, and Protein/DNA-Binding Properties of Sulfate-Coordinated Ni(II) Complex with Pyridoxal-Semicarbazone (PLSC) Ligand. *Inorganics* **2024**, *12*, 280. [CrossRef]

https://doi.org/10.3390/inorganics14030075

27. Spackman, M.A.; Byrom, P.G. A Novel Definition of a Molecule in a Crystal. *Chem. Phys. Lett.* **1997**, *267*, 215–220. [CrossRef]
28. Spackman, M.A.; Jayatilaka, D. Hirshfeld Surface Analysis. *CrystEngComm* **2009**, *11*, 19–32. [CrossRef]
29. Becke, A.D. Density-Functional Thermochemistry. III. The Role of Exact Exchange. *J. Chem. Phys.* **1993**, *98*, 5648. [CrossRef]
30. Dunning, T.H. Gaussian Basis Sets for Use in Correlated Molecular Calculations. I. The Atoms Boron through Neon and Hydrogen. *J. Chem. Phys.* **1989**, *90*, 1007. [CrossRef]
31. Hay, P.J.; Wadt, W.R. Ab Initio Effective Core Potentials for Molecular Calculations. Potentials for K to Au Including the Outermost Core Orbitale. *J. Chem. Phys.* **1985**, *82*, 299–310. [CrossRef]
32. Weigend, F. Accurate Coulomb-Fitting Basis Sets for H to Rn. *Phys. Chem. Chem. Phys.* **2006**, *8*, 1057. [CrossRef]
33. Frisch, M.J.; Trucks, G.W.; Schlegel, H.B.; Scuseria, G.E.; Robb, M.A.; Cheeseman, J.R.; Scalmani, G.; Barone, V.; Mennucci, B.; Petersson, G.A.; et al. *Gaussian 09*; Gaussian, Inc.: Wallingford, CT, USA, 2009.
34. Bader, R.F.W. Atoms in Molecules. *Acc. Chem. Res.* **1985**, *18*, 9–15. [CrossRef]
35. Murillo, M.I.; Mejia, C.F.; Restrepo-Acevedo, A.; Barraud, B.; Orjuela, A.L.; Flores-Alamo, M.; Toscano, R.A.; Alí-Torres, J.; Ryabov, A.D.; Le Lagadec, R. DNA Binding with Dipyrromethene Ruthenium(II) Complexes. *Inorganics* **2025**, *13*, 198. [CrossRef]
36. Roy, L.E.; Hay, P.J.; Martin, R.L. Revised Basis Sets for the LANL Effective Core Potentials. *J. Chem. Theory Comput.* **2008**, *4*, 1029–1031. [CrossRef] [PubMed]
37. De Matos, P.A.; Do Carmo, M.E.G.; Parussulo, A.L.A.; Silva, C.M.F.; Lino, R.C.; Toma, H.E.; Beletti, M.E.; De Oliveira Júnior, R.J.; Patrocinio, A.O.D.T.; Matias, T.A.; et al. Ligand Rigidity and π-Surface Modulate Biomolecular Interactions and Cytotoxicity in Ru(II) Polypyridyl Complexes. *Inorganics* **2026**, *14*, 63. [CrossRef]
38. Thota, S.; Rodrigues, D.A.; Crans, D.C.; Barreiro, E.J. Ru(II) Compounds: Next-Generation Anticancer Metallotherapeutics? *J. Med. Chem.* **2018**, *61*, 5805–5821. [CrossRef]
39. Alessio, E. Thirty Years of the Drug Candidate NAMI-A and the Myths in the Field of Ruthenium Anticancer Compounds: A Personal. *Eur. J. Inorg. Chem.* **2017**, *55*, 1549–1560. [CrossRef]
40. Simović, A.R.; Masnikosa, R.; Bratsos, I.; Alessio, E. Chemistry and Reactivity of Ruthenium(II) Complexes: DNA/Protein Binding Mode and Anticancer Activity Are Related to the Complex Structure. *Coord. Chem. Rev.* **2019**, *398*, 113011. [CrossRef]
41. Milutinović, M.M.; Rilak, A.; Bratsos, I.; Klisurić, O.; Vraneš, M.; Gligorijević, N.; Radulović, S.; Bugarčić, Ž.D. New 4′-(4-Chlorophenyl)-2,2′:6′,2″-Terpyridine Ruthenium(II) Complexes: Synthesis, Characterization, Interaction with DNA/BSA and Cytotoxicity Studies. *J. Inorg. Biochem.* **2017**, *169*, 1–12. [CrossRef]
42. Čanović, P.; Simović, A.R.; Radisavljević, S.; Bratsos, I.; Demitri, N.; Mitrović, M.; Zelen, I.; Bugarčić, Ž.D. Impact of Aromaticity on Anticancer Activity of Polypyridyl Ruthenium(II) Complexes: Synthesis, Structure, DNA/Protein Binding, Lipophilicity and Anticancer Activity. *J. Biol. Inorg. Chem.* **2017**, *22*, 1007–1028. [CrossRef] [PubMed]
43. Akın-Polat, Z.; Şahin, N.; Hkiri, S.; Ly, B.M.T.; Özdemir, İ.; Sémeril, D. In Vitro Evaluation of Silver-NHC Complexes Against a Clinical Isolate of Acanthamoeba Castellanii: Time- and Dose-Dependent Effects. *Inorganics* **2025**, *13*, 204. [CrossRef]
44. Geisen, S.; Fiore-Donno, A.M.; Walochnik, J.; Bonkowski, M. Acanthamoeba Everywhere: High Diversity of Acanthamoeba in Soils. *Parasitol. Res.* **2014**, *113*, 3151–3158. [CrossRef]
45. Schuster, F.L.; Visvesvara, G.S. Free-Living Amoebae as Opportunistic and Non-Opportunistic Pathogens of Humans and Animals. *Int. J. Parasitol.* **2004**, *34*, 1001–1027. [CrossRef]
46. Halevas, E.; Mavroidi, B.; Varna, D.; Zahariou, G.; Litsardakis, G.; Pelecanou, M.; Hatzidimitriou, A.G. Structurally Characterized Cobalt and Nickel Complexes of Flavonoid Chrysin as Potential Radical Scavenging Compounds. *Inorganics* **2025**, *13*, 230. [CrossRef]
47. Stompor-Gorący, M.; Bajek-Bil, A.; Machaczka, M. Chrysin: Perspectives on Contemporary Status and Future Possibilities as Pro-Health Agent. *Nutrients* **2021**, *13*, 2038. [CrossRef]
48. Chan, C.W.; Deadman, B.J.; Manley-Harris, M.; Wilkins, A.L.; Alber, D.G.; Harry, E. Analysis of the Flavonoid Component of Bioactive New Zealand Mānuka (*Leptospermum scoparium*) Honey and the Isolation, Characterisation and Synthesis of an Unusual Pyrrole. *Food Chem.* **2013**, *141*, 1772–1781. [CrossRef]
49. Woźniak, M.; Mrówczyńska, L.; Kwaśniewska-Sip, P.; Waśkiewicz, A.; Nowak, P.; Ratajczak, I. Effect of the Solvent on Propolis Phenolic Profile and Its Antifungal, Antioxidant, and In Vitro Cytoprotective Activity in Human Erythrocytes Under Oxidative Stress. *Molecules* **2020**, *25*, 4266. [CrossRef]
50. Halevas, E.; Mitrakas, A.; Mavroidi, B.; Athanasiou, D.; Gkika, P.; Antoniou, K.; Samaras, G.; Lialiaris, E.; Hatzidimitriou, A.; Pantazaki, A.; et al. Structurally Characterized Copper-Chrysin Complexes Display Genotoxic and Cytotoxic Activity in Human Cells. *Inorganica Chim. Acta* **2021**, *515*, 120062. [CrossRef]
51. Halevas, E.; Mavroidi, B.; Antonoglou, O.; Hatzidimitriou, A.; Sagnou, M.; Pantazaki, A.A.; Litsardakis, G.; Pelecanou, M. Structurally Characterized Gallium–Chrysin Complexes with Anticancer Potential. *Dalton Trans.* **2020**, *49*, 2734–2746. [CrossRef]
52. Tsiliou, S.; Kefala, L.-A.; Hatzidimitriou, A.G.; Kessissoglou, D.P.; Perdih, F.; Papadopoulos, A.N.; Turel, I.; Psomas, G. Cobalt(II) Complexes with Non-Steroidal Anti-Inflammatory Drugs and α-Diimines. *J. Inorg. Biochem.* **2016**, *160*, 125–139. [CrossRef] [PubMed]

https://doi.org/10.3390/inorganics14030075

53. Mohammed, E.H.; Mohammed, E.R.; Yahya, E.M.; Alsultan, M. Synthesis, Characterisation, DFT Study and Biological Evaluatio of Complexes Derived from Transition Metal and Mixed Ligands. *Inorganics* **2025**, *13*, 334. [CrossRef]
54. Kowalewska, A.; Majewska-Smolarek, K. Eugenol-Based Polymeric Materials—Antibacterial Activity and Applications. *Antib otics* **2023**, *12*, 1570. [CrossRef]
55. Kanwal, A.; Afzal, U.; Zubair, M.; Imran, M.; Rasool, N. Synthesis of Anti-Depressant Molecules via Metal-Catalyzed Reaction A Review. *RSC Adv.* **2024**, *14*, 6948–6971. [CrossRef] [PubMed]
56. Refat, M.S. Synthesis and Characterization of Ligational Behavior of Curcumin Drug towards Some Transition Metal Ion Chelation Effect on Their Thermal Stability and Biological Activity. *Spectrochim. Acta Part A Mol. Biomol. Spectrosc.* **2013**, *10* 326–337. [CrossRef]
57. Prasad, S.; DuBourdieu, D.; Srivastava, A.; Kumar, P.; Lall, R. Metal–Curcumin Complexes in Therapeutics: An Approach t Enhance Pharmacological Effects of Curcumin. *Int. J. Mol. Sci.* **2021**, *22*, 7094. [CrossRef]

https://doi.org/10.3390/inorganics14030075

inorganics MDPI

Review

Metal Complexes with Hydroxyflavones: A Study of Anticancer and Antimicrobial Activities

Ljiljana E. Mihajlović [1], Monica Trif [2] and Marijana B. Živković [3]

[1] Innovation Center of the Faculty of Chemistry, Studentski trg 12–16, 11158 Belgrade, Serbia; ljiljanam@chem.bg.ac.rs
[2] Centre for Innovative Process Engineering (CENTIV) GmbH, Boschstr. 8–10, 28857 Syke, Germany; monica_trif@hotmail.com
[3] Department of Chemistry, Institute of Chemistry, Technology and Metallurgy, National Institute of the Republic of Serbia, University of Belgrade, Studentski trg 12–16, 11000 Belgrade, Serbia
* Correspondence: marijana.zivkovic@ihtm.bg.ac.rs

Abstract: Metal chelation to bioactive small molecules is a well-established strategy to enhance the biological activity of the resulting complexes. Among the widely explored structural motifs, the combination of prominent metal centers with naturally inspired derivatives has attracted considerable attention. One such promising platform is the flavone scaffold, derived from flavonoids and studied since ancient times. Flavones are plant-derived compounds known for their diverse biological activities and health benefits. They exhibit significant structural variability, primarily through backbone modifications such as hydroxylation. Importantly, coordination of metal ions to hydroxylated flavone cores often improves their natural bioactivities, including anticancer and antimicrobial effects. In this review, we summarize transition metal complexes incorporating hydroxyflavone (OH–F) ligands reported over the past 15 years. We provide a concise overview of synthetic approaches and structural characterization, with a particular emphasis on coordination modes (e.g., maltol-type, acetylacetonate-type, catechol-type, and others). Furthermore, we discuss biological evaluation results, especially anticancer and antimicrobial studies, to highlight the therapeutic potential of these complexes. Finally, we suggest directions for the future development of metal-based agents bearing hydroxyflavone moieties through several critical points in terms of the accuracy, reproducibility, and relevance of biological studies involving metal-based compounds.

Keywords: transition metal complexes; flavonoid metal chelates; coordination mode; biological activity; cytotoxicity; antimicrobial

1. Introduction

According to the World Health Organization (WHO), cancer remains one of the leading causes of mortality, with nearly 10 million deaths reported worldwide in 2020, accounting for approximately one in six deaths [1]. The rapid mutation of cancer cells and the development of resistance to existing therapies represent major obstacles to effective treatment [2,3]. Furthermore, anticancer drugs often exhibit a limited selectivity for malignant cells over normal cells, resulting in systemic toxicity, adverse side effects, and damage to healthy tissues and organs.

Another pressing global health concern is the growing resistance to antimicrobial agents. The uncontrolled and excessive use of antibiotics may potentially return us to a time when infectious diseases were frequently fatal due to the absence of effective treatment options [4].

 https://doi.org/10.3390/inorganics13080250

One promising strategy to address both anticancer and antimicrobial resistance issues involves the rational design of novel compounds through the combination of different molecular fragments. Such an approach can lead to synergistic effects, improving therapeutic selectivity while minimizing toxicity. Notably, a substantial proportion of recently approved drugs are inspired by natural products and their structural motifs [5]. Evidence from the literature indicates that researchers have been investigating the therapeutic potential of metal complexes incorporating nature-inspired ligands, such as flavones, for several decades [6,7].

Flavones are natural compounds composed of three six-membered rings: two benzene rings (A and B) and a γ-pyrone ring (C). Benzene ring A is fused to the heterocyclic oxygen-containing ring C, while benzene ring B is attached to ring C at position C-2 (Figure 1).

Figure 1. Origin of hydroxyflavones, their roles in plants, and associated biological activities.

Hydroxyflavones (OH–F) are characterized by the presence of one or more hydroxyl groups within their molecular structure. Figure 2 illustrates representative OH–F compounds selected from the scientific literature that will be examined in the present review.

OH–F are biosynthesized through the shikimate pathway, which ultimately leads to the formation of the aromatic amino acid phenylalanine (Figure 3). Phenylalanine is subsequently deaminated by phenylalanine ammonia-lyase (PAL) to yield cinnamic acid, which is hydroxylated by cinnamate-4-hydroxylase (C4H) to form p-coumaric acid. The activation of p-coumaric acid by 4-coumarate-CoA ligase (4CL) results in the formation of p-coumaroyl-CoA. This intermediate undergoes condensation with three molecules of malonyl-CoA, catalyzed by chalcone synthase (CHS), to generate chalcones. Chalcones are then isomerized to flavanones via chalcone isomerase (CHI), which are further oxidized by flavone synthase (FNS) to produce flavones. Subsequent hydroxylation of flavones at specific positions, mediated by cytochrome P450-dependent monooxygenases, leads to the formation of hydroxyflavones. A comprehensive overview of this biosynthetic pathway is provided in the review by Lou et al. [8].

OH–F play several essential roles in the plants that synthesize them. They function as effective antioxidants, safeguarding plant tissues from oxidative stress induced by UV radiation and environmental stressors [10]. Specific OH–F, such as chrysin and 3-hydroxyflavone, contribute to UV–B absorption, thus serving as natural sunscreens [11]. Furthermore, they are involved in plant defense mechanisms by exhibiting antimicrobial activity against bacterial and fungal pathogens [12]. Some hydroxyflavones also modulate plant–microbe interactions and influence auxin transport, thereby affecting plant growth and development [13].

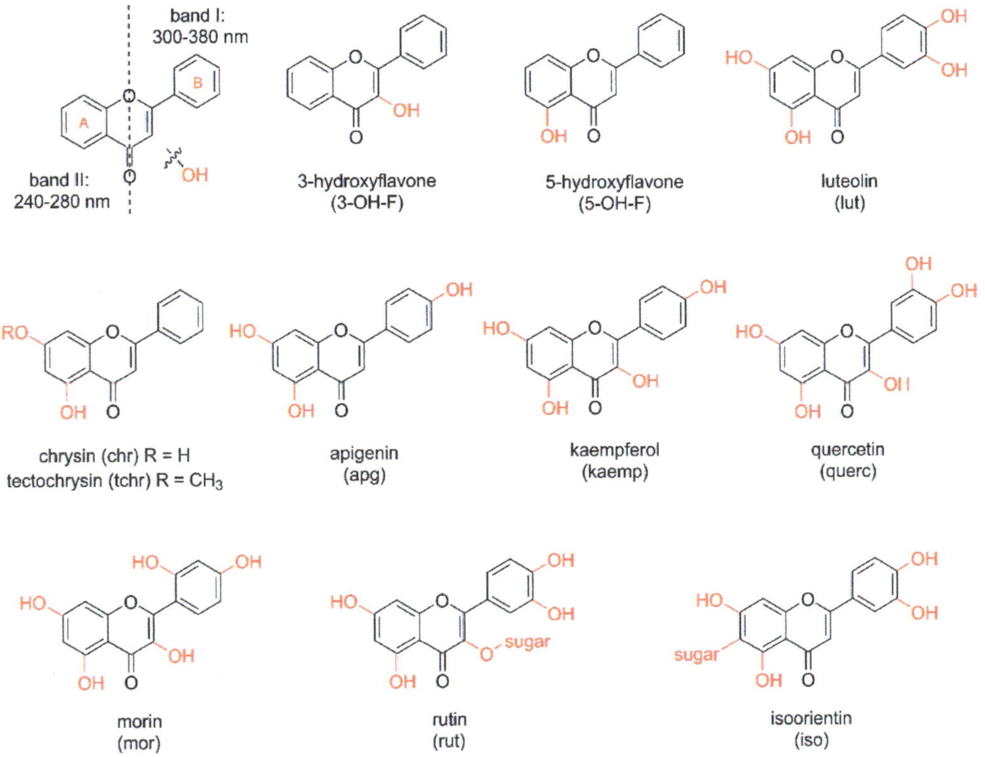

Figure 2. Characteristic UV–Vis absorption bands of hydroxyflavones [9], along with representative hydroxyflavone structures reported in the analyzed literature.

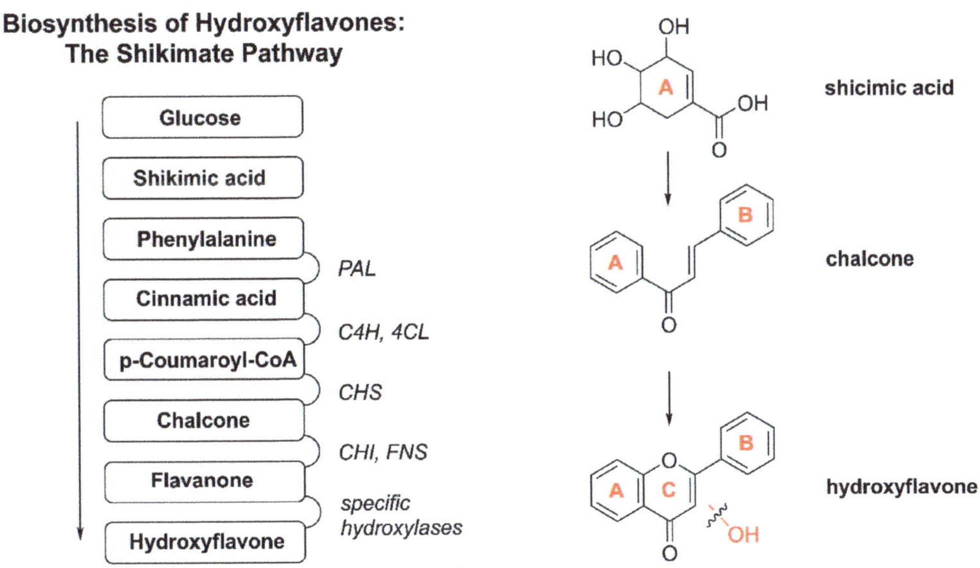

Figure 3. Biosynthetic pathway of hydroxyflavones derived from the shikimic acid route [8].

Over the past three decades, various biological activities of OH–F and its derivatives have been reported, including antioxidant [14–18], cytotoxic effects with distinct mechanisms of action [19–24], antimicrobial [18,25], antiviral [18], antidiabetic [26,27], vasorelaxant [28], antidepressant [29], antiparasitic [30], antiallergenic [31], and cholinesterase inhibitory activity [32].

This review focuses on transition metal complexes [M(OH–F)] featuring OH–F structural motifs, covering research published over the last 15 years. From a strictly chemical perspective, we outline common synthetic approaches used to obtain such compounds, highlight key techniques employed for their structural characterization, and discuss physicochemical properties relevant to biological applications, including stability and lipophilicity. The metal complexes are broadly categorized based on coordination modes, including maltol-type, acetylacetonate (acac)-type, catechol-type, mixed-type, and linear-type coordination.

Furthermore, their anticancer and antimicrobial activities are reviewed in the context of biological profiling. Ultimately, the aim of this review is to provide a comprehensive overview of recent advancements in this field and to underline how these findings may guide the development of next-generation anticancer and antimicrobial agents.

2. Metal–Hydroxyflavone Complexes and Their Biological Properties

In the coordination chemistry of OH–F, several commonly recognized bidentate (O,O) coordination modes to metal ions have been identified, including: (i) coordination via the 4-carbonyl and 3-hydroxyl groups (maltol-type, forming a five-membered chelate ring, see Section 2.1), (ii) coordination via the 4-carbonyl and 5-hydroxyl groups (acac-type, forming a six-membered chelate ring, see Section 2.2), (iii) coordination through two adjacent hydroxyl groups (catechol-type, five-membered chelate, see Section 2.3), and (iv) bridging coordination involving both acac- and catechol-type donor sites (see Section 2.4) (Figure 4). In addition to these bidentate modes, hydroxyflavones may also coordinate in a monodentate manner, as observed in some gold complexes (linear-type coordination, see Section 2.5) [33,34].

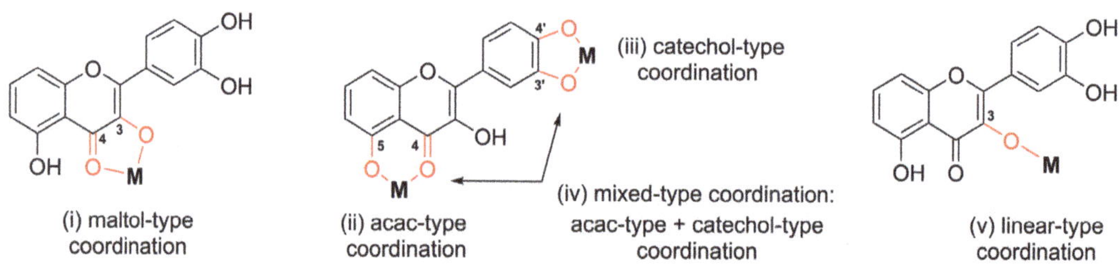

Figure 4. Representative coordination modes of OH–F to metal ions, including binding via: (i) the 4-carbonyl and 3-hydroxyl groups (maltol-type; forming a 5-membered chelate ring), (ii) the 4-carbonyl and 5-hydroxyl groups (acac-type; forming a 6-membered chelate ring), (iii) two adjacent hydroxyl groups (catechol-type; 5-membered chelate), (iv) mixed coordination involving both acac- and catechol-type donor sites, and (iv) only one hydroxyl group (linear-type).

Besides mononuclear complexes, which contain only one central metal ion, bi- and polynuclear complexes with hydroxyflavones have also been reported. However, recent studies have not identified any of these multinuclear complexes exhibiting anticancer or antimicrobial activity. Only a few examples of these complexes with mixed coordination and demonstrated biological activity will be discussed [35,36].

2.1. Maltol-Type Coordination

In metal complexes based on OH–F coordinated in a maltol-type manner, the ligand typically functions as a bidentate chelator, binding the metal center through two adjacent oxygen atoms located on the C-ring, namely the hydroxyl group at position C-3 and the carbonyl group at position C-4. This (O,O)-coordination mode results in the formation of a stable five-membered, envelope-like metallocycle, which critically influences the structural and electronic characteristics of the resulting complexes.

Coordination induces pronounced changes in the spectroscopic profiles of hydroxyflavones, which can be used to confirm a complex formation. In ultraviolet–visible (UV–Vis) spectra, band I, associated with a $\pi \rightarrow \pi^*$ transition involving the B-ring, typically undergoes a bathochromic shift of 10–30 nm upon metal coordination. In infrared (IR) spectra, the C=O stretching vibration (originally observed at 1650–1665 cm^{-1} in the free ligand) shifts to lower wavenumbers (around 1610–1630 cm^{-1}), while the broad O–H stretching band (3200–3500 cm^{-1}) diminishes or disappears, indicating deprotonation and coordination via the C-3 hydroxyl group. Additionally, new bands appear in the 400–600 cm^{-1} region, corresponding to metal–ligand vibrations. In ^{1}H NMR spectra, the disappearance of the characteristic signal for the hydroxyl proton at ~10–12 ppm further supports coordination, while the ^{13}C nuclear magnetic resonance (NMR) signal of the carbonyl carbon shifts downfield by a few ppm, consistent with metal binding.

2.1.1. (O,O)-Coordinated Complexes: Simple Metal–Hydroxyflavone Chelates

In 2016, Dell'Anna et al. synthesized three square-planar Pt(II) complexes bearing triphenylphosphine ligands and (O,O) bidentate donor ligands: 3-hydroxyflavone, quercetin, and ethyl gallate [37]. The complex containing ethyl gallate is not included in the present study. The complex with quercetin, coordinated in a catechol-type manner, is described in the following section, while complex **1** with 3-hydroxyflavone, exhibiting maltol-type coordination, is discussed here and shown in Figure 5. A concentrated methanolic solution of KOH was added dropwise to a dichloromethane solution containing equimolar amounts of *cis*-$[PtCl_2(PPh_3)_2]$ and 3-hydroxyflavone under continuous stirring at room temperature. The reaction mixture was stirred overnight at an ambient temperature. The formed precipitate of KCl was removed by filtration, and the resulting filtrate was concentrated. Ethanol was added to the concentrated solution, followed by the addition of three equivalents of *n*-pentane, which induced the precipitation of an orange solid. The solid was isolated by filtration, washed with cold *n*-pentane, and dried under a vacuum.

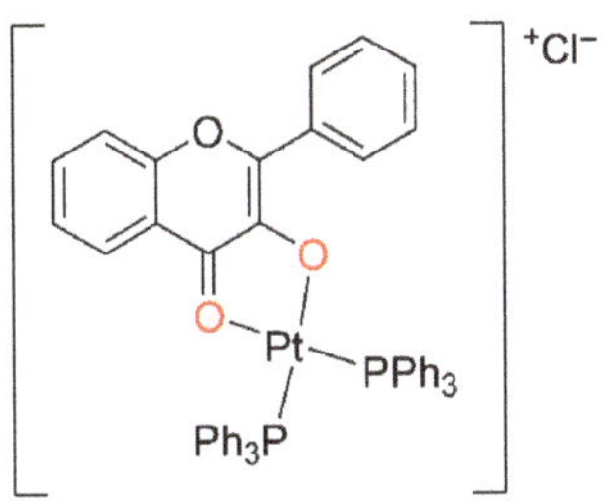

1

Figure 5. Proposed structure of the platinum (II) complex with 3-hydroxyflavon (**1**), *cis*-$[Pt(PPh_3)_2(3\text{-}OH\text{–}F)]Cl$, as suggested by Dell'Anna et al. [37].

The structure of complex **1** was proposed based on data obtained from NMR and IR spectroscopy, high-resolution mass spectrometry (HRMS), and elemental analysis (EA). The coordination mode in the solution was elucidated by NMR spectroscopy, while in the solid state it was determined using single-crystal X-Ray diffraction (XRD) analysis. The HRMS data supported the proposed molecular composition and structural formulation of the complex.

The single crystal obtained by slow diffusion of *n*-pentane into the tetrahydrofuran (THF) reaction solution was not of sufficient quality to allow complete structural refinement, likely due to the presence of photodegradation products. Upon standing in air, the complex gradually underwent photodegradation, yielding ortho-benzoyl-salicylic acid after five days, and subsequently, benzoic and salicylic acids upon prolonged exposure.

The cytotoxicity of complex **1** was evaluated using the 3-(4,5-dimethylthiazol-2-yl)-2,5-diphenyltetrazolium bromide (MTT) assay after 72 h of incubation against two tumor cell lines, U87 (human glioblastoma astrocytoma cell line), and MCF-7 (human breast adenocarcinoma cell line), without including any non-cancerous cell line to allow for the assessment of selectivity (Table 1). Cisplatin was not tested under the same experimental conditions; instead, the literature values were used for comparison. The cytotoxicity of free 3-OH–F and its platinum (II) complex **1** was found to be similar on the U87 cell line, indicating only a minor synergistic effect upon complexation. In contrast, on the MCF-7 cell line, the platinum complex exhibited approximately twice the cytotoxic activity of the free ligand. Overall, the cytotoxicity of the complex on both cell lines was significantly lower than that of cisplatin; however, due to the reliance on data from the literature for cisplatin, the validity of this comparison remains uncertain. Theoretical calculations support the observation that the platinum complex readily releases its coordinated bioactive ligand in the solution, implying that the biological activity of the complex may largely depend on the cytotoxicity of its leaving groups. Antimicrobial activity of the synthesized complex was not evaluated.

Table 1. Cytotoxicity of Pt(II) complex **1** determined by MTT assay after 72 h treatment against U87 (glioblastoma) and MCF-7 (breast) cancer cell lines. Data are mean values ± SD (n = 3) [37].

Compd.	IC_{50} (μM)	
	U87	MCF-7
cis-[$PtCl_2(PPh_3)2$]	>200	>200
3-OH–F	27.5 ± 2.3	108.1 ± 3.5
1	26.3 ± 2.1	55.2 ± 1.7
Cisplatin [a]	1.76 ± 0.22	14 ± 3

[a] Reported from the literature.

In 2016, Raza and coworkers synthesized an octahedral complex of Fe(II) with quercetin, compound **2**, as presented in Figure 6 [38]. $FeSO_4 \cdot 7H_2O$ was refluxed and being stirred for 6 h with a double stoichiometric amount of quercetin in methanol. The reaction mixture was cooled, washed, and vacuum dried. The blackish-brown precipitate stable at room temperature was obtained.

The authors did not succeed in obtaining a crystal suitable for XRD analysis; therefore, the complex was characterized using other experimental techniques. Based on conductometric analysis (CA), complex **2** behaves as a nonelectrolyte, indicating its neutral nature. Two bidentately coordinated quercetin molecules and two water molecules are bound to a single Fe(II) ion. Upon addition of an Fe(II) salt to a methanolic solution of quercetin, a bathochromic shift in the UV–Vis absorption bands was observed, indicating a complex formation. The main changes in the IR spectrum upon complexation include a shift in the carbonyl stretching band from 1661 to 1646 cm^{-1}, and the appearance of a new band at 630 cm^{-1}, corresponding to Fe–O stretching. The presence of two water molecules in the complex was confirmed by both EA and thermogravimetric analyses (TGA). The absence of the NMR signal corresponding to the hydroxyl group at the C-3 position of quercetin indicates a maltol-type coordination mode of quercetin to the Fe(II) ion. Additionally, scanning electron microscopy (SEM) of complex **2** revealed a uniform matrix morphology with ice rock-like surface structures, and the average particle size was approximately 5 μm.

The interaction of the quercetin–iron (II) complex **2** with deoxyribonucleic acid (DNA) was investigated using UV–Vis spectroscopy, fluorescence spectroscopy (FS), and agarose gel electrophoresis. The results indicated that the complex moderately intercalates into DNA, effectively quenches the fluorescence of the classical intercalator ethidium bromide (EB), and competes for intercalative binding sites. In addition, the complex was shown to cleave DNA via an oxidative mechanism. These findings suggest that the complex is capable of directly interacting with genetic material, which may underlie its cytotoxic effects and support its potential as an anticancer agent.

The antibacterial activity of the Fe(II) complex **2** was evaluated using the well diffusion method after 24 h of incubation against two bacterial strains: *Staphylococcus aureus* (Gram-positive, G+) and *Escherichia coli* (Gram-negative, G−) (Table 2). Penicillin sodium was used as the reference compound. The complex exhibited significantly stronger antibacterial activity compared to the free quercetin, as evidenced by larger inhibition zones, which may be explained by chelation theory [39]. The antibacterial effect of the complex was shown to be concentration dependent. The inhibition zones produced by complex **2** were 4–8 times wider than those observed for free quercetin, although still 2–3 times narrower than those produced by the reference compound.

Table 2. Antibacterial activity of the Fe(II) complex **2** evaluated using the well diffusion method. Zones of inhibition were measured in millimeters after 24 h of incubation [38].

Compd.	Zone of Inhibition (mm)			
	Staphylococcus aureus (G+)		*Escherichia coli* (G−)	
	1 mM	2 mM	1 mM	2 mM
Quercetin	0	1	1	3
2	5	7	8	11
Penicilin Sodium	14	15	16	19

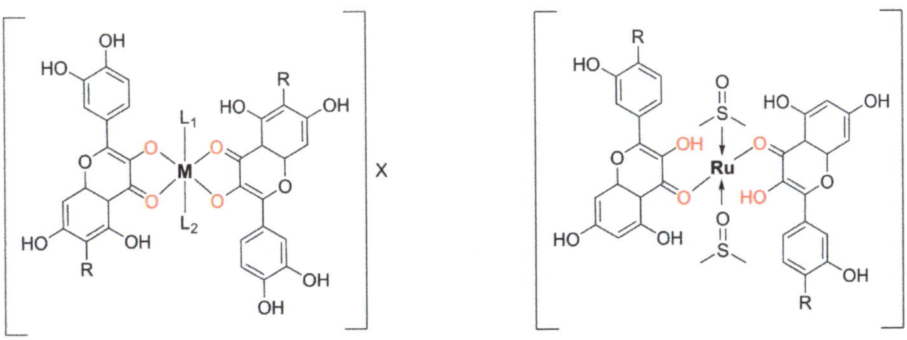

Figure 6. Proposed structures of Fe(II), Rh(III), and Ru(II) complexes with quercetin and its derivatives, as reported by Raza et al., Sahyon et al., and Singh et al., respectively [38,40,41].

In 2022, Sahyon et al. synthesized the Rh(III) complex **3** presented in Figure 3 by reacting quercetin with $RhCl_3 \cdot 3H_2O$ in an ethanolic NaOH solution under stirring at room temperature [40]. After the solvent was evaporated, a dark red microcrystalline complex precipitated. The product was then filtered, washed with an ethanol/diethyl ether (1:1) mixture, and dried under anhydrous conditions.

CA revealed that the quercetin-based Rh(III) complex **3** behaves as a non-electrolyte in the solution. TGA confirmed the presence of one water molecule in the coordination sphere of the complex, in agreement with the results of EA. The coordination mode was elucidated by comparing the IR spectra of free quercetin and the synthesized Rh(III) complex **3**. A shift in the carbonyl stretching frequency from 1664 cm^{-1} in free quercetin to 1620 cm^{-1} in complex **3** indicates the involvement of the carbonyl group in metal coordination. Among the hydroxyl groups, the C-3 hydroxyl was identified as the primary donor site due to its increased acidity attributed to its proximity to the carbonyl group and position within an oxygen-containing heterocyclic ring.

Rhodium(III) is a d^6 transition metal with an ^1A_1g ground state. In the synthesized complex **3**, the Rh(III) center adopts a hexa-coordinated, low-spin octahedral geometry, as evidenced by UV–Vis spectroscopy and magnetic susceptibility measurements. The diamagnetic character of the complex confirmed a ${t_{2g}}^6$ electron configuration, consistent with d^2sp^3 hybridization. Powder XRD (PXRD) analysis further supported this structure, showing similar bond lengths and bond angles close to 90°, indicating a quasi-ideal octahedral geometry. Good agreement was observed between the experimental PXRD data and theoretical calculations. The complex binds strongly to calf thymus DNA via a surface interaction with the phosphate backbone, inducing secondary structure changes ($K_b = 2.76 \times 10^6$ M^{-1}), which may underlie its antiproliferative effect.

The cytotoxic potential of complex **3** was assessed using the MTT assay against five human cancer cell lines: cervical adenocarcinoma cell line (HeLa), hepatocellular carcinoma cell line (HepG2), colorectal adenocarcinoma cell line (Caco-2), prostate adenocarcinoma cell line (PC-3), and MCF-7 (breast), as well as one normal human lung fibroblast cell line (WI-38). Cisplatin was used as the reference drug (Table 3). In all tumor cell lines, complex **3** exhibited significantly higher cytotoxicity than free quercetin (4- to 15-fold); however, quercetin itself demonstrated markedly greater selectivity toward cancer cells. Compared to cisplatin, the complex was 2–4 times less cytotoxic but demonstrated an approximately 6-fold greater selectivity for cancer cells over normal cells. The highest selectivity indices (SI), calculated as the ratio of IC_{50} values for normal to tumor cells, were observed for HepG-2 (liver) and HeLa (cervix) cell lines, with values of 4.8 and 4.3, respectively.

Table 3. Cytotoxicity ($IC_{50} \pm$ SEM) of quercetin, the Rh(III) complex **3**, and cisplatin determined by MTT assay after 48 h of incubation [40].

Compd.	IC_{50} (μM)					
	HepG-2	**HeLa**	**MCF-7**	**PC-3**	**Caco-2**	**WI-38**
Quercetin	129.28 ± 13.5	68.78 ± 1.39	61.72 ± 3.29	91.92 ± 7.32	80.92 ± 9.04	454.5 ± 48.1
3	8.49 ± 0.40	9.40 ± 0.46	16.32 ± 0.81	17.71 ± 0.87	21.73 ± 1.13	40.58 ± 2.80
Cisplatin	4.50 ± 0.12	5.57 ± 0.23	4.17 ± 0.12	8.870 ± 0.635	12.49 ± 0.64	6.72 ± 0.29

Mechanistic studies suggest that complex **3** inhibits HeLa cell proliferation by inducing apoptosis, evidenced by cell cycle arrest in the pre-G1 phase, most likely due to inhibited DNA replication. This arrest may occur through the activation of the p53 tumor suppressor gene, as indicated by decreased levels of the anti-apoptotic Bcl-2 protein. Additionally, a reduction in MMP-9 levels was observed, supporting the complex's antiproliferative and anti-metastatic effects. Based on these findings, a mechanistic hypothesis can be proposed: the complex **3** binds to the DNA surface, preventing replication in cancer cells and promoting p53 expression. Increased p53 levels can lead to the inhibition of both Bcl-2 and MMP-9, followed by the activation of caspase-9. Caspase-9 then activates caspase-3, initiating apoptosis. This apoptotic process results in the pre-G1 cell cycle arrest and decreased proliferation of HeLa cells, as reflected in the reduced proportion of cells in the G2/M phase.

To the best of our knowledge the antimicrobial activity of complex **3** was not investigated.

In 2017, Singh et al. synthesized a series of quercetin derivatives bearing para-substituted B-rings and subsequently prepared their $[Ru^{II}(DMSO)_2(querc)_2]$ complexes (**4**–**7**, Figure 6) [41]. The distorted octahedral geometry structure of the complexes was proposed based on spectroscopic data (IR, NMR, and HRMS) and supported by density functional theory (DFT) calculations. In the resulting complexes **4**–**7**, two dimethyl sulfoxide (DMSO) ligands were coordinated monodentately, one via the sulfur atom and the other via the oxygen atom, reflecting the ambidentate nature of DMSO.

The cytotoxicity of the synthesized complexes **4**–**7** was evaluated using the MTT assay after 24 h of exposure against MCF-7 (breast) cell line (Table 4). Among the tested compounds, complex **4**, featuring a methoxy substituent on the B-ring, exhibited the highest cytotoxic activity against the MCF-7 cancer cell line (IC_{50} = 16 μM). In ligand–complex pairs, the IC_{50} values were consistently lower for the metal complexes **4**–**7** compared to their corresponding quercetin-based ligands, suggesting a potential synergistic effect between the metal center and the coordinated ligand.

Table 4. IC_{50} values of the synthesized complexes **4**–**7** determined by the MTT assay after 24 h of exposure [41].

Compd.	IC_{50} (μM)	Compd.	IC_{50} (μM)	Compd.	IC_{50} (μM)	Compd.	IC_{50} (μM)
	MCF7		MCF7		MCF7		MCF7
L4	17.2	**L5**	29.5	**L6**	38.4	**L7**	35.4
4	16.0	**5**	28.0	**6**	36.2	**7**	32.1

L4–**L7** are hydroxyflavone derivatives used as ligands in the corresponding complexes.

In 2010, Prajapati and coworkers synthesized two Ru(II) complexes with DMSO [42]. One of these complexes featured a chalcone ligand coordinated in a bidentate (O,O) fashion and is therefore not considered in this study. The other complex, compound **8**, which incorporates a 3-OH–F derivative, is relevant to the present discussion and is shown in Figure 7.

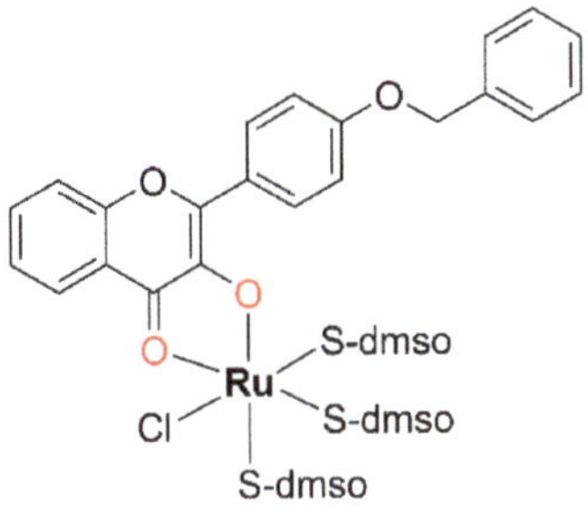

8

Figure 7. Structure of the Ru(II) complex **8**, *fac*-$[RuCl(S\text{-}dmso)_3(3\text{-}OH\text{–}F)]$, containing a 3-OH–F derivative and coordinated DMSO ligands, as reported by Prajapati et al. [42].

The complex **8** was synthesized by the dropwise addition of a methanolic solution of *cis*-[$Ru^{II}Cl_2(dmso)_4$] to a methanolic solution of the 3-OH–F derivative containing an equimolar amount of Et_3N, adjusted to pH around 9.0. The resulting reaction mixture was stirred on a steam bath for 30 min. Subsequently, the solution was concentrated to one fourth of its original volume and left to stand at room temperature for 24 h. The resulting dark reddish-brown crystalline solid was collected by filtration, washed sequentially with methanol and diethyl ether, and finally dried under a vacuum. According to the authors, complex **8** is reported to be air-stable. CA indicates that it is a neutral species. The evidence for the complex **8** formation includes the appearance of a characteristic S=O stretching band around 1050 cm^{-1} and a Ru–S vibration band at approximately 430 cm^{-1}. The proposed structure of complex **8** was further confirmed by XRD analysis, which revealed a monoclinic crystal system containing four discrete complex molecules arranged in a chain, with no solvent-accessible voids in the crystal packing.

Preliminary cytotoxicity screening of complex **8** was performed using the MTT assay after 48 and 72 h of incubation against Dalton's lymphoma cell line, a murine transplantable T-cell lymphoma (DL) (Table 5). The IC_{50} value of complex **8** was found to be in the nanomolar range, indicating significantly higher cytotoxicity compared to the corresponding free OH–F ligand **L8**. This enhanced activity of complex **8** may suggest a synergistic effect between the coordinated ligand **L8** and the Ru(II) metal center in mediating the cytotoxic response.

Table 5. Cytotoxicity of complex **8** determined by MTT assay after 48 and 72 h of incubation against DL (lymphoma) cells [42].

Compd.	IC_{50} (μM)	
	48 h	72 h
DMSO control a	none	none
L8	>5	0.054
8	0.816	0.042

[a] DMSO was used as a vehicle control at 0.01% concentration. According to the original source, DL cells incubated with 0.01% DMSO alone did not produce any cytotoxicity.

Ruthenium–DMSO complexes have attracted considerable interest due to their promising selectivity toward solid tumor metastases and relatively low toxicity to healthy tissues [43], which has led some of these compounds to enter clinical trials [44].

In 2020, Gantsho and coworkers synthesized 18 Re(I) carbonyl complexes incorporating tropolone, 3-OH–F, and three differently substituted quinolones as bidentate ligands [45]. Since tropolone and quinolones are not members of the OH–F family, only five Re(I) tricarbonyl complexes **9–13** containing 3-OH–F will be discussed here (Figure 8). Starting from $[NEt_4]_2[Re(CO)_3Br_3]$, the addition of $AgNO_3$ at pH 2.2 yielded the triaqua complex *fac*-$[Re(CO)_3(H_2O)_3]^+$. Subsequent substitution of two labile aqua ligands with the bidentate 3-OH–F resulted in complex **9**. In methanolic solution, one of the coordinated water molecules is replaced by methanol, resulting in the complex **10**. Complexes **11–13** were synthesized by introducing one equivalent of a monodentate phosphine ligand. All complexes **9–13** were isolated as yellow precipitates or thin, needle-like crystals, which were unsuitable for XRD analysis. Structural characterization was performed using EA along with IR, UV–Vis, and NMR spectroscopy.

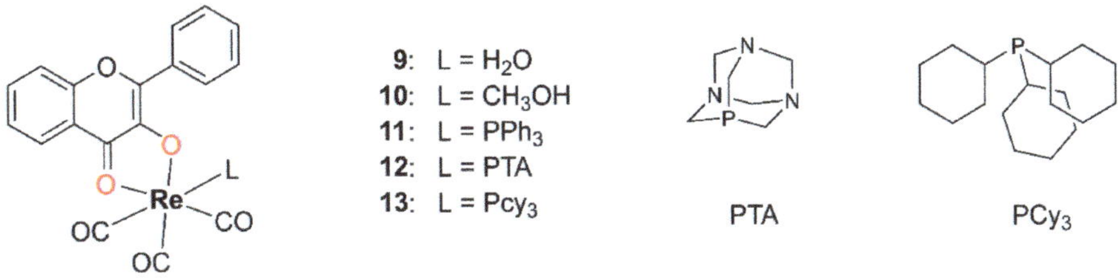

Figure 8. Proposed structure of the Re(I) tricarbonyl complexes **9–13** with 3-OH-F, as reported by Gantsho et al. [45].

Complexes **9** and **10** exhibit a high tendency toward the substitution of the coordinated water or methanol ligand. In complexes **11–13**, elevated temperatures and an excess of the phosphine ligand promote the efficient substitution of the axially positioned carbonyl ligand trans to the phosphine. The kinetics of the substitution of complex **10** with triphenylphosphine (PPh_3), 1,3,5-triaza-7-phosphaadamantane (PTA), and tricyclohexylphosphine (PCy_3) were studied, and all reactions proceeded via a single-step, first-order process, with rates directly dependent on the concentration of the entering phosphine ligand.

A cytotoxicity assessment was performed exclusively for complex **11** due to the limited solubility of the other complexes in this series. The evaluation employed a fluorometric cell viability assay based on resazurin after 48 h of incubation with HeLa (cervix) as the tumor cell line, and human retinal pigment epithelial cells immortalized with hTERT (RPE-1) as the normal cell line. Cisplatin served as a positive control (Table 6). The cytotoxicity of complex **11** against HeLa cells was comparable to that of cisplatin. The SI of the complex, calculated as the ratio of IC_{50} values for RPE-1 and HeLa cells, was 1.5, indicating a limited selectivity toward tumor cells under the tested conditions. In contrast, cisplatin exhibited an SI of approximately five. The authors suggest that the moderate cytotoxicity observed for complex **11** may be attributed to its high kinetic stability.

Table 6. Cytotoxicity of Re(I) complex **11**, *fac*-[Re(CO)$_3$(3-OH–F)(PPh_3)], determined by a fluorometric cell viability assay using resazurin after 48 h incubation. Results are expressed as IC_{50} values (µM). Cisplatin was used as a positive control [45].

Compd.	IC_{50} (µM)	
	HeLa	RPE-1
11	12.21 ± 0.17	18.41 ± 3.16
cisplatin	8.02 ± 0.57	39.07 ± 0.45

Kurzwernhart et al. synthesized 28 distinct metal complexes **14–41** derived from 3-OH–F derivatives with Ru(II), Os(II), and Rh(III) as metals [46–49]. The complexes adopt a pseudo-octahedral "piano-stool" geometry, characterized by slight variations in M–O bond lengths and a moderately twisted phenyl substituent on the ligand. The general structure of the complexes is presented in Figure 9, while the individual ligand structures and corresponding cytotoxicity data are summarized in Table 7. The cytotoxicity of complexes **14–41** was evaluated by an MTT assay after 96 h of exposure using three human cancer cell lines: the ovarian carcinoma cell line (CH1), the colorectal adenocarcinoma cell line (SW480), and the lung carcinoma epithelial cell line (A549). The positive control value for cisplatin was obtained from the literature.

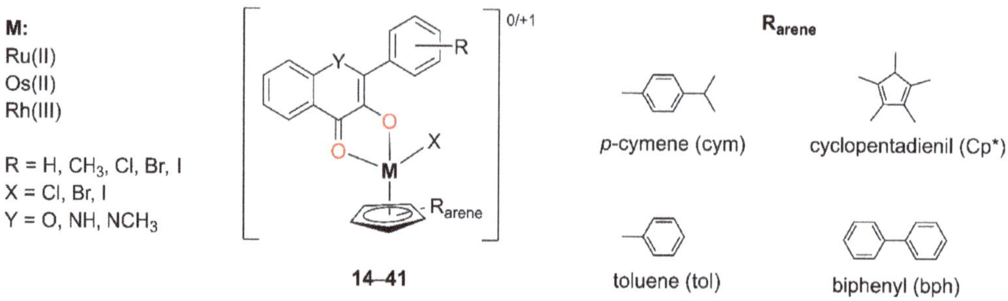

14–41

Figure 9. General structure of complexes **14–41** bearing 3-OH–F derivatives, as reported by Kurzwernhart et al. [46–49].

Table 7. Ligands of complexes **14–41** and their IC_{50} cytotoxicity values in CH1 (ovarium), SW480 (colon), and A549 (lung) cancer cell lines after 96 h exposure, determined by MTT assay (mean ± SD, n = 3) [46–49].

Compd.	X	Y	R	R_{arene}	M	IC_{50} ± SD (μM)			Ref. No.
						CH1 [a]	SW480	A549	
14	Cl	O	H	cym	Ru^{II}	2.1 ± 0.2	9.6 ± 1.5	20 ± 2	[46]
15	Cl	O	p-CH_3	cym	Ru^{II}	1.8 ± 0.2	7.2 ± 0.5	17 ± 2	[46]
16	Cl	O	p-F	cym	Ru^{II}	1.7 ± 0.4	7.9 ± 2.1	18 ± 1	[46]
17	Cl	O	p-Cl	cym	Ru^{II}	0.86 ± 0.06	3.8 ± 0.5	9.5 ± 0.5	[46]
18	Cl	O	m-F	cym	Ru^{II}	1.5 ± 0.1	7.0 ± 1.0	15 ± 1	[47]
19	Cl	O	o-F	cym	Ru^{II}	4.0 ± 0.8	24 ± 3	30 ± 1	[47]
20	Cl	O	m-Cl	cym	Ru^{II}	1.0 ± 0.1	7.0 ± 0.7	12 ±2	[47]
21	Cl	O	o-Cl	cym	Ru^{II}	7.9 ± 0.6	26 ± 1	51 ± 5	[47]
22	Cl	O	p-Br	cym	Ru^{II}	1.2 ± 0.2	3.4 ± 0.1	8.6 ± 0.7	[47]
23	Cl	O	m-Br	cym	Ru^{II}	2.3 ± 0.7	7.2 ± 0.4	17 ± 3	[47]
24	Br	O	H	cym	Ru^{II}	2.8 ± 0.4	12 ± 1	27 ± 4	[48]
25	Br	O	p-Cl	cym	Ru^{II}	0.86 ± 0.04	3.4 ± 0.4	7.9 ± 0.6	[48]
26	I	O	H	cym	Ru^{II}	1.6 ± 0.2	9.6 ± 1.5	16 ± 1	[48]
27	I	O	p-Cl	cym	Ru^{II}	1.2 ± 0.3	4.7 ± 0.9	8.9 ± 0.8	[48]
28	Cl	O	H	tol	Ru^{II}	3.2 ± 0.1	12 ± 3	19 ± 1	[48]
29	Cl	O	p-Cl	tol	Ru^{II}	0.88 ± 0.17	4.7 ± 0.6	7.8 ± 2.5	[48]
30	Cl	O	H	bph	Ru^{II}	5.5 ± 1.2	9.2 ± 1.9	28 ± 5	[48]
31	Cl	O	p-Cl	bph	Ru^{II}	6.3 ± 1.1	21 ± 4	59 ± 1	[48]
32	Cl	N-H	H	cym	Ru^{II}	4.0 ± 0.2	14 ± 1	17 ± 2	[48]
33	Cl	N-CH_3	H	cym	Ru^{II}	5.3 ± 0.2	12 ± 2	19 ± 1	[48]
34	Cl	O	H	cym	Os^{II}	2.5 ± 0.2	12 ± 1	16 ± 1	[49]
35	Cl	O	p-CH_3	cym	Os^{II}	0.58 ± 0.06	6.6 ± 0.2	4.8 ± 1.0	[49]
36	Cl	O	p-F	cym	Os^{II}	1.7 ± 0.2	7.6 ± 0.1	16 ± 2	[49]
37	Cl	O	p-Cl	cym	Os^{II}	0.90 ± 0.06	4.2 ± 0.3	8.0 ± 0.5	[49]
38	Cl	O	H	Cp*	Rh^{III}	3.1 ± 0.3	7.9 ± 0.8	15 ± 3	[49]
39	Cl	O	p-CH_3	Cp*	Rh^{III}	2.0 ± 0.2	6.3 ± 0.8	11 ± 3	[49]
40	Cl	O	p-F	Cp*	Rh^{III}	2.0 ± 0.1	7.6 ± 0.6	13 ± 4	[49]
41	Cl	O	p-Cl	Cp*	Rh^{III}	1.0 ± 0.0	2.5 ± 0.8	4.3 ± 1.8	[49]
CDDP [b]	-	-	-	-	-	0.14 ± 0.03	3.3 ± 0.4	1.3 ± 0.4	[46]

[a] It was the CH1/PA-1 cell line for compounds **34–41**. [b] Literature-reported data.

Complexes **14–23**, with the general formula $[Ru^{II}(Cl)(cym)(3\text{-OH–F})]$, differ in both the nature and position of substituents on the B-ring of the 3-OH–F scaffold. For several of these complexes (**15**, **17**, **18**, and **21**), single-crystal XRD data were obtained. Complexes bearing para-substituted B-rings exhibited the highest cytotoxic activity. In contrast, the reduced activity observed for ortho-substituted derivatives may be attributed to steric effects, as the phenyl ring in these compounds adopts a significantly more twisted conformation. This structural distortion is likely to impair molecular recognition and diminish the binding affinity to biological targets. Notably, all complexes showed a stronger inhibition of topoisomerase IIα compared to the corresponding free OH–F derivatives, which was attributed to their multitargeted mechanism of action. Furthermore, the extent of topoisomerase IIα inhibition correlated well with the observed cytotoxicity.

In a subsequent set of compounds (**24–33**, $[Ru^{II}(X)(R_{arene})(3\text{-OH–F})]$), the aromatic ligand coordinated to the metal center (R_{arene}) was varied. The results showed that changes in lipophilicity were more strongly influenced by the 3-OH–F ligand than by the R_{arene} moiety [40]. Finally, in complexes **34–41**($[Os^{II}/Rh^{III}(Cl)(R_{arene})(3\text{-OH–F})]^{0/+1}$), ruthenium (II) was replaced by osmium (II) or rhodium (III) as the central metal ion, but this did not significantly alter the observed cytotoxicity [41].

The complexes **14–41** synthesized by Kurzwernhart et al. exhibited the highest cytotoxic activity against the CH1 (ovarian) cell line, while lower activity was observed against the SW480 (colon) and the lowest toward A549 (lung) cells, which are generally more resistant. Most IC_{50} values for the CH1 cell line were in the low micromolar range ($\leq$7.9 μM), with compounds **17**, **25**, **29**, **35**, and **37** showing nanomolar potency. These five compounds share structural features including a chloride leaving group and a para-substituted B-ring of the OH–F moiety bearing either a chlorine or methyl substituent. These results suggest that both the nature and position of substituents on the B-ring, as well as the identity of the leaving group, play a critical role in enhancing cytotoxicity.

2.1.2. Mixed-Ligand Systems: Incorporating Both (O,O) and (N,N) Donor Atoms

Starting from *cis*-$[Ru^{II}(bpy)_2Cl_2]$ or $[Ru^{II}(phen)_2(CO_3)]$, Zahirović et al. synthesized a series of octahedral cationic ruthenium (II) complexes **42–47** with OH–F ligands [50]. In these complexes, the diimine ligands, 2,2′-bipyridine (bpy) or 1,10-phenanthroline (phen), are coordinated in a bidentate manner through nitrogen atoms, while the OH–F ligand is coordinated via two oxygen atoms in a maltol-type binding mode (Figure 10). The resulting complexes are low-spin, diamagnetic species consistent with a $t_{2g}{}^6$ electronic configuration of the Ru(II) center.

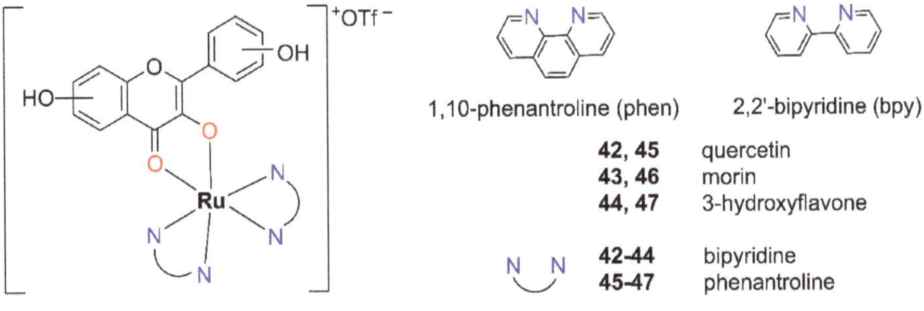

Figure 10. Proposed structures of the synthesized heteroleptic $[Ru^{II}(phen/bpy)_2(OH\text{–}F)]OTf{\cdot}nH_2O$ complexes **42–47**, as reported by Zahirović et al. [50].

Given that the synthesized complexes **42–47** lack good leaving groups, the most probable mode of DNA interaction is intercalation. This is supported by their cationic nature and the presence of planar π-extended aromatic systems. On the other hand, the efficiency of π–π stacking interactions with nucleobases tends to decrease with the increasing number of hydroxyl groups on the OH–F ligand.

The cytotoxicity of these types of Ru(II) complexes **42–47** was evaluated using the MTT assay after 72 h of incubation on four human cancer cell lines: colorectal adenocarcinoma cell line (SW620), HepG2 (liver), MCF-7 (breast), and HeLa (cervix) (Table 8). Unfortunately, the measurement errors were sufficiently high enough to raise concerns regarding the reliability and significance of the obtained results. However, two results stood out as significant: the bipyridine-based complex **44** exhibited nanomolar cytotoxicity against SW620 cells (IC_{50} = 0.75 μM), while the phenanthroline-based complex **47** showed micromolar activity against the MCF-7 cell line (IC_{50} = 8.32 μM). Both of these complexes are derivatives of 3-OH–F.

Table 8. Cytotoxicity of complexes **42–47** determined by MTT assay after 72 h treatment expressed as (mean ± SD, n = 3) [50].

Compd.	**IC_{50}**			
	SW620	**HepG-2**	**MCF-7**	**HeLa**
quercetin	>100	59.54 ± 20.48	>100	93.22 ± 22.97
morin	>100	>100	>100	>100
3-OH–F	50.73 ± 22.29	8.88 ± 17.68	42.06 ± 21.08	5.44 ± 31.22
***cis*-[Ru(bpy)$_2$Cl$_2$]·2H$_2$O**	4.53 ± 60.11	>100	2.10 ± 70.64	>100
[Ru(phen)$_2$(CO$_3$)]·2H$_2$O	>100	>100	92.38 ± 44.00	>100
42	>100	>100	0.39 ± 70.64	>100
43	>100	>100	85.87 ± 70.64	>100
44	0.75 ± 0.15	2.51 ± 0.67	0.52 ± 0.38	0.78 ± 0.20
45	>100	>100	>100	>100
46	>100	>100	7.64 ± 70.43	>100
47	8.23 ± 46.41	11.42 ± 66.02	8.32 ± 0.86	19.32 ± 65.89

Antimicrobial activity was assessed by the disk diffusion method against four Gram-positive bacterial strains (*Staphylococcus aureus* ATCC 25923, *Enterococcus faecalis* ATCC 19433, *Streptococcus* β-hemolytic group A, and methicillin-resistant *Staphylococcus aureus* (MRSA)). Vancomycin was used as the reference drug for all Gram-positive strains, except for MRSA, where gentamicin served as the control. Four Gram-negative bacterial strains were also tested (*Klebsiella pneumoniae* ATCC 1705, *Acinetobacter baumannii* ATCC BAA-747, *Pseudomonas aeruginosa*, and *Escherichia coli*), with gentamicin as the reference drug. Additionally, the fungal strain *Candida albicans* was included, with nystatin used as the control. The results are presented in Table 9. Interestingly, complex **44** exhibited zones of inhibition against Gram-positive bacteria comparable in width to that of vancomycin. In the case of MRSA, the zone of inhibition was significantly wider than that of gentamicin. Against the tested fungal strain, the inhibition zone observed for **44** was identical to that measured for nystatin. Comparison of the inhibition zones of the synthesized metal complexes with those of the free hydroxyflavone ligands indicates a synergistic effect of coordination, enhancing antimicrobial activity against both Gram-positive bacteria and the fungal strain *Candida albicans*. Based on the obtained results, a significant improvement in the antimicrobial efficacy of complex **44** was observed against the tested Gram-positive bacteria and *Candida albicans* strains, relative to the uncoordinated 3-hydroxyflavone ligand.

Table 9. Antimicrobial activity of free 3-OH–F and its ruthenium (II) complex $[Ru(bpy)_2(3\text{-}OH\text{-}F)](OTf)\cdot nH_2O$ **44**, evaluated by the disk diffusion method and expressed as inhibition zone diameters (mm) [50].

Compd.	*Staphylococcus aureus* ATCC 25923	*Enterococcus faecalis* ATCC 19433	*Streptococcus* Beta-Hemolytic Group A	Methicillin-Resistant *Staphylococcus aureus*	*Klebsiella pneumoniae* ATCC 1705	*Acinetobacter baumannii* ATCC—BAA 747	*Pseudomonas aeruginosa*	*Escherichia coli*	*Candida albicans*
	G+	G+	G+	G+	G−	G−	G−	G−	Fungus
	Diameter of Inhibition Zone (mm)								
3-OH–F	*	*	*	12	*	16.5	*	*	18.5
44	25	20	20	26	*	16	13	*	28
vancomycin	27	26	35	-	-	-	-	-	-
gentamicin	-	-	-	20	25	35	36	21	-
nystatin	-	-	-	-	-	-	-	-	28

* No activity, - not tested, G+ = Gram-positive bacteria, and G− = Gram-negative bacterial strain.

Gençkal et al. synthesized heteroleptic complexes of Co(II), Ni(II), and Cu(II) with quercetin and diimine (phen or bpy) ligands (**48–51**, Figure 11) [51]. A methanolic solution of metal (II) chloride x-hydrate was added dropwise to a methanolic solution of quercetin and sodium hydroxide under constant stirring at room temperature. The reaction mixture was stirred for an additional hour, after which a methanolic solution of phen or bpy was added. The mixture was then refluxed for approximately 4 h. Upon completion, it was allowed to cool to room temperature, and the resulting solid was filtered, washed with methanol, and dried under ambient conditions.

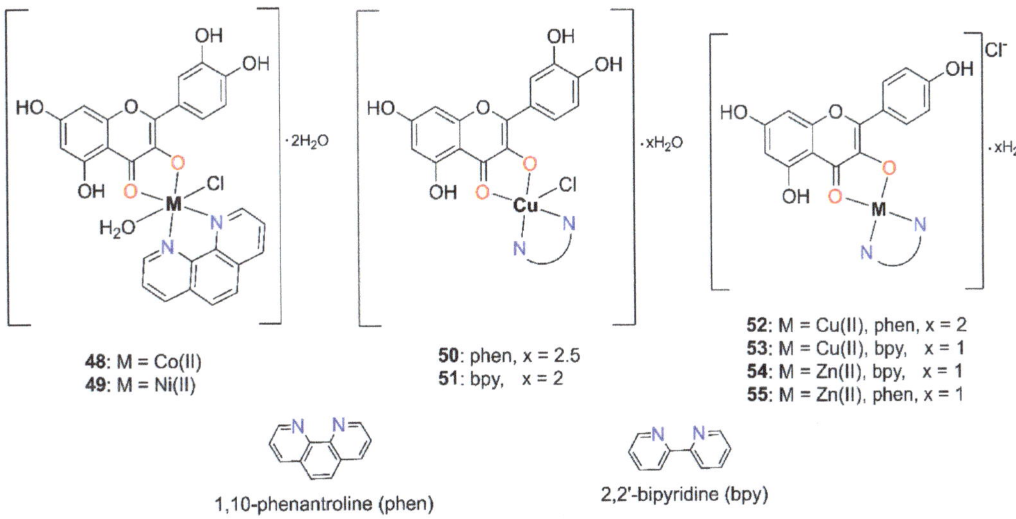

Figure 11. Proposed structures of M(II) complexes with quercetin (**48–51**) or kaempferol (**52–55**), as suggested by Gençkal et al. and Wang et al., respectively [51,52].

According to experimental measurements, two distinct types of complexes were obtained: an octahedral $[Co^{II}/Ni^{II}(queH\text{-}1)Cl(phen)(H_2O)]\cdot 2H_2O$ (**48–49**), and a square-pyramidal geometry for $[Cu^{II}(queH\text{-}1)Cl(phen/bpy)]\cdot xH_2O$ (**50–51**), as illustrated in Figure 11.

EA confirms the proposed molecular formulas of the complexes **48–51**. CA indicates that the complexes are neutral in nature and thus behave as non-electrolytes. Their stability was verified by repeating the CA after 24 and 48 h, with consistent results. The magnetic susceptibility (χ) values for the Ni(II) and Co(II) complexes are consistent with mononuclear octahedral coordination, while those for the Cu(II) complexes support the assignment of a mononuclear square-pyramidal geometry for complexes **50** and **51**. In the UV–Vis spectrum, the characteristic absorption band of free quercetin at 371 nm, assigned to the B-ring, undergoes a bathochromic shift of approximately 50 nm upon complexation, supporting a maltol-type coordination mode of the quercetin ligand. The band maximum corresponding to the carbonyl stretching vibration shifted to lower wavenumbers in all metal complexes, indicating coordination through the adjacent hydroxyl group. Since the hydroxyl group at the C3 position is significantly more acidic than the one at C5, it is expected to preferentially participate in metal coordination. Furthermore, the IR spectra reveal new bands in the 433–417 cm^{-1} and 652–643 cm^{-1} regions, corresponding to M–N and M–O bond vibrations, respectively, confirming the formation of metal–ligand bonds through nitrogen and oxygen donor atoms.

The cytotoxic activity of the compounds **48–51** was initially evaluated using Sulforhodamine B (SRB) and ATP-based cell viability assays after 48 h of incubation against four different human cancer cell lines: A549 (lung), PC-3 (prostate), HeLa (cervix), and MCF-7 (breast). Improved results were obtained using the ATP assay, which is generally more sensitive than the SRB assay. Among them, the Cu(II) complexes **50** and **51** exhibited the most pronounced cytotoxicity toward the MCF-7 breast cancer cell line. Consequently, an additional breast cancer cell line, namely the human triple-negative breast cancer cell line (MDA-MB-231), was selected for further evaluation. At higher concentrations, some cytotoxic activity was also observed for the free ligands, quercetin, phen, and bpy, in breast cancer cell lines, although this activity was significantly lower compared to that of the Cu(II) complexes. Complex **50**, containing phen, showed high cytotoxicity after 48 h of exposure in both MCF-7 and MDA-MB-231 cell lines, with IC_{50} values in the low micromolar range. One of the mechanisms underlying the cytotoxic effect of complex **50** was identified as apoptosis induction, as evidenced by an approximately 40% increase in apoptotic cells, mediated through the Caspase-3/7-dependent pathway. These findings identify complex **50** as a promising candidate for further development as an antitumor agent targeting breast cancer.

In 2014, Wang et al. synthesized four M(II) complexes **52–55** (M = Cu(II) or Zn(II)) incorporating the OH–F kaempferol as an (O,O) donor ligand, along with bpy or phen as (N,N) donor ligands, as depicted in Figure 11 [52]. A methanolic solution of MCl_2 was added to a methanolic solution containing equimolar amounts of kaempferol and NaOH at an ambient temperature. Subsequently, bpy or phen, dissolved in methanol, was added dropwise. The reaction mixture was filtered, and the resulting yellow precipitate was washed with ethanol and dried under a vacuum. Since single crystals suitable for XRD analysis could not be obtained, the structures of the complexes were proposed based on EA, HRMS, UV–Vis, IR, and NMR spectroscopy. A fluorometric competitive experiment using EB as a probe revealed moderate intercalation by the complexes. The increased viscosity of CT-DNA in the presence of the metal complexes **52–55** further supported these findings. Notably, the complexes **52–55**, especially complex **52**, were able to induce the cleavage of plasmid DNA (pUC19) in a dose-dependent manner.

Kaempferol and its complexes (**52–55**) were tested for cytotoxic activity against the MDA-MB-231 (breast) cancer cell line using the MTT assay after 24 h of incubation (Table 10). All complexes showed significantly improved cytotoxicity compared to the free ligand, indicating a potential synergistic interaction between kaempferol and the coordinated metal ions.

Table 10. Cytotoxic activity of metal complexes **52–55** with kaempferol against MDA-MB-231 cells, determined by MTT assay after 24 h of treatment. Data are expressed as mean ± SD (n = 3) [52].

Compd.	IC_{50} (μM)
	MDA-MB-231
kaempferol	39.76 ± 1.06
52	17.39 ± 1.42
53	23.71 ± 1.12
54	29.89 ± 1.19
55	27.49 ± 1.42

In 2021, Kozsup et al. reported the synthesis and characterization of a series of sixteen octahedral Co(III) complexes containing derivatized OH–F and tetradentate nitrogen ligands [53], namely tris(2-aminoethyl)amine (tren) or tris(2-pyridylmethyl)amine (tpa). Thirteen compounds of these (**56–68**, $[Co^{III}(4N)(OH–F)]^{2+}$) were obtained by maltol-type coordination (Figure 12, Table 11), using variously substituted OH–F-ligands (3-hydroxyflavone (flavH), 6-F-3′-NO_2-3-hydroxyflavone (NO_2FflavH), 3′-NO_2-3-hydroxyflavone (NO_2flavH), 4′-Me-3-hydroxyflavone (MeflavH), 6-Cl-4′-OMe-3-hydroxyflavone (ClO-MeflavH), 6-Br-3-hydroxyflavone (BrflavH), and 4′-iPr-3-hydroxyflavone (iPrflavH). Two additional complexes (**69** and **70**, Figure 13) were obtained by the acac method and will be discussed in the following Section 2.2. One complex, ($[Co(tren)(nar)](Cl)(ClO_4)$), which contains naringenin (nar) from the flavanone family, will not be a subject of this review.

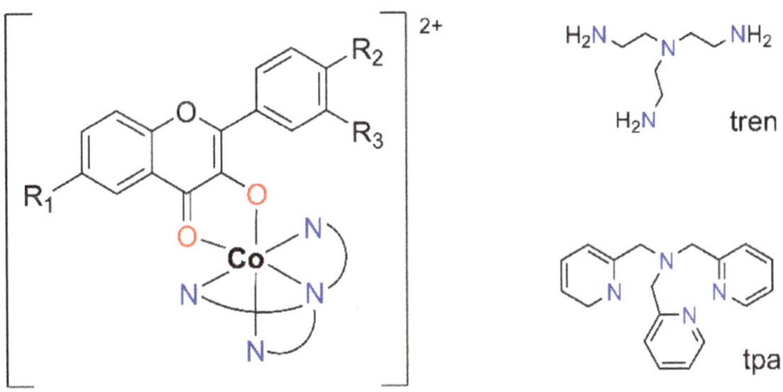

56–68

Figure 12. The chemical structures of Co(III) complexes **56–68** with differently substituted OH–F obtained by maltol-type coordination, as reported by Kozsup et al. [53].

The $[Co(4N)(flav)]^{2+}$ type of compounds (**56–70**) was obtained in an overnight reaction by mixing the stoichiometric amount of the metal precursor ($[Co(4N)Cl_2]Cl$ (4N = tren or tpa)) and the ligand in MeOH at 60 °C with an equivalent of NaOH as a base. Most of the complexes were crystallized from water solutions after the addition of a counterion, ClO_4^-.

Table 11. Numeration of Co(III) complexes **56–68** bearing different OH–F derivatives [53].

Compd.	Abbreviation	4N	R_1	R_2	R_3
$[Co(tren)(flav)](ClO_4)_2$	**56**	tren	-	-	-
$[Co(tpa)(flav)](ClO_4)_2$	**57**	tpa	-	-	-
$[Co(tren)(Brflav)](ClO_4)_2$	**58**	tren	Br	-	-
$[Co(tpa)(Brflav)](ClO_4)_2$	**59**	tpa	Br	-	-
$[Co(tren)(Meflav)](ClO_4)_2$	**60**	tren	-	Me	-
$[Co(tpa)(Meflav)](ClO_4)_2$	**61**	tpa	-	Me	-
$[Co(tren)(NO_2Fflav)](ClO_4)_2$	**62**	tren	F	-	NO_2
$[Co(tpa)(NO_2Fflav)](ClO_4)_2$	**63**	tpa	F	-	NO_2
$[Co(tren)(NO_2flav)](Cl)(ClO_4)$	**64**	tren	-	-	NO_2
$[Co(tpa)(NO_2flav)](ClO_4)_2$	**65**	tpa	-	-	NO_2
$[Co(tren)(ClOMeflav)](ClO_4)_2$	**66**	tren	Cl	OMe	-
$[Co(tren)(iPrflav)](ClO_4)_2$	**67**	tren	-	iPr	-
$[Co(tpa)(iPrflav)](ClO_4)_2$	**68**	tpa	-	iPr	-

The full characterization of a series of compounds **56–70** included NMR, IR, and HRMS measurements, as well as crystal structure and a cyclic voltammetric (CV) analysis. Characterization data confirmed acac coordination mode (O,O) of the OH–F ligand to the metal ion and octahedral geometry in all cases. Prior to biological study, the Co(III) species were tested for their stability in a D_2O:DMSO = 5:1 (v/v) mixture revealing no spectral changes in ambient conditions. Additionally, their electrochemical profile was determined using CV. The redox study indicated the influence of the substituents on the values of reduction potential as the presence of electron-donating substituents on the OH–F further shifted the cobalt complex's reduction potential toward more negative values. The nature of the substituents demonstrated the greatest impact on the cytotoxic activity of the compounds (Table 11).

The cytotoxicity of most complexes (**56–65**), as well as the free flavone ligands, was evaluated after 72 h of incubation against two human cancer cell lines, epithelial carcinoma cell line (A431) and A549 (lung), under both normoxic (21% O_2) and hypoxic (1% O_2) conditions and end-point SRB assay (Table 12). The A549 cell line was generally more sensitive to the tested compounds than A431, with the exception of compound **60**, for which the EC_{50} values were lower in the A431 cells. Among the free ligands, MeflavH exhibited no cytotoxicity against either cell line, while flavH showed slight activity only against A549 under normoxic conditions. The ligands NO_2flavH, NO_2FflavH, and BrflavH were also exclusively active against A549, but their cytotoxic effects in this cell line were significant. Overall, the results indicate that the Co(III) species derived from the $[Co(tpa)Cl_2]Cl$ precursor exhibit significantly stronger cytotoxic effects compared to those obtained from $[Co(tren)Cl_2]Cl$, highlighting the influence of substituent variation in OH–F-based ligands.

Table 12. Studied Co(III) complexes **56–65** obtained by maltol-type coordination and their cytotoxicity against A549 and A431 human cancer cell lines under normoxic and hypoxic conditions after 72 h incubation using end-point SRB assay. EC_{50} and 95% confidence interval (CI) values are presented in μM as a mean of three independent experiments [53].

Compd.	EC_{50}(μM)			
	A431		A549	
	Normoxia	Hypoxia	Normoxia	Hypoxia
$[Co(tren)Cl_2]Cl$	>100	>100	>100	>100
$[Co(tpa)Cl_2]Cl$	>100	>100	>100	>100
flavH	>100	>100	36	>100

Table 12. *Cont.*

Compd.	EC_{50}(μM)			
	A431		A549	
	Normoxia	Hypoxia	Normoxia	Hypoxia
MeflavH	>100	>100	>100	>100
BrflavH	84 [+29, −19]	>100	7.0 [+1, −1]	12 [+7, −4]
NO_2flavH	64 [+7, −6]	>100	15 [+3, −2]	17 15 [+3, −3]
NO_2FflavH	>100	>100	16 15 [+2, −2]	15 15 [+3, −2]
56	>100	>100	>100	>100
57	45 [+4, −3]	47 [+7, −6]	8.3 [+1.2, −1.1]	21 [+3, −4]
58	55 [+11, −9]	91 [+39, −23]	89 [+12, −12]	73 [+20, −16]
59	13 [+2, −2]	17 [+2, −2]	6.8 [+1.0, −0.8]	8.9 [+3.6, −2.9]
60	44 [+5, −4]	88 [+15, −12]	>100	>100
61	17 [+3, −2]	26 [+4, −4]	14 [+1, −1]	17 [+11, −7]
62	>100	>100	72 [+15, −12]	48 [+8, −7]
63	15 [+3, −2]	22 [+4, −3]	5.2 [+0.7, −0.6]	6.9 [+1.2, −1.1]
64	n.e.	n.e.	n.e.	n.e.
65	14 [+4, −3]	23 [+5, −4]	6.3 [+1.0, −0.9]	7.3 [+1.3, −1.4]
cisplatin	3.2 [+0.7, −0.6]	5.6 [+1.3, −1.1]	1.8 [+0.2, −0.2]	2.4 [+0.7, −0.6]

n.e.—not evaluated; the original study did not provide a reason.

2.2. Acac-Type Coordination

The second part of our review study will focus on the acetylacetone (2,4-pentanedione, acac) type of coordination using the common acac ligand platform in coordination chemistry. Acac's ubiquity originates from its stability, easy synthesis, and the chelate effect. Typical acac chemistry relies on two types of reactions: (a) conversions with Lewis bases when acac coordinates via both oxygens; (b) electrophilic substitution due to the nucleophilic central carbon atom forming 3-substituted acac derivatives. Besides their significant role in coordination chemistry, they are key precursors in rapidly developing fields such as nanoparticle research, polymer science, and catalysis (both homo- and heterogeneous) [54].

In hydroxyflavone-based metal complexes exhibiting a 4-carbonyl/5-hydroxyl coordination mode, the ligand acts as a bidentate chelator, coordinating the metal center through two oxygen atoms located on adjacent aromatic rings, namely the carbonyl group at position C-4 on the C-ring and the hydroxyl group at position C-5 on the A-ring. This (O,O)-coordination mode results in the formation of a six-membered chelate ring.

The spectroscopic features associated with this coordination pattern largely mirror those observed in maltol-type (3-hydroxy-4-keto) complexes. However, the key distinction lies in the ^{1}H NMR spectrum, where the characteristic signal of the C-5 hydroxyl proton disappears, providing direct evidence of its involvement in metal coordination. In contrast to 3-OH–F complexes, where coordination primarily involves the C-ring and affects band I, the 4-carbonyl/5-hydroxyl coordination mode may also induce minor spectral shifts in band II, associated with electronic transitions in the A-ring. This subtle bathochromic or hypsochromic shift further supports the involvement of the A-ring 5-hydroxyl group in metal coordination.

The series of Co(III) complexes **56**–**68** synthesized by Kozsup et al. (Table 11), previously reviewed in Section 2.1, has been expanded with two additional complexes, **69** and **70** (Figure 13) featuring a chrysin moiety capable of acac-type coordination (Table 13) [45]. The coordination chemistry is well established, as seen in their maltol-type analogs, and is supported by NMR, IR, and HRMS characterization data. The crystal structure analysis of complex **69** confirmed the proposed coordination mode. While chrysin, the metal precursors, and **69** exhibited no cytotoxicity against A431 and A549 cell lines, complex **70**, obtained with [Co(tpa)Cl_2]Cl, demonstrated improved activity in both cell lines under the tested conditions. A previously spotted trend for more active compounds with tpa moiety was confirmed for acac complexes **69** and **70** as well.

Table 13. Numeration of Co(III) complexes **69** and **70** obtained by acac-type coordination and their cytotoxicity against A431 and A549 human cancer cell lines under normoxic and hypoxic conditions after 72 h incubation using end-point SRB assay. EC_{50} and 95% confidence interval (CI) values are presented in μM as a mean of three independent experiments [53].

Compd.	Abbr.	EC_{50}(μM)			
		A431		A549	
		Normoxia	Hypoxia	Normoxia	Hypoxia
chrysH	–	>100	>100	>100	>100
[Co(tren)Cl_2]Cl	–	>100	>100	>100	>100
[Co(tpa)Cl_2]Cl	–	>100	>100	>100	>100
[Co(tren)(chrys)](Cl)(ClO_4)	69	>100	>100	>100	>100
[Co(tpa)(chrys)]$(ClO_4)_2$	70	68 [+17, −13]	64 [+15, −12]	34 [+4, −3]	33 [+5, −4]
cisplatin	–	3.2 [+0.7, −0.6]	5.6 [+1.3, −1.1]	1.8 [+0.2, −0.2]	2.4 [+0.7, −0.6]

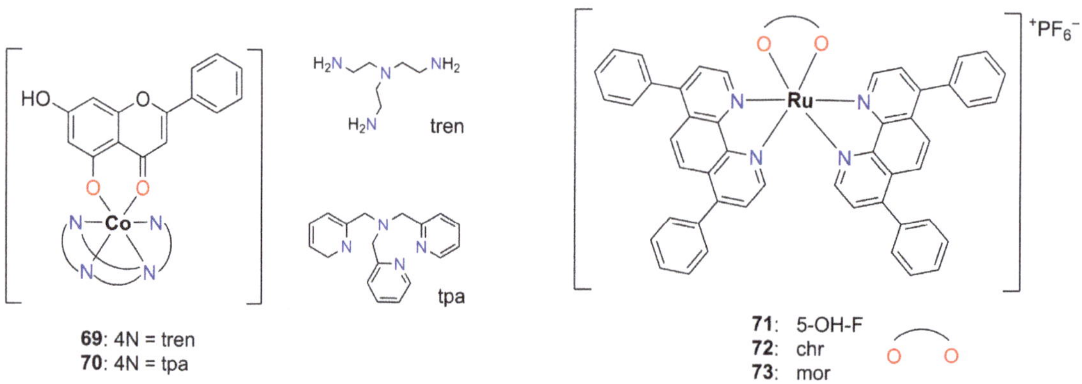

Figure 13. The chemical structures of Co(III) complexes **69** and **70**, as reported by Kozsup et al. [53], and Ru(II) representatives (**71–73**), as reported by Gasser's group [55].

In the same year, the research group of Gasser published a series of monocationic Ru(II) polypyridyl complexes **71–73** of the type [Ru$(DIP)_2$(OH–F)]X (Figure 13), containing 4,7-diphenyl-1,10-phenanthroline (DIP), OH–F (5-hydroxyflavone, chrysin, or morin) and X as the counterion, PF_6^-, and OTf^- [55]. These complexes were obtained by different synthetic routes which can be summarized as the following: all complexes were obtained starting from the metal precursor, Ru$(DIP)_2Cl_2$, in EtOH under N_2 and protected from light. Compound **71**, [Ru$(DIP)_2$(5-OHF)](PF_6), carrying 5-hydroxyflavone, involved direct coordination in the presence of NaOH. For the rest of the two complexes with chrysin (**72**, [Ru$(DIP)_2$(chr)](PF_6)), or morin (**73**, [Ru$(DIP)_2$(mor)](PF_6)), respectively, silver triflate was used prior to the addition of ligands. In the case of morin-based complex **73**, the ligand was silylated with tetramethylsilane (TMS) bromide in THF before complexation, though it later appeared to be unnecessary as TMS protecting groups were hydrolyzed during the complexation. The reactions for all compounds **71–73** involved refluxing (1.5–2 h), followed by cooling, solvent removal under a vacuum, and purification via DCM extraction, NH_4PF_6 precipitation, or Celite filtration, and preparative thin-layer chromatography (TLC) as needed. The final isolation steps included sequential washes with Et_2O and heptane, sonication, centrifugation, and vacuum drying to obtain the purified Ru(II) complexes in a form of a racemic mixture of Δ and Λ enantiomers. The chemical characterization included HRMS and NMR spectroscopy while their purity was checked using EA. Prior to the cytotoxicity study, the stability of **71–73** was confirmed by NMR in DMSO for 5 days.

The biological profile of Ru(II) complexes **71–73** was evaluated using a fluorometric cell viability assay on four human cancer cell lines MCF-7 (breast), hypopharyngeal squamous cell carcinoma (FaDU), metastatic breast cancer cell line (MDA-MB-435S), and U87 (glioblastoma), as well as on two non-cancerous cell lines, RPE-1 (retina) and embryonic kidney (HEK293) cells. Cisplatin and doxorubicin were used as positive controls (Table 14). Among the complexes' precursors, only chrysin was found to be active mainly against HEK293 (IC_{50} = 26.80 ± 2.79), while $[Ru(DIP)_2]Cl_2$ showed good activity on the RPE-1 cell line with IC_{50} = 3.13 ± 0.28. However, the coordination via Ru(II) significantly improved starting IC_{50} values for the majority of **71–73** complexes. None of the complexes **71–73** showed activity against the MCF-7 cell line. While they exhibited some activity against other tested cell lines, no selectivity for cancerous over normal cell lines was observed.

Table 14. Cytotoxicity results for reviewed Ru(II) **71–73** complexes obtained by acac-type coordination and their precursors after 48 h treatment using the fluorometric cell viability assay on tested human cell lines [55].

Compd.	IC_{50} (μM)					
	MCF-7	**FaDU**	**MDA-MB-435S**	**U87**	**RPE-1**	**HEK293**
5-OH–F	>100	>100	>100	>100	>100	>100
chr	62.59 ± 3.23	95.06 ± 11.55	79.37 ± 8.13	91.14 ± 13.76	>100	26.80 ± 2.79
mor	>100	>100	>100	>100	>100	>100
CDDP	19.69 ± 1.63	5.17 ± 0.21	17.62 ± 0.54	6.94 ± 0.46	39.9 ± 9.14	2.27 ± 0.67
Dox. *	9.39 ± 1.37	1.55 ± 0.18	5.55 ± 1.37	0.59 ± 0.03	14.90 ± 1.31	0.21 ± 0.03
prec. *	>50	>50	27.73 ± 5.33	25.59 ± 0.29	3.13 ± 0.28	12.11 ± 1.30
71	>50	38.21 ± 5.22	24.48 ± 1.92	30.72 ± 1.48	19.72 ± 8.23	24.46 ± 3.20
72	>50	>50	27.73 ± 5.33	25.59 ± 0.29	23.21 ± 8.08	33.02 ± 3.25
73	>50	>50	>50	>50	>50	>50

* Dox. doxorubicin; prec. precursor $[Ru(DIP)_2Cl_2]$.

Munteanu et al. studied four mononuclear lanthanide-containing complexes, **74–77**, with 5-hydroxyflavone and phen (Figure 14) [56]. A series of complexes containing Ln(III) metal ions were synthesized by reacting the corresponding metal chlorides, $LnCl_3$, with deprotonated 5-OH–F in the presence of triethylamine and an ethanolic solution of phen. The reactions were carried out under reflux, yielding yellow products with the general formula $[Ln(OH)_2(5\text{-OH–F})(phen)]\cdot nH_2O$, where n = 4, 3.5, 2, and 3 for Sm(III), Eu(III), Gd(III), and Tb(III), respectively. The physicochemical characterization of newly synthesized compounds included UV–Vis, IR, HRMS, EA, and TA. According to a DFT study, authors suggested a six-coordinated distorted octahedron geometry for all four complexes. The next step included the evaluation of the cytotoxic effects for both ligands and complexes on five human cancer cell lines (HeLa (cervix), colorectal adenocarcinoma (HT-29), colon adenocarcinoma (LoVo), MCF-7 (breast), and ovarian adenocarcinoma (SK-OV-3)) using cisplatin and doxorubicin as positive controls. Cell viability tests were performed by MTS assay within the incubation period of 48 h. Interestingly, obtained IC_{50} values indicated a relatively good cytotoxic activity for reported complexes, **74–77**, particularly against MCF-7 cancer cells (Table 15). In this case, IC_{50} values for **74–77** were remarkably close to doxorubicin, highlighting the Eu(III)-complex **75** as the most active one. Regarding the cytotoxic profile of the precursor compounds, 5-HO–F showed an enhanced cytotoxicity in HeLa cells in comparison to Ln(III) complexes, while the Sm(III)-complex **74** demonstrated a lower IC_{50} in comparison to cisplatin (13.78 vs. 31.55 μM). In parallel, 5-HO–F was not active on HT-29 cells in contrast to **74–77** which were more potent than cisplatin. Moreover, in the cases of LoVo and SK-OV-3, the investigated complexes **74–77** displayed better activity in comparison to free ligands and cisplatin, especially against LoVo. Therefore, the authors chose HeLa and LoVo cell lines to proceed with the biological profiling that involved apoptotic effects, DNA-binding, and serum-protein-binding studies.

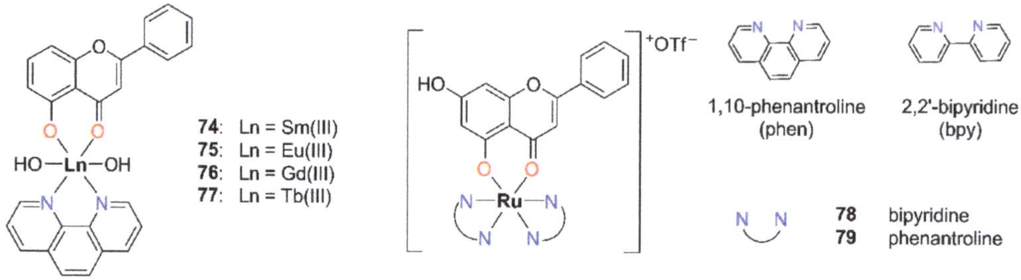

Figure 14. The proposed structures of Ln(III) complexes **74–77**, as reported by Munteanu et al. [56], and Ru(II) complexes **78–79**, as reported by Zahirović et al. [50].

Table 15. Cytotoxicity results for reviewed $[Ln^{III}(OH)_2(5\text{-}OH\text{-}F)(phen)]\cdot nH_2O$ complexes **74–77** obtained by acac-type coordination and their precursors after 48 h treatment using MTS assay on five cancer human cell lines: HeLa, HT-29, LoVo, MCF-7, and SK-OV-3 [56].

Compd.	Ln(III)	IC_{50} (μM)				
		HeLa	HT-29	LoVo	MCF-7	SK-OV-3
5-HO–F	–	5.94 ± 1.77	118.81 ± 16.93	33.67 ± 6.46	30.43 ± 5.85	60.27 ± 5.03
phen	–	50.99 ± 6.51	7.81 ± 1.89	33.86 ± 5.15	7.59 ± 3.25	60.11 ± 7.97
74	**Sm**	13.78 ±1.51	24.71 ± 4.52	8.22 ± 2.01	7.62 ± 3.68	36.28 ± 3.73
75	**Eu**	36.55 ± 3.30	17.24 ± 2.41	11.32 ± 2.44	2.50 ± 2.18	43.44 ± 3.69
76	**Gd**	31.75 ± 2.74	35.18 ± 6.44	17.63 ± 3.84	3.35 ± 0.61	41.03 ± 3.99
77	**Tb**	31.55 ± 2.76	20.47 ± 2.75	27.55 ± 3.84	3.60 ± 0.46	31.13 ± 2.93
cisplatin	–	24.06 ± 2.51	53.62 ± 4.68	30.47 ± 5.36	n.e.	50.85 ± 4.32
doxorubicin	–	n.e.	n.e.	n.e.	3.27 ± 0.45	n.e.

n.e.—not evaluated

Moreover, all synthesized complexes demonstrated potential as DNA intercalators. This claim was supported by UV–Vis spectroscopy and various fluorescence-based assays (EB displacement assay, SYBR Gold® fluorescence quenching assay, and Alexa488 fluorescence quenching assay) and viscosimetric studies. These findings provided insights into binding dynamics, confirming effective DNA binding at micromolar concentrations highlighting the Sm(III)-complex **74** as the compound with the fastest binding kinetics and the highest activity. Molecular docking simulations further supported the intercalative binding mode. In addition to DNA interaction, the lanthanide complexes displayed effective binding to serum transport proteins, with a generally higher affinity for human serum albumin (HSA) compared to transferrin (Tf). Importantly, the authors underlined the significant role of structural flexibility in macromolecular interactions, as binding to DNA or proteins was accompanied by distinct conformational rearrangements. Overall, this study introduces four novel Ln(III) complexes capable of targeting multiple cellular components indicating how the integration of multiple (hetero)cyclic motifs and the adaptable coordination environment of trivalent lanthanide ions present a promising platform for the development of multimodal anticancer agents.

As previously discussed in the section on maltol-type coordination, Zahirović et al. synthesized mixed-ligand complexes with Ru(II) [50]. This time, complexes **78** and **79**, incorporating the hydroxyflavone chrysin, which lacks a 3-hydroxyl group, formed six-membered chelate rings (Figure 14).

Cytotoxicity was evaluated using the MTT assay after 72 h of treatment, and the results are presented in Table 16. Complex **78**, containing bpy, exhibited the best activity against the MCF-7 cell line, while complex **79**, bearing phen, showed the highest cytotoxicity toward the HeLa cell line.

Table 16. Cytotoxicity of complexes **78–79** determined by MTT assay after 72 h treatment expressed as (mean ± SD, n = 3) [50].

Compd.	**IC_{50} (μM)**			
	SW620	**HepG-2**	**MCF-7**	**HeLa**
chrysin	43.01 ± 16.34	85.81 ± 51.93	63.81 ± 25.45	33.82 ± 9.95
***cis*-[Ru(bpy)$_2$Cl$_2$]·2H$_2$O**	>100	4.53 ± 60.11	2.10 ± 70.64	>100
[Ru(phen)$_2$(CO$_3$)]·2H$_2$O	>100	>100	92.38 ± 44.00	>100
78	62.78 ± 19.36	45.02 ± 6.67	12.55 ± 18.18	26.80 ± 1.21
79	53.39 ± 36.52	43.26 ± 21.31	43.67 ± 37.06	8.42 ± 20.23

The antimicrobial activity of complex **78** is approximately two to three times lower than that of standard antibiotics, except in the case of MRSA, where its efficacy is comparable to that of gentamicin (Table 17). Compound **79** exhibits roughly half the antibacterial activity of gentamicin against *Acinetobacter baumannii* ATCC BAA-747, yet demonstrates slightly higher activity than the parent flavonoid, chrysin.

Table 17. Antimicrobial activity of free chrysin and its ruthenium (II) complexes [Ru(bpy/phen)$_2$(chrys)](OTf)·nH$_2$O, evaluated using the disk diffusion method and expressed as inhibition zone diameters (mm) [50].

Compd.	*Staphylococcus aureus* ATCC 25923	*Enterococcus faecalis* ATCC 19433	*Streptococcus* Beta-Hemolytic Group A	Methicillin-Resistant *Staphylococcus aureus*	*Klebsiella pneumoniae* ATCC 1705	*Acinetobacter baumannii* ATCC—BAA 747	*Pseudomonas aeruginosa*	*Escherichia coli*	*Candida albicans*
	G+	G+	G+	G+	G−	G−	G−	G−	Fungus
	Diameter of Inhibition Zone (mm)								
chrysin	*	*	*	*	*	14	*	*	14
***cis*-[Ru(bpy)$_2$Cl$_2$]**	*	*	*	*	*	15	*	*	14
[Ru(phen)$_2$CO$_3$]	*	*	*	*	*	15	11	*	15
78	15	*	15	16	*	13	*	*	17
79	*	*	*	*	*	17	*	*	*
vancomycin	27	26	35	-	-	-	-	-	-
gentamicin	-	-	-	20	25	35	36	21	-
nystatin	-	-	-	-	-	-	-	-	28

* No activity, - not tested, G+ = Gram-positive bacteria, and G− = Gram-negative bacterial strain.

In 2019, Marques et al. contributed to the field of Ru(II) metal complexes with OH–F motifs coordinated in acac mode by reporting two Ru(II) trithiacyclononane ([9]aneS$_3$) complexes, **80–81**, presented in Figure 15 [57]. Ru(II) complex **80** was synthesized by reacting chrysin with one equivalent of [Ru(II)([9]aneS$_3$)Cl$_2$(S-DMSO)] and half an equivalent of tetrabutylammonium hydroxide. The light yellow solution was refluxed in methanol for 24 h. Upon evaporation to half its volume, the mixture yielded both complex **80** and an unreacted metal precursor. The pure form of **80** was obtained by filtration of the supernatant, followed by volumization with an additional portion of diethyl ether. After one week at 4 °C, complex **80** ([Ru(II)([9]aneS$_3$)(chrys)(*S*-DMSO)]Cl) was isolated as a dark orange solid, washed with cold ethanol and diethyl ether, and dried. Prior to the synthesis of the second complex (**81**), the ligand precursor 5-hydroxy-7-methoxyflavone (tectochrysin, tchr) was prepared by reacting chrysin with potassium carbonate (1:3 molar ratio), followed by the addition of dimethyl sulfate. The reaction mixture was refluxed in acetone for one hour, yielding a yellow precipitate that was purified via column chromatography and recrystallization from a dichloromethane/methanol mixture. In the following synthetic step, complex **81**, also featuring acac-type coordination, was synthesized by reacting tchr with [Ru(II)([9]aneS_3)Cl_2(S-DMSO)] in a 1:1 molar ratio under a 24 h reflux in methanol, with the addition of tetrabutylammonium hydroxide. After cooling, diethyl ether and dichloromethane were added to isolate the dark orange oily product. To obtain a solid form of **81**, several recrystallization steps were performed in dichloromethane, followed by washing with diethyl ether. Unlike complex **80**, which was obtained in relatively low yields (23%, respectively), **2** was synthesized in a slightly improved yield of 43%. Both compounds (**80** and **81**) were characterized by IR, ^{1}H, and ^{13}C NMR spectroscopy, HRMS, and EA.

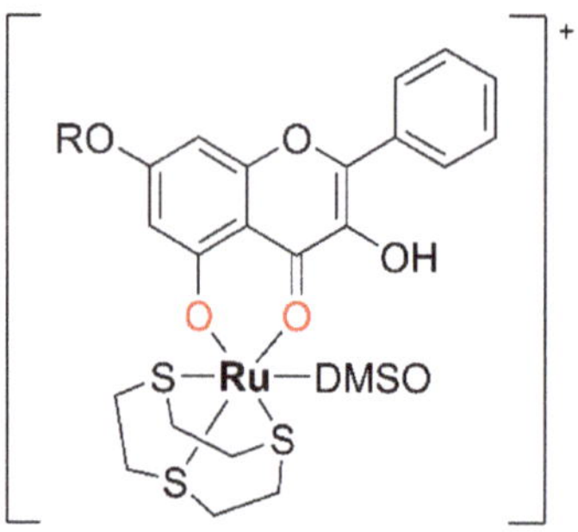

80: R = H
81: R = CH_3

Figure 15. The proposed chemical structures of Ru(II) trythiacyclononane complexes **80–81**, featuring acac-type coordination, as reported by Marques et al. [57].

The cytotoxic effects of Ru(II) trithiacyclononane ([9]aneS_3) complexes **80–81** and their precursor compounds (chr and tchr) were evaluated against three cancer types using four human cancer cell lines: osteosarcoma cell line (MG-63), PC-3 (prostate), MCF-7 (breast), and MDA-MB-231 (breast) (Table 18). Cell viability was assessed using the MTT assay after a 72 h incubation period, with cisplatin as a positive control. Interestingly, the pure ligands exhibited stronger cytotoxicity, particularly against osteosarcoma cells, with the 5-methoxy derivative of chr showing the most promising effect. This finding is especially valuable given the limited availability of effective chemotherapeutic options for osteosarcoma. Moreover, OH–F compounds are generally associated with milder side effects, adding further therapeutic relevance to these results. Simultaneously, complexes **80** and **81** displayed mild cytotoxic activity, which is rather modest compared to the numerous previous studies reporting enhanced bioactivity upon metal complexation [58].

Table 18. The cytotoxicty of reviewed Ru(II) complexes **80–81** obtained by acac-type coordination against MG-63, PC-3, MCF-7, and MDA-MB-231 after 72 h of incubation using MTT assay and cisplatin as a positive control [57].

Compd.	IC_{50} (μM)			
	MG-63	PC-3	MCF-7	MDA-MB-231
chr	41.0	15.7	29.5	17.1
80	>200 [a]	146.2	>200 [a]	180.6
tchr	67.9	32.8	113.5	44.8
81	>200 [a]	>200 [a]	>200 [a]	>200 [a]
cisplatin	4.5	6.6	13.8	9.5

[a] Growth inhibition at 200 μM, the highest concentration tested, was lower than 50%.

2.3. Catechol-Type Coordination

Wherever two vicinal hydroxyl groups are present within the OH–F molecule, catechol-type coordination to a metal center may occur, leading to the formation of a five-membered chelate ring. Upon coordination, the deprotonation of the hydroxyl groups is expected, which can be observed as the disappearance of the corresponding signals in the ^{1}H NMR spectrum, along with characteristic changes in the UV–Vis spectrum.

A representative example of catechol-type coordination is described in the work of Naso et al., which reports an oxidovanadium(IV) complex with luteolin, $[VO(lut)(H_2O)_2]Na{\cdot}3H_2O$ (compound **82**; Figure 16) [59]. The background behind designing such a complex originates from both the favorable biological profile of luteolin, known for its antioxidant, anti-inflammatory, antimicrobial, and anticancer activities [60], and the well-documented therapeutic potential of oxidovanadium(IV) complexes with other flavonoids [61–63]. Complex **82** was synthesized by reacting $VOCl_2$ with luteolin in a 1:2 molar ratio in ethanol, with the pH of the reaction mixture adjusted to five using 1 M NaOH. After 3 h of reflux, a dark green precipitate formed, which was isolated by filtration, washed with ethanol, and dried to yield the pure compound. The proposed structure of **82** was based on EA, UV–Vis spectroscopy, and TGA. Electron paramagnetic resonance (EPR) spectroscopy revealed the presence of two coordinated water molecules and two phenolate (ArO^-) groups in the metal's coordination sphere. Moreover, EPR data indicated the formation of chains involving adjacent VO units, as supported by calculated exchange pathways. Prior to biological evaluation, the stability of **82** was confirmed by monitoring its electronic absorption spectra in DMSO under N_2 atmosphere, which remained unchanged over time, indicating good solution stability.

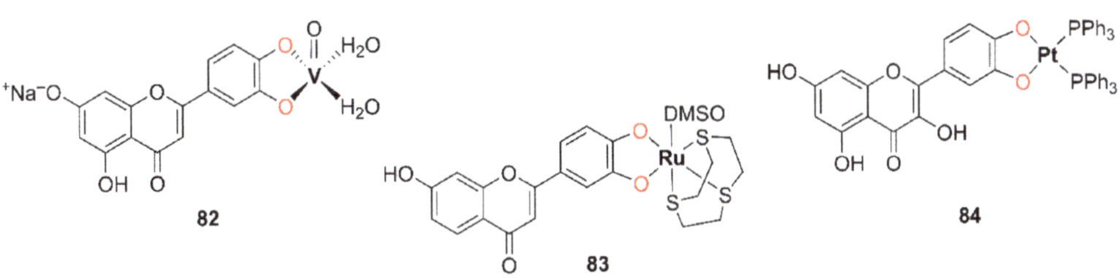

Figure 16. The proposed chemical structures of catechol-type coordinated hydroxyflavones: oxidovanadium(IV) complex **82** with luteolin, as reported by Naso et al. [59], and Ru(II) complex **83** with 7,3′,4′-OH–Fand trythiacyclononane, as reported by Marques et al. [57], and Pt(II)complex **84** with quercetin and triphenylphosphine (PPh_3), as reported by Dell'Anna et al. [37].

Cell viability assays were conducted using the crystal violet assay for A549 (lung) cells and the MTT assay for MDA-MB-231 (breast) cells (Table 19). For the A549 cell line, cells were incubated at 37 °C in a growth medium for 24 h, after which the medium was replaced with fresh solutions of luteolin and complex **82**. Following additional 24 h incubation, cells were stained with crystal violet. The cytotoxicity evaluation revealed comparable IC_{50} values for both compounds, 66.3 μM for luteolin, and 60.5 μM for complex **82**. It is worth noting that significantly lower IC_{50} values have been reported in other studies [64].

Table 19. Studied oxidovanadium(IV) complex **82** obtained by catechol-type coordination and its cytotoxicity against two cell lines (A549 using crystal violet assay and MDA-MB-231 using MTT assay) [59].

Compd.	IC_{50} (μM)	
	A549	**MDA-MB-231**
luteolin	66.3	88.3
82	60.5	17.0

In the next step, the MTT assay was applied to MDA-MB-231 cells treated with varying concentrations (0–100 μM) of luteolin and complex **82**. After 24 h of incubation, luteolin exhibited an IC_{50} value of 88.3 μM, consistent with previously reported data. In contrast, its oxidovanadium(IV) complex, **82**, demonstrated significantly enhanced cytotoxicity, with an IC_{50} of 17.0 μM. According to the authors of this study, the obtained results underscore the role of metal coordination in enhancing the anticancer activity of luteolin, as previously reported for vanadium complexes [65].

In addition to inducing cellular reactive oxygen species (ROS) generation and mitotic arrest, both luteolin and its oxidovanadium(IV) complex (VOlut) cause damage to the cytoplasmic and nuclear membranes. Their binding to bovine serum albumin (BSA) involves a combined quenching mechanism, with static quenching being the dominant process. Thermodynamic parameters, derived from the Van't Hoff equation, revealed that the interactions are spontaneous. The luteolin–BSA complex is primarily stabilized by electrostatic forces, whereas the VOlut–BSA interaction is mainly driven by hydrogen bonding and van der Waals forces. Given the confirmed cytotoxic effects of both luteolin and **82**, the complex shows potential as a promising candidate for further in vivo evaluations in cancer treatment studies.

Beside two acac-type coordinated complexes, Marques et al. (reference [57] in previous Section) prepared a catechol-type Ru(II) complex **83** (Figure 16). It was synthesized by reacting 7,3′,4′-OH–F with the metal precursor [Ru([9]aneS_3)Cl_2(DMSO)] in a 1:1 molar ratio, in the presence of one equivalent of tetrabutylammonium hydroxide in ethanol. The reaction mixture was refluxed for 48 h, during which the solution changed color from light yellow to orange and finally to the dark green solid product. Complex **83** was isolated by washing with ethanol and diethyl ether, then dried. According to the cytotoxicity test, complex **83** was inactive, likely due to its limited water solubility (Table 20).

Table 20. The cytotoxicty of reviewed Ru(II) complex obtained by catechol-type coordination against MG-63, PC-3, MCF-7, and MDA-MB-231 after 72 h of incubation using MTT assay and cisplatin as a positive control [57].

Compd.	IC_{50} (μM)			
	MG-63	**PC-3**	**MCF-7**	**MDA-MB-231**
7,3′,4′-OH–F	38.1	23.2	36.9	24.8
83	>18.5 [a]	>18.5 [a]	>18.5 [a]	>18.5 [a]
cisplatin	4.5	6.6	13.8	9.5

[a] An amount of 18.5 was the highest concentration tested for **84** due to its low solubility.

As previously mentioned in Section 2.1, Dell'Anna et al. [37] synthesized Pt(II) complexes with OH–F and PPh_3. One of the complexes (**84**), shown in Figure 16, features quercetin coordinated in a catechol-type fashion. The complex was synthesized using the same procedure as described for the maltol-type complex, with the exception that the precipitate was obtained using *n*-hexane instead of *n*-pentane. In this case, slow diffusion of *n*-pentane into a THF solution containing the respective complex over a period of two weeks yielded crystals suitable for XRD analysis. The complex crystallizes in the orthorhombic crystal system. Probably due to the multiple possible coordination modes of quercetin, complex **84** was obtained with approximately 5% impurities, which the authors were unable to separate. Consequently, the cytotoxicity of the complex was not evaluated.

2.4. Mixed-Type Coordination

The combination of both acac- and catechol-type coordination modes within the same complex molecule can lead to so-called mixed coordination. The observed spectral changes will reflect a combination of features characteristic of both acac- and catechol-type coordinated complexes.

One of the earliest examples of mixed-type coordination was reported by Ikeda et al., who described the synthesis and characterization of a rutin–zinc (II) flavonoid–metal complex (Figure 17) [35]. The rutin–Zn(II) complex **85** was synthesized by the slow addition of an aqueous solution of $[Zn(CH_3COO)_2]\cdot 2H_2O$ to a methanolic solution of dehydrated rutin. The reaction mixture was stirred at temperatures up to 40 °C for 24 h, after which the resulting complex **85** was isolated by filtration, rinsed with methanol, and dried. Structural characterization of compound **85** was carried out using IR spectroscopy, ^1H NMR, UV–Vis spectroscopy, and EA. The proposed structure (Figure 17) features a mixed coordination mode, involving both catechol-type coordination via the 3′,4′-dihydroxy groups of the B-ring and acac-type coordination via the 4-keto and 5-hydroxy groups of the C-ring.

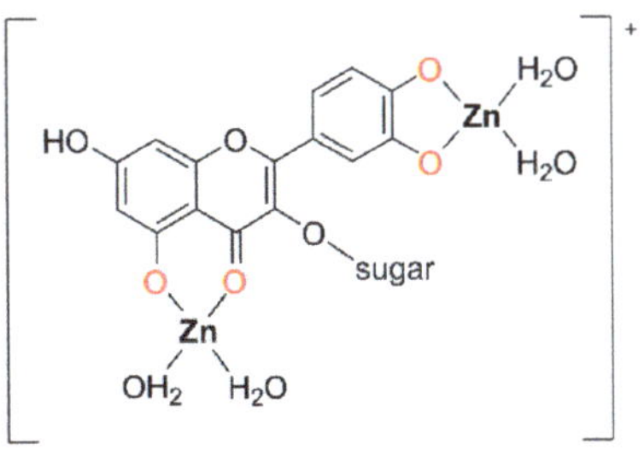

85

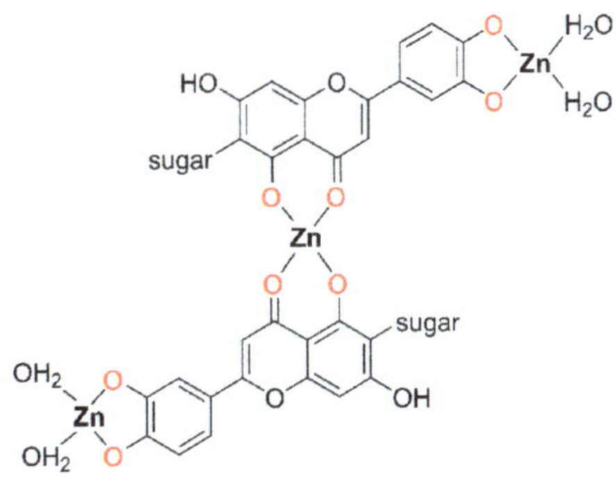

86

Figure 17. The proposed chemical structures of Zn(II) complexes with mixed-type coordination mode: the cationic complex **85** with rutin, reported by Ikeda et al. [35], and the neutral complex **86** with isoorientin, reported by Wang et al. [36].

This dual chelation stabilizes the complex and is typical of flavonoid–metal interactions.

The cytotoxicity study was conducted on a panel of cancer cell lines, including murine melanoma (B16F10, derived from a C57BL/6J mouse), human skin melanoma (SK-MEL-28), acute myelogenous leukemia (KG-1), multiple myeloma (RPMI 8226), T-cell leukemia (Jurkat), and chronic myelogenous leukemia (K562), as well as on normal human cell lines, namely fibroblasts and umbilical vein endothelial cells (HUVECs). Cells were treated with compound **85** and its precursor rutin over a concentration range of 17.2–275.6 μM for 24 h.

Cell viability was then assessed using the MTT assay (Table 21). While rutin alone exhibited no cytotoxic effects on either cancerous or healthy cells, and was therefore not included in the IC_{50} analysis, compound **85** showed significant cytotoxic activity against all tested tumor cell lines.

Table 21. Studied Zn(II) complex **85** obtained by mixed-type coordination and its cytotoxicity assessed by MTT assay against eight cell lines after 24 h incubation [35].

Cell Line	IC_{50} (μM)
B16F10	160.7
SK-MEL-28	194.0
KG-1	91.4
RPMI 8226	196.6
Jurkat	150.2
K562	173.2
Fibroblasts	>275.6
HUVEC	>275.6

Importantly, **85** displayed favorable selectivity, as no cytotoxic effect was observed on normal (non-cancerous) cells. Although the compound demonstrated its strongest anticancer activity against the KG-1 leukemia cell line, it is worth noting that the study did not include a comparison with a standard positive control, which limits the interpretation of its relative potency.

In addition, Ikeda et al. reported in vivo toxicological studies in female mice. Both rutin and **85** were administered intraperitoneally (i.p.) for 7, 14 and 21 days. The toxicity was observed for 30, 60, and 90 days after the initial treatment following the parameters such as body weight measurement, hematological and biochemical analyses, and histological examination of the liver and kidneys. After 21 days of treatment, mice treated with rutin and **85** showed increased platelet, red blood cell, and white blood cell counts, which returned to normal after 90 days. A histopathological analysis revealed no abnormalities in the liver, kidneys, or lungs compared to controls. Mitochondrial membrane potential ($\Delta\psi m$) was evaluated by flow cytometry using Rhodamine 123. Significant loss of $\Delta\psi m$ was observed in Ehrlich ascites carcinoma (EAC) cells treated with paclitaxel or its combinations with rutin and Rutin–zinc (II), although the single compounds had milder effects.

In vivo studies in EAC-bearing mice showed that **85** induced stronger DNA fragmentation in tumor cells than rutin alone. When combined with paclitaxel, both compounds significantly enhanced DNA fragmentation. These effects were associated with the activation of caspase-dependent apoptosis, reduced cyclin D1 and vascular endothelial growth factor (VEGF) expression, and upregulation of caspases-8 and -3. Importantly, the combination of the Rutin–zinc (II) complex, **85**, with paclitaxel exhibited synergistic antitumor activity, reduced anemia and myelosuppression, common side effects of chemotherapy, and increased the survival of tumor-bearing mice.

In one of their recent studies, Wang et al. synthesized and characterized a trinuclear Zn(II) complex **86** (Figure 17) using OH–F isoorientin (iso) as a ligand [36]. This natural bioactive flavonoid C-glycoside was selected due to its broad biological activity, including anticancer and antioxidant properties [66], as well as antibacterial activity [67]. The complex was synthesized by reacting the iso ligand in ethanol with $Zn(CH_3COO)_2{\cdot}2H_2O$. After adjusting the pH to approximately eight, the reaction mixture was stirred and refluxed for 8 h. Upon cooling, a brown-yellow precipitate formed, which was isolated, washed with ethanol, dried, and stored at 4 °C. Comprehensive characterization of the complex was performed using ^{1}H NMR and IR spectroscopy, EA, and TGA. To further support the proposed structure, UV–Vis spectroscopy, inductively coupled plasma atomic emission spectrometry (ICP-AES), and SEM were also employed.

The biological profiling of **86** was initiated with antibacterial activity assays, which included measurements of the bacteriostatic zone diameter, bacteriostatic rate, and bacterial growth curves. Antibacterial effects were evaluated against *Escherichia coli* and *Staphylococcus aureus* using the disk diffusion method. Initial results revealed stronger antibacterial effects of both **86** and its precursor iso against *S. aureus* compared to *E. coli*, across the same concentration range (0, 200, 400, 600, and 800 μg/mL, Table 22). Notably, **86** exhibited higher bacteriostatic rates against both bacterial strains compared to the free ligand. Additionally, a concentration-dependent increase in the diameter of the bacteriostatic zone was observed for both compounds, with **86** consistently outperforming iso (Table 23). These findings suggest that coordination with Zn(II) enhances the antibacterial efficacy of isoorientin, resulting in a stronger bacteriostatic effect. The next phase of the antibacterial study assessed the effects of iso and **86** on bacterial cell membrane integrity using SEM and transmission electron microscopy (TEM) techniques. Morphological and ultrastructural analyses revealed visible alterations in cell shape and clear signs of membrane damage, primarily induced by compound **86** and, to a lesser extent, iso. The main indication of this membrane disruption was the leakage of cytoplasmic contents, accompanied by an increase in electrical conductivity in the surrounding medium. This effect was attributed to the release of proteins and nucleic acids, as well as the continuous efflux of intracellular electrolytes such as K^+ and Na^+ ions. The observed damage was more pronounced in cells treated with complex **86**, suggesting its superior antibacterial efficacy. This enhancement is likely due to structural and property changes upon Zn(II) coordination, which may increase the compound's ability to interact with bacterial proteins and DNA. Finally, the authors suggested the influence of a synergistic effect, meaning that the combination of the ligand (iso) and the metal (**Zn**) works better than either component alone.

Table 22. Summarized inhibition zone diameters and inhibition rates of isoorientin (iso) and complex **86** on *E. coli* and *S. aureus* [36].

Compd.	Concentration (μg/mL)	*E. coli*		*S. aureus*	
		Diameter of Bacteriostatic Circle (mm)	Bacteriostatic Rate (%)	Diameter of Bacteriostatic Circle (mm)	Bacteriostatic Rate (%)
iso	0	6.03 ± 0.23	0.00	6.06 ± 0.12	0.00
	200	6.08 ± 0.15	0.82	8.29 ± 0.31	26.90
	400	8.10 ± 0.24	25.56	10.83 ± 0.29	44.04
	600	11.13 ± 0.27	45.82	12.62 ± 0.14	51.98
	800	13.29 ± 0.25	54.62	14.16 ± 0.35	57.20
86	0	6.04 ± 0.28	0.00	6.07 ± 0.17	0.00
	200	7.69 ± 0.16	21.46	10.66 ± 0.32	43.06
	400	9.46 ± 0.18	36.62	12.24 ± 0.27	50.41
	600	11.97 ± 0.39	49.54	13.91 ± 0.36	56.36
	800	15.37 ± 0.44	60.70	16.59 ± 0.21	63.41

The cytotoxicity evaluation included both a hemolysis assay and an MTT assay for cell viability. The hemolysis assay was performed following the protocol described by Liu et al. in 2009 [68], using a microplate reader for absorbance measurements. The results provided clear evidence that both **86** and iso inhibited erythrocyte hemolysis. As lipid peroxidation of the erythrocyte membrane is typically induced by hydrogen peroxide (H_2O_2), the authors observed that **86** exhibited low cytotoxicity even after H_2O_2 exposure, indicating its favorable safety profile for potential therapeutic applications. This observation was further supported in the final stage of **86** biological profiling, which involved assessing cell viability using the MTT assay, following a slightly modified protocol based on Yuan et al. [69]. The experiments were conducted on normal human liver cells (HL-7702) and the results showed no significant difference in viability between the **86**-treated group and the untreated control, indicating low cytotoxicity toward normal hepatocytes. Overall, the findings of this study classify the **86**-type of compounds as primarily antibacterial agents rather than for anticancer compounds.

Table 23. MIC values of studied compounds iso and **86** against *E. coli* and *S. aureus* [36].

Compd.	Bacteria	Concentration (μg/mL)					
		0	50	100	200	400	800
iso	*E. coli*	++	++	+	+	-	-
	S. aureus	++	+	+	-	-	-
86	*E. coli*	++	++	+	-	-	-
	S. aureus	++	+	-	-	-	-

+ Indicates small amount of bacterial growth. ++ Indicates mass bacterial growth. - Indicates no bacterial growth.

2.5. Linear-Type Coordination

A characteristic feature of this type of complex is coordination through a single hydroxyl group of the OH–F ligand.

In 2019, Mármol et al. described the synthesis and a full characterization of flavone-based ligands bearing propargyl ether group and their corresponding phosphane gold(I) complexes (Figure 18) as promising antiproliferative agents for colon cancer [33].

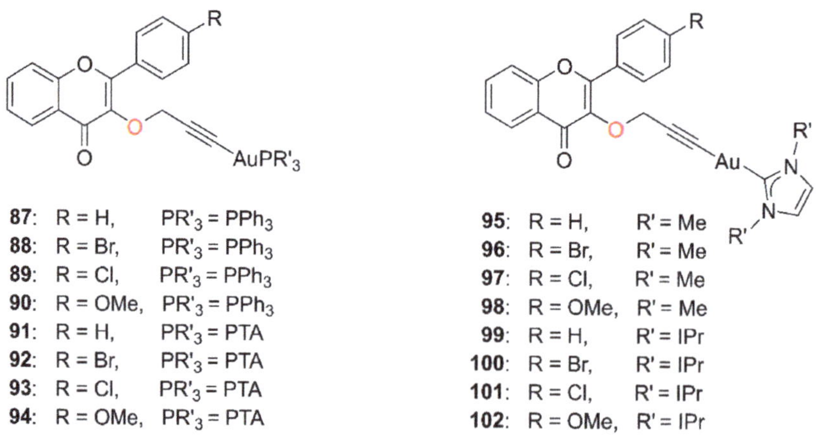

Figure 18. Chemical structures of Au(I) hydroxyflavone complexes **87–94** bearing phosphane ligands, and complexes **95–102** featuring N-heterocyclic carbene derivatives, as reported by Mármol et al. [33,34].

A two-step synthesis of 3-OH–F-derivatives followed a slightly changed synthetic path of Gunduz et al. [70] when 2-hydroxyacetophenone was reacted with benzaldehyde or the different para-substituted benzaldehydes in a 1:1 molar ratio under base conditions and a 3 h reflux. The obtained derivatives were further derivatized with propargyl bromide to obtain four new ligands bearing a propargyl ether group. Au(I) complexes **87–94** were synthesized in MeOH after 20 h of stirring obtained ligands with the metal precursor, [AuCl(PR′$_3$)] (PR′$_3$ = PPh$_3$, PTA), at room temperature. The characterization of novel compounds included NMR, IR, and XRD analysis for the OH–F derivative-bearing OMe group, while their purity was confirmed by EA. Their stability under physiological conditions (in phosphate-buffered saline (PBS) buffer at pH = 7.4 and 37 °C) was confirmed by UV–Vis absorption spectroscopy. The structural optimization of these flavonoid-inspired compounds, known for their pronounced lipophilicity, involved the introduction of a propargyl ether unit followed by coordination with a Au–phosphane moiety. This modification resulted in a well-balanced lipophilic–hydrophilic profile, with improved lipophilicity (log P) values ranging from 0.23 to 1.2.

Prior to the biological study, authors evaluated the interaction of Au(I) species with BSA adding the increasing amounts of **87–94** to the BSA solution. Using fluorescence emission spectroscopy, no emission was noticed in the range from 310 to 400 nm for **87–94**. In addition, the binding constants and the number of binding sites for each experiment were calculated from fluorescence intensity data, identifying the hydrophobic cavity of site I in subdomain IIA of BSA as the most probable binding location.

Cell culture experiments were performed on the human colorectal adenocarcinoma subline (Caco-2/TC7), as well as on the MCF-7 (breast), HepG2 (liver), and differentiated Caco-2 cells (noncancerous colon model), in the concentration range of 0–100 μM, using auranofin and cisplatin as positive controls. As the literature reports that flavones-based compounds can influence MTT assay, Mármol et al. primarily tested the toxicity of 3-OH–F and its derivatives on undifferentiated Caco-2/TC7 cells using both MTT and SRB assays. Mármol's study combined IC_{50} for four different 3-OH–F ligands obtained by MTT and SRB indicating that IC_{50} values from SRB were 1.2 to 5 times lower in comparison to those obtained by MTT. In addition, the SRB results were more reproducible, making this technique more reliable. In contrast, this type of behavior was not observed for eight Au(I) complexes **87–94** at concentrations below 20 mM with the exception of **90** (Table 24). MTT and SRB tests resulted in similar dose–response curves and comparable IC_{50} values. Since all Au(I) complexes **87–94** showed a reduced cell proliferation for all cell lines, the study clearly demonstrated the positive impact of metal coordination. Therefore, complexes **88** and **93**, with the strongest anticancer potential and selectivity, emerged as the most promising candidates for further biological evaluations. The first step included the measurement of cyclooxygenase-1 and cyclooxygenase-2 (COX-1/2) activity, keeping in mind the flavones affinity toward cyclooxygenase [71], and the analysis of redox enzymes thioredoxin reductase (TrxR) and glutathione reductase (GR) inhibition. In both experiments, it was confirmed that Au(I) species interact with COX-1/2 and TrxR and GR redox enzymes, proceeding to ROS increase on CaCo-2 cell lines. At the stage when biological targets were identified, the type of a cell death was examined for both complexes **88** and **93**. The experiments revealed that cell death occurred through the intrinsic apoptotic pathway and was accompanied by changes in cell cycle progression. Toxicity evaluation on differentiated Caco-2 cells, used as a model for noncancerous tissue, suggested that flavone-inspired gold complexes **88** and **93** may preferentially target cancer cells over normal differentiated cells, making them promising candidates for cancer treatment.

The same research group expanded the series of Au(I) complexes bearing N-heterocyclic carbene (NHC) derivatives and an OH–F moiety functionalized with a propargyl ether group (Figure 18) [34]. Mármol et al. reported eight [AuCl(NHC)]-type complexes (**95–102**), where NHC stands for 1,3-bis(2,6-diisopropylphenyl)imidazolidin-2-ylidene (IPr; $C_{27}H_{38}N_2$) or 1,3-dimethylimidazolidin-2-ylidene (IMe; $C_5H_{10}N_2$). The compounds **95–102** were synthesized by reacting four different previously obtained 3-OH–F derivatives with [AuCl(NHC)] in methanol in the presence of two equivalents of KOH as the base. The reaction mixture was stirred for 8 h at room temperature, after which the solvent was evaporated. The pure products were obtained as white, stable solids following a work-up procedure that included the addition of dichloromethane and diethyl ether. The next step involved a comprehensive characterization of complexes **95–102** using NMR, IR, HRMS, and EA. Subsequently, the shake-flask method combined with UV–Vis spectroscopy was used to determine their distribution coefficients. Based on the results (log P: 0.34–1.96 for IMe; 0.42–3.4 for IPr), the complexes were classified as lipophilic species. Given that chemical stability under physiological conditions is essential for upcoming biological studies, Mármol et al. conducted stability tests in PBS at pH 7.4 and 37 °C over a 24 h period. UV–Vis monitoring of the solutions revealed no spectral changes in the main absorption bands (λ = 210–350 nm), nor any noticeable red or blue shifts. These findings confirmed that complexes **95–102** remain stable under the tested experimental conditions.

Table 24. Cytotoxicity results for reviewed Au(I) **87–94** complexes obtained by linear-type coordination after 72 h treatment using MTT assay on human cell lines undifferentiated Caco-2/TC7, MCF-7 and HepG2 cancer cell lines, and differentiated Caco-2 cells (noncancerous model) compared with auranofin and cisplatin [33].

Compd.	R	PR'_3	IC_{50} (μM)			
			Caco-2/TC7	MCF-7	HepG2	Dif Caco-2
87	H	PPh_3	3.81 ± 1.18	2.08 ± 0.17	32.34 ±4.27	131.30 ± 38.54
88	Br	PPh_3	1.52 ± 0.91	13.87 ± 0.78	3.38 ± 0.07	43.89 ± 0.02
89	Cl	PPh_3	5.34 ± 0.05	8.99 ± 3.89	34.14 ± 5.88	114.38 ± 9.24
90	OMe	PPh_3	4.78 ± 0.62	3.40 ± 0.85	47.97 ± 5.32	75.45 ± 11.28
91	H	PTA	7.68 ± 1.74	18.49 ± 0.90	10.84 ± 0.67	18.25 ± 1.34
92	Br	PTA	6.42 ± 0.35	8.80 ± 3.14	11.25 ± 0.63	24.39 ± 0.86
93	Cl	PTA	2.33 ± 1.26	7.57 ± 0.08	5.88 ± 0.04	25.46 ± 0.66
94	OMe	PTA	5.22 ± 0.43	9.19 ± 2.89	10.70 ± 1.35	25.29 ± 0.60
cisplatin	–	–	37.24 ± 5.15	41.82 ± 0.07	49.85 ± 6.66	151.13 ± 58.12
auranofin	–	–	1.80 ± 0.10	0.77 ± 0.05	0.92 ± 0.08	6.21 ± 0.44

The antiproliferative activity of compounds **95–102** was evaluated on Caco-2 (colon) cells after a 72 h incubation period using the MTT assay in a concentration range of 0–20 μM, with auranofin as a reference drug. Despite their structural similarity to the previously reported series of Au(I) complexes **87–94** with notable anticancer properties, the new series (**95–102**) exhibited only mild to low activity in comparison to auranofin (Table 25). Nevertheless, the presence of the IMe ligand appeared to have a favorable effect, as evidenced by a significant increase in the antiproliferative activity in the corresponding complexes. Overall, the IMe-containing compounds demonstrated a higher cytotoxicity than their IPr analogs, with the exception of complexes **95** and **99**. However, the antiproliferative results as a whole did not indicate sufficient potential to continue the study.

Selected compounds from the **95–102** series exhibited enhanced biological activity against both Gram-positive and Gram-negative bacterial strains (Table 26). Their antimicrobial properties were evaluated using the disk diffusion assay against six strains: *Escherichia coli, Pseudomonas aeruginosa, Salmonella enterica, Enterococcus faecalis, Listeria monocytogenes, Staphylococcus epidermidis*, and *Staphylococcus aureus*. The assay was conducted following the guidelines of the Clinical and Laboratory Standards Institute (CLSI) [72]. Stock solutions of the compounds were prepared in concentrations ranging from 0 to 100 μM, and bacterial cultures were incubated for 24 h at 37 °C prior to assessing inhibition zones.

Table 25. Cytotoxicity results for reviewed Au(I) **95–102** complexes obtained by linear-type coordination and their precursors after 72 h treatment using MTT assay on Caco-2 cells compared to auranofin [34].

Compd.	R	R′	IC_{50} (μM)
[AuCl(IMe)]	–	–	>100
[AuCl(IPr)]	–	–	9.09 ± 3.22
95	H	Me	43.34 ± 5.06
96	Br	Me	23.01 ± 2.42
97	Cl	Me	15.05 ± 3.04
98	OMe	Me	16.33 ± 1.04
99	H	IPr	16.34 ± 2.04
100	Br	IPr	53.85 ± 22.72
101	Cl	IPr	46.15 ± 22.54
102	OMe	IPr	48.22 ± 14.56
auranofin	–	–	1.80 ± 0.10

Table 26. Antimicrobial activity of selected Au(I) complexes bearing N-heterocyclic carbene derivatives after 24 h incubation obtained by disk diffusion assay using phenol as positive control [34].

Bacterial Strain	Inhibition Zone (mm)				
	96	**98**	**100**	**101**	**Phenol**
E. coli 25922	9.04 ± 2.17	-	-	-	23.39 ± 1.26
E. coli 8739	11.10 ± 0.08	-	8.97 ± 0.49	-	22.71 ± 1.37
E. coli 13216	10.95 ± 0.64	-	8.10 ± 0.60	-	20.70 ± 2.97
P. aeruginosa	-	-	-	-	30.48 ± 1.45
S. enterica 25928	8.90 ± 1.56	-	-	-	19.20 ± 0.42
S. enterica 14028	9.39 ± 1.26	-	-	-	27.79 ± 1.11
E. faecalis	12.50 ± 1.41	9.47 ± 1.67	12.10 ± 0.99	10.25 ± 1.41	13.5 ± 2.12
L. monocytogenes	18.50 ± 1.41	14.25 ± 1.06	17.17 ± 0.24	15.33 ± 0.47	16.58 ± 0.12
S. epidermidis	25.05 ± 1.48	17.23 ± 0.87	20.95 ± 0.91	18.06 ± 2.04	22.67 ± 4.49
S. aureus 2593	13.76 ± 0.54	9.70 ± 0.02	12.79 ± 1.11	10.49 ± 1.13	12.05 ± 0.73
S. aureus 13565	16.58 ± 1.63	11.86 ± 0.20	14.03 ± 0.17	12.19 ± 0.88	17.25 ± 3.38

Two compounds, **96** and **100**, were distinguished in the series, demonstrating greater activity in comparison to phenol. Additionally, the authors evaluated the mechanism of action for **96**, relying on estimations of TrxR and dihydrofolate reductase (DHFR) activities, and further supported by SEM. In the case of *Escherichia coli* strain, no inhibition of the TrxR was spotted, but DHFR activity was altered in the presence of **96**. This result insinuated that DHFR had a crucial role in the final antibacterial activity of **96**.

3. Conclusions

3.1. Retrospective View of the Authors

In the current era of medicinal chemistry, naturally inspired compounds play a crucial role in the discovery and design of new therapeutics. Notably, nearly 50% of FDA-approved drugs are structurally related to natural products [5]. In this context, hydroxyflavone (OH–F) scaffolds combined with metal centers have gained attention as promising platforms for the development of anticancer and antimicrobial agents. Across the majority of published studies, a common feature of OH–F was recognized: its broad spectrum of biological activities, ranging from anticancer and antimicrobial to anti-inflammatory, antiviral, antimutagenic, and vasodilatory effects [73].

In our present study, we reviewed more than twenty studies published between 2010 and 2024 that investigated the biological activity of [M(OH–F)] coordination compounds, focusing on their anticancer and/or antimicrobial effects. In total, 102 complexes with various transition metals were described and classified based on their coordination mode (e.g., maltol-like, acetylacetonate, catecholate, and mixed-ligand types).

Importantly, nineteen out of twenty-two studies reported that coordination with metal centers enhanced the biological activity compared to the free ligands, justifying the incorporation of metals into OH–F-based pharmacophores. Only three studies deviated from this trend.

Despite these promising results of [M(OH–F)], their pharmaceutical application remains limited due to poor aqueous solubility. To address this challenge, we propose two key strategies. First, structural modification through the incorporation of nitrogen-containing heterocycles could enhance solubility and ligand versatility, given the prevalence of such moieties in approved drugs. Second, systematic derivatization of the OH–F platform could improve water compatibility and stability under physiological conditions, thus increasing the potential for clinical translation.

In parallel, there is a growing interest in coupling these bioactive metal complexes with biocompatible carriers and delivery matrices (e.g., chitosan, alginate, and gelatin, as well as liposomes, solid lipid nanoparticles (SLNs), and poly(lactic-co-glycolic acid) (PLGA) nanoparticles) to overcome solubility barriers and improve therapeutic outcomes. Natural polymers, lipid-based systems, and hybrid nanostructures are being actively explored for their ability to stabilize coordination compounds and modulate their release under physiological conditions. Incorporating [M(OH–F)] complexes into such delivery systems could significantly expand their biomedical utility, enabling targeted delivery, reduced toxicity, and enhanced bioavailability. These directions represent a promising frontier in medicinal chemistry, where the convergence of natural product-inspired scaffolds and metal coordination continues to unlock new therapeutic opportunities. These systems offer advantages such as controlled release, enhanced bioavailability, and targeted delivery, making them ideal candidates for the administration of coordination compounds. Combining the therapeutic potential of [M(OH–F)] scaffolds with modern drug delivery strategies has great promise for translating these complexes into clinically viable treatments.

The final part of our review addresses several common shortcomings identified in the chemical characterization and biological evaluation of the studied compounds. Some studies lacked comprehensive structural characterization of the newly synthesized complexes, with missing data such as NMR, HRMS, EA, or XRD crystallography. Others showed deviations in the biological testing, including the absence of reference compounds, lack of solubility or stability assessments, or insufficient experimental controls. These gaps can compromise the reliability of the reported biological activity and may lead to misleading conclusions at early stages of drug development and pre-clinical evaluation.

In light of the abovementioned conclusions, we would like to finalize this review by emphasizing the significant biological potential of [M(OH–F)] coordination compounds. With proper structural optimization and fine-tuning, particularly regarding solubility, stability, and target selectivity, these complexes represent promising candidates for further development into clinically relevant anticancer and antimicrobial agents.

3.2. Conclusions and Future Prospects

In this concluding section of our review, we would like to highlight several critical considerations that we believe are essential for ensuring the accuracy, reproducibility, and relevance of biological studies involving metal-based compounds.

From a strictly chemical perspective, the comprehensive structural characterization of newly synthesized compounds is strongly advised. This should definitely include techniques such as NMR spectroscopy (^{1}H and ^{13}C), high-resolution mass spectrometry

(HRMS), infrared spectroscopy (IR), elemental analysis (EA), and X-Ray crystallography (XRD). Only the combined usage of these methods can reliably confirm the chemical structure. Despite the well-established nature of this practice, we still observe cases in the literature where characterization is incomplete, and compounds are directly subjected to biological evaluation.

Furthermore, the stability of the complexes in the culture medium and in the solvent used for cytotoxicity assays must be thoroughly assessed, ideally through ^{1}H NMR monitoring over the incubation period. This is a crucial step, as interactions with biological media and stability under physiological conditions significantly impact the therapeutic potential and reproducibility of results.

Cytotoxicity screening should, where possible, be performed on at least five cancerous and one non-cancerous cell line. This enables the calculation of the selectivity index, which provides valuable insight into the compound's tumor-targeting ability. The choice of a positive control must correspond to the specific tumor type under investigation, and its cytotoxicity should be evaluated under identical experimental conditions rather than cited from the literature to ensure a meaningful comparison.

Therefore, we hope this review offers constructive recommendations that will contribute to more rigorous and informative future studies on metal-based compounds with promising potential in drug development.

Funding: The authors gratefully acknowledge financial support from the European Union HORIZON-MSCA through the MET-EFFECT project under grant No. 101086373, PROMISEANG under grant agreement 101036768 funded from the Bio Based Industries Joint Undertaking (JU), and the Ministry of Science, Technological Development and Innovation of the Republic of Serbia, grant Nos. 451-03-136/2025-03/200288 and 451-03-136/2025-03/200026. The JU-CBE receives support from the European Union's Horizon 2020 research and innovation programme and the Bio Based Industries Consortium.

Data Availability Statement: No new data were created or analyzed in this study.

Conflicts of Interest: Author Ljiljana E. Mihajlović was employed by the company Innovation Center of the Faculty of Chemistry. Author Monica Trif was employed by the company Centre for Innovative Process Engineering (CENTIV) GmbH. The remaining authors declare that the research was conducted in the absence of any commercial or financial relationships that could be construed as a potential conflict of interest.

Abbreviations

3-OH-F	3-hydroxyflavone
4CL	4-coumarate-CoA ligase
5-OH-F	5-hydroxyflavone
A431	human epithelial carcinoma cell line
A549	human lung carcinoma epithelial cells
ATP	adenosine triphosphate
apg	apigenin
Bcl-2	B-cell lymphoma 2
bpy	2,2′-bipyridine
B16F10	murine melanoma, c57bl/6j mouse
BSA	bovine serum albumin
CA	conductometric analysis
Caco-2	human colorectal adenocarcinoma cell line
Caco-2/TC7	human colorectal adenocarcinoma cell subline

C4H	cinnamate-4-hydroxylase
CHI	chalcone isomerase
CHS	chalcone synthase
CDDP	cisplatin, cis-diamminedichloroplatinum(II)
chr	chrysin
CH1	human ovarian carcinoma cell line
COX-1/2	cyclooxygenase-1 and cyclooxygenase-2
CV	cyclic voltammetry
Compd.	compound
DFT	density functional theory
DHFR	dihydrofolate reductase
DIP	4,7-diphenyl-1,10-phenanthroline
DL	Dalton's lymphoma, murine transplantable t-cell lymphoma
DMSO	dimethyl sulfoxide
DNA	deoxyribonucleic acid
EA	elemental analysis
EAC	Ehrlich ascites carcinoma
EB	ethidium bromide
EPR	electron paramagnetic resonance
FaDU	human hypopharyngeal squamous cell carcinoma
FNS	flavone synthase
FS	fluorescence spectroscopy
GR	glutathione reductase
HL-7702	normal human liver cells
HEK293	non-cancerous human embryonic kidney cells
HeLa	human cervical adenocarcinoma cell line
HepG2	human hepatocellular carcinoma cell line
HRMS	high resolution mass spectrometry
HSA	human serum albumin
HT-29	human colorectal adenocarcinoma
HUVECs	human umbilical vein endothelial cells
ICP-AES	coupled plasma atomic emission spectrometry
IMe	1,3-dimethylimidazolidin-2-ylidene
IPr	1,3-bis(2,6-diisopropylphenyl)imidazolidin-2-ylidene
i.p.	intraperitoneally
IR	infrared
iso	isoorientin
Jurkat	human T-lymphocyte
kaemp	kaempferol
K562	human myelogenous leukemia
KG-1	human acute myelogenous leukemia
L	ligand
log P	lipophilicity
LoVo	human colon adenocarcinoma
lut	luteolin
NMR	nuclear magnetic resonance
MCF-7	human hormone-dependant breast adenocarcinoma cell line
MDA-MB-231	human triple-negative breast cancer cell line, highly metastatic
MDA-MB-435S	human metastatic breast cancer cell line
MG63	human osteosarcoma cell line
MMP-9	matrix metalloproteinase-9
mor	morin
MTT	3-(4,5-dimethylthiazol-2-yl)-2,5-diphenyltetrazolium bromide

OH-F	hydroxyflavone
p53	tumor suppressor protein involved in cell cycle regulation and apoptosis
PBS	phosphate-buffered saline
phen	1,10-phenanthroline
PAL	phenylalanine ammonia-lyase
PC3	human prostate adenocarcinoma cell line
PCy3	tricyclohexylphosphine
PPh3	triphenylphosphine
PTA	1,3,5-triaza-7-phosphaadamantane
PXRD	powder X-Ray diffraction analyses
querc	quercetin
rut	rutin
ROS	reactive oxygen species
RPE-1	non-cancerous human retinal pigment epithelial cells immortalized with hTERT
RPMI8226	human myeloma
SEM	scanning electron microscopy
SEM	standard error of the mean
SD	standard deviation
SI	selectivity index
SK-MEL-28	human skin melanoma
SK-OV-3	human ovarian adenocarcinoma
SRB	sulforhodamine B
SW480	human colorectal adenocarcinoma cell line
SW620	human colorectal adenocarcinoma cell line
TA	thermal analysis
tchr	tectochrysin
TEM	transmission electron microscopy
TGA	thermogravimetric analysis
Tf	transferrin
THF	tetrahydrofuran
TLC	thin layer chromatography
TMS	tetramethylsilane
tpa	tris(2-pyridylmethyl)amine
tren	tris(2-aminoethyl)amine
TrxR	thioredoxin reductase
WHO	world health organization
U87	human glioblastoma astrocytoma cell line
UV-Vis	ultraviolet-visible spectroscopy
VEGF	vascular endothelial growth factor
WI-38	human normal lung fibroblast cell line (derived from embryonic lung tissue)
XRD	X-Ray diffraction analysis
χ	magnetic susceptibility

References

1. Cancer. Available online: https://www.who.int/news-room/fact-sheets/detail/cancer (accessed on 30 June 2025).
2. Preisler, H.D. Resistence to cytotoxic therapy: A speculative overview. *Ann. Oncol.* **1995**, *6*, 651–657. [CrossRef] [PubMed]
3. Mansoori, B.; Mohammadi, A.; Davudian, S.; Shirjang, S.; Baradaran, B. The different mechanisms of cancer drug resistance: A brief review. *Adv. Pharm. Bull.* **2017**, *7*, 339–348. [CrossRef] [PubMed]
4. Davies, J.; Davies, D. Origins and evolution of antibiotic resistance. *Microbiol. Mol. Biol. Rev.* **2010**, *74*, 417–433. [CrossRef] [PubMed]
5. Newman, D.J.; Cragg, G.M. Natural products as sources of new drugs over the nearly four decades from 01/1981 to 09/2019. *J. Nat. Prod.* **2020**, *83*, 770–803. [CrossRef] [PubMed]
6. Malešev, D.; Kuntić, V. Investigation of metal–flavonoid chelates and the determination of flavonoids via metal-flavonoid complexing reactions. *J. Serb. Chem. Soc.* **2007**, *72*, 921–939. [CrossRef]

7. Guo, M.; Perez, C.; Wei, Y.; Rapoza, E.; Su, G.; Bou-Abdallah, F.; Chasteen, N.D. Iron-binding properties of plant polyphenols and their biological significance. *Dalton Trans.* **2007**, *2007*, 4951–4961. [CrossRef] [PubMed]
8. Lou, H.; Hu, L.; Lu, H.; Wei, T.; Chen, Q. Metabolic Engineering of Microbial Cell Factories for Biosynthesis of Flavonoids: A Review. *Molecules* **2021**, *26*, 4522. [CrossRef] [PubMed]
9. Kasprzak, M.M.; Erxleben, A.; Ochocki, J. Properties and applications of flavonoid metal complexes. *RSC Adv.* **2015**, *5*, 45853–45877. [CrossRef]
10. Agati, G.; Azzarello, E.; Pollastri, S.; Tattini, M. Flavonoids as antioxidants in plants: Location and functional significance. *Plant Sci.* **2012**, *196*, 67–76. [CrossRef] [PubMed]
11. Treutter, D. Significance of flavonoids in plant resistance: A review. *Plant Biol.* **2005**, *7*, 581–591. [CrossRef] [PubMed]
12. Hernández, I.; Alegre, L.; Van Breusegem, F.; Munné-Bosch, S. How relevant are flavonoids as antioxidants in plants? *Trends Plant Sci.* **2009**, *14*, 125–132. [CrossRef] [PubMed]
13. Buer, C.S.; Imin, N.; Djordjevic, M.A. Flavonoids: New roles for old molecules. *J. Integr. Plant Biol.* **2010**, *52*, 98–111. [CrossRef] [PubMed]
14. Cotelle, N.; Bernier, J.-L.; Catteau, J.-P.; Pommery, J.; Wallet, J.-C.; Gaydou, E.M. Antioxidant properties of hydroxy-flavones. *Free Radic. Biol. Med.* **1996**, *20*, 35–43. [CrossRef] [PubMed]
15. Magnani, L.; Gaydou, E.M.; Hubaud, J.C. Spectrophotometric measurement of antioxidant properties of flavones and flavonols against superoxide anion. *Anal. Chim. Acta* **2000**, *411*, 209–216. [CrossRef]
16. Lemanska, K.; Szymusiak, H.; Tyrakowska, B.; Zielinsky, R.; Soffers, E.M.F.; Rietjens, I.M.C.M. The influence of pH on antioxidant properties and the mechanism of antioxidant action of hydrxiflavones. *Free Radic Biol. Med.* **2001**, *31*, 869–881. [CrossRef]
17. de Souza, R.F.V.; Sussuchi, E.M.; De Giovani, W.F. Synthesis, Electrochemical, Spectral, and Antioxidant Properties of Complexes of Flavonoids with Metal Ions. *Synth. React. Inorg. Met.-Org. Chem.* **2003**, *33*, 1125–1144. [CrossRef]
18. Lewandowski, W.; Lewandowska, H.; Golonko, A.; Świderski, G.; Świsłocka, R.; Kalinowska, M. Correlations between molecular structure and biological activity in "logical series" of dietary chromone derivatives. *PLoS ONE* **2020**, *15*, e0229477. [CrossRef] [PubMed]
19. Leonte, D.; Ungureanu, D.; Zaharia, V. Flavones and Related Compounds: Synthesis and Biological Activity. *Molecules* **2023**, *28*, 6528. [CrossRef] [PubMed]
20. Fotsis, T.; Pepper, M.S.; Montesano, R.; Aktas, E.; Breit, S.; Schweigerer, L.; Rasku, S.; Wahala, K.; Adlercreutz, H. Phytoestrogens and inhibition of angiogenesis. *Baillieres Clin. Endocrinol. Metab.* **1998**, *12*, 649–666. [CrossRef] [PubMed]
21. Kim, M.H. Flavonoids Inhibit VEGF/bFGF-Induced Angiogenesis In Vitro by Inhibiting the Matrix-Degrading Proteases. *J. Cell. Biochem.* **2003**, *89*, 529–538. [CrossRef] [PubMed]
22. Sanderson, J.T.; Hordijk, J.; Denison, M.S.; Springsteel, M.F.; Nantz, M.H.; van den Berg, M. Induction and Inhibition of Aromatase (CYP19) Activity by Natural and Synthetic Flavonoid Compounds in H295R Human Adrenocortical Carcinoma Cells. *Toxicol. Sci.* **2004**, *82*, 70–79. [CrossRef] [PubMed]
23. Holder, S.; Zemskova, M.; Zhang, C.; Tabrizizad, M.; Bremer, R.; Neidigh, J.W.; Lilly, M.B. Characterization of a potent and selective small-molecule inhibitor of the PIM1 kinase. *Mol. Cancer Ther.* **2007**, *6*, 163–172. [CrossRef] [PubMed]
24. Loa, J.; Chow, P.; Zhang, K. Studies of structure–activity relationship on plant polyphenol-induced suppression of human liver cancer cells. *Cancer Chemother. Pharmacol.* **2009**, *63*, 1007–1016. [CrossRef] [PubMed]
25. Math, H.H.; Kumar, R.S.; Chakraborty, B.; Almansour, A.I.; Perumal, K.; Kantli, G.B.; Nayaka, S. Antimicrobial Efficacy of 7-hydroxyflavone Derived from Amycolatopsis sp. HSN-02 and Its Biocontrol Potential on Cercospora Leaf Spot Disease in Tomato Plants. *Antibiotics* **2023**, *12*, 1175. [CrossRef] [PubMed]
26. Torres-Piedra, M.; Ortiz-Andrade, R.; Villalobos-Molina, R.; Singh, N.; Medina-Franco, J.L.; Webster, S.P.; Binnie, M.; Navarrete-Vázquez, G.; Estrada-Soto, S. A comparative study of flavonoid analogues on streptozotocinenicotinamide induced diabetic rats: Quercetin as a potential antidiabetic agent acting via 11b–Hydroxysteroid dehydrogenase type 1 inhibition. *Eur. J. Med. Chem.* **2010**, *45*, 2606–2612. [CrossRef] [PubMed]
27. Pillai, S.I.; Subramanian, S.P.; Kandaswamy, M. A novel insulin mimetic vanadiumeflavonol complex: Synthesis, characterization and in vivo evaluation in STZ-induced rats. *Eur. J. Med. Chem.* **2013**, *63*, 109–117. [CrossRef] [PubMed]
28. Calderone, V.; Chericoni, S.; Martinelli, C.; Testai, L.; Nardi, A.; Morelli, I.; Breschi, M.C.; Martinotti, E. Vasorelaxing effects of flavonoids: Investigation on the possible involvement of potassium channels. *Naunyn-Schmiedeberg's Arch. Pharmacol.* **2004**, *370*, 290–298. [CrossRef] [PubMed]
29. Liu, X.; Chan, C.-B.; Jang, S.-W.; Pradoldej, S.; Huang, J.; He, K.; Phun, L.H.; France, S.; Xiao, G.; Jia, Y.; et al. A Synthetic 7,8-hihydroxyflavone Derivative Promotes Neurogenesis and Exhibits Potent Antidepressant Effect. *J. Med. Chem.* **2010**, *53*, 8274–8286. [CrossRef] [PubMed]
30. Tasdemir, D.; Kaiser, M.; Brun, R.; Yardley, V.; Schmidt, T.J.; Tosun, F.; Rüedi, P. Antitrypanosomal and Antileishmanial Activities of Flavonoids and Their Analogues: In Vitro, In Vivo, Structure-Activity Relationship, and Quantitative Structure-Activity Relationship Studies. *Antimicrob. Agents Chemother.* **2006**, *50*, 1352–1364. [CrossRef] [PubMed]

31. Grazul, M.; Budzisz, E. Biological activity of metal ions complexes of chromones, coumarins and flavones. *Coord. Chem. Rev.* **2009**, *253*, 2588–2598. [CrossRef]
32. Yang, A.; Liu, C.; Zhang, H.; Wu, J.; Shen, R.; Kou, X. A multifunctional anti-AD approach: Design, synthesis, X-ray crystal structure, biological evaluation and molecular docking of chrysin derivatives. *Eur. J. Med. Chem.* **2022**, *233*, 114216–114232. [CrossRef] [PubMed]
33. Mármol, I.; Castellnou, P.; Alvarez, R.; Gimeno, M.C.; Rodríguez-Yoldi, M.J.; Cerrada, E. Alkynyl gold(I) complexes derived from 3-hydroxyflavones as multi-targeted drugs against colon cancer. *Eur. J. Med. Chem.* **2019**, *183*, 111661. [CrossRef] [PubMed]
34. Mármol, I.; Quero, J.; Azcárate, P.; Atrián-Blasco, E.; Ramos, C.; Santos, J.; Gimeno, M.C.; Rodríguez-Yoldi, M.J.; Cerrada, E. Biological activity of NHC-gold-alkynyl complexes derived from 3-hydroxyflavones. *Pharmaceutics* **2022**, *14*, 2064. [CrossRef] [PubMed]
35. Andrades Ikeda, N.E.; Novak, E.M.; Maria, D.A.; Velosa, A.S.; Silva Pereira, R.M. Synthesis, characterization and biological evaluation of rutin–zinc(II) flavonoid-metal complex. *Chem.-Biol. Interact.* **2015**, *239*, 184–191. [CrossRef] [PubMed]
36. Wang, X.; He, S.; Yuan, L.; Deng, H.; Zhang, Z. Synthesis, structure characterization, and antioxidant and antibacterial activity study of iso-orientin–zinc complex. *J. Agric. Food Chem.* **2021**, *69*, 3952–3964. [CrossRef] [PubMed]
37. Dell'Anna, M.M.; Censi, V.; Carrozzini, B.; Caliandro, R.; Denora, N.; Franco, M.; Veclani, D.; Melchior, A.; Tolazzi, M.; Mastrorilli, P. Triphenylphosphane Pt(II) complexes containing biologically active natural polyphenols: Synthesis, crystal structure, molecular modeling and cytotoxic studies. *J. Inorg. Biochem.* **2016**, *163*, 346–361. [CrossRef] [PubMed]
38. Raza, A.; Xu, X.; Xia, L.; Xia, C.; Tang, J.; Ouyang, Z. Quercetin-Iron Complex: Synthesis, Characterization, Antioxidant, DNA Binding, DNA Cleavage, and Antibacterial Activity Studies. *J. Fluoresc.* **2016**, *26*, 2023–2031. [CrossRef] [PubMed]
39. Repac Antić, D.; Parčina, M.; Gobin, I.; Petković Didović, M. Chelation in Antibacterial Drugs: From Nitroxoline to Cefiderocol and Beyond. *Antibiotics* **2022**, *11*, 1105. [CrossRef] [PubMed]
40. Sahyon, H.A.; Althobaiti, F.; Ramadan, A.E.-M.M.; Fathy, A.M. Quercetin—Based rhodium(III) complex: Synthesis, characterization and diverse biological potentials. *J. Mol. Struct.* **2022**, *1257*, 132584. [CrossRef]
41. Singh, A.K.; Saxena, G.; Sahabjada; Arshad, M. Synthesis, characterization and biological evaluation of ruthenium flavanol complexes against breast cancer. *Spectrochim. Acta Part A Mol. Biomol. Spectrosc.* **2017**, *180*, 97–104. [CrossRef] [PubMed]
42. Prajapati, R.; Kumar Dubey, S.; Gaur, R.; Koiri, R.K.; Maurya, B.K.; Trigun, S.K.; Mishra, L. Structural characterization and cytotoxicity studies of ruthenium(II)–dmso–chloro complexes of chalcone and flavone derivatives. *Polyhedron* **2010**, *29*, 1055–1061. [CrossRef]
43. Sava, G.; Bergamo, A. Ruthenium-based compounds and tumour growth control (review). *Int. J. Oncol.* **2000**, *17*, 353–365. [CrossRef] [PubMed]
44. Alessio, E.; Balducci, G.; Lutman, A.; Mestroni, G.; Calligaris, M.; Attia, W.M. Synthesis and characterization of two new classes of ruthenium(III)-sulfoxide complexes with nitrogen donor ligands (L): Na[trans-$RuCl_4(R_2SO)(L)$] and mer, cis-$RuCl_3(R_2SO)(R_2SO)(L)$. The crystal structure of Na[trans-$RuCl_4(DMSO)(NH_3)$] · 2DMSO, Na[trans-$RuCl_4$(DMSO)(Im)] · H_2O, Me_2CO (Im = imidazole) and mer, cis-$RuCl_3(DMSO)(DMSO)(NH_3)$. *Inorg. Chim. Acta* **1993**, *203*, 205–217. [CrossRef]
45. Gantsho, V.L.; Dotou, M.; Jakubaszek, M.; Goud, B.; Gasser, G.; Vissera, H.G.; Schutte-Smith, M. Synthesis, characterization, kinetic investigation and biological evaluation of Re(I) di- and tricarbonyl complexes with tertiary phosphine ligands. *Dalton Trans.* **2020**, *49*, 35–46. [CrossRef] [PubMed]
46. Kurzwernhart, A.; Kandioller, W.; Bartel, C.; Bächler, S.; Trondl, R.; Mühlgassner, G.; Jakupec, M.A.; Arion, V.B.; Marko, M.; Keppler, B.K.; et al. Targeting the DNA-topoisomerase complex in a double-strike approach with a topoisomerase inhibiting moiety and covalent DNA binder. *Chem. Commun.* **2012**, *48*, 4839–4841. [CrossRef] [PubMed]
47. Kurzwernhart, A.; Kandioller, W.; Bächler, S.; Bartel, C.; Martic, S.; Buczkowska, M.; Mühlgassner, G.; Jakupec, M.A.; Kraatz, H.-B.; Bednarski, P.J.; et al. Structure-Activity Relationships of Targeted $Ru^{II}(\eta^6$-p-Cymene) Anticancer Complexes with Flavonol-Derived Ligands. *J. Med. Chem.* **2012**, *55*, 10512–10522. [CrossRef] [PubMed]
48. Kurzwernhart, A.; Kandioller, W.; Enyedy, É.A.; Novak, M.; Jakupec, M.A.; Keppler, B.K.; Hartinger, C.G. 3-Hydroxyflavones vs. 3-hydroxyquinolinones: Structure–activity relationships and stability studies on Ru^{II}(arene) anticancer complexes with biologically active ligands. *Dalton Trans.* **2013**, *42*, 6193–6202. [CrossRef] [PubMed]
49. Kurzwernhart, A.; Mokesch, S.; Klapproth, E.; Adib-Razavi, M.S.; Jakupec, M.A.; Hartinger, C.G.; Kandioller, W.; Keppler, B.K. Flavonoid-based organometallics with different metal centers—Investigations of the effects on reactivity and cytotoxicity. *Eur. J. Inorg. Chem.* **2016**, *2016*, 240–246. [CrossRef]
50. Zahirović, A.; Kahrović, E.; Cindrić, M.; Kraljević Pavelić, S.; Hukić, M.; Harej, A.; Turkušić, E. Heteroleptic ruthenium bioflavonoid complexes: From synthesis to in vitro biological activity. *J. Coord. Chem.* **2017**, *70*, 4030–4053. [CrossRef]
51. Gençkal, H.M.; Erkisa, M.; Alper, P.; Sahin, S.; Ulukaya, E.; Ari, F. Mixed ligand complexes of Co(II), Ni(II) and Cu(II) with quercetin and diimine ligands: Synthesis, characterization, anti-cancer and anti-oxidant activity. *J. Biol. Inorg. Chem.* **2020**, *25*, 161–177. [CrossRef] [PubMed]

52. Wang, Q.; Huang, M.; Huang, Y.; Zhang, J.-S.; Zhou, G.-F.; Zeng, R.-Q.; Yang, X.-B. Synthesis, characterization, DNA interaction, and antitumor activities of mixed-ligand metal complexes of kaempferol and 1,10-phenanthroline/2,2′-bipyridine. *Med. Chem. Res.* **2014**, *23*, 2659–2666. [CrossRef]
53. Kozsup, M.; Zhou, X.Q.; Farkas, E.; Bényei, A.C.; Bonnet, S.; Patonay, T. (the late); Kónya, K.; Buglyó, P. Synthesis, characterization and cytotoxicity studies of Co(III)-flavonolato complexes. *J. Inorg. Biochem.* **2021**, *217*, 111382. [CrossRef] [PubMed]
54. Sodhi, R.K.; Paul, S. An overview of metal acetylacetonates: Developing areas/routes to new materials and applications in organic syntheses. *Catal. Surv. Asia* **2018**, *22*, 31–62. [CrossRef]
55. Munteanu, A.-C.; Notaro, A.; Jakubaszek, M.; Cowell, J.; Tharaud, M.; Goud, B.; Uivarosi, V.; Gasser, G. Synthesis, characterization, cytotoxic activity, and metabolic studies of ruthenium(II) polypyridyl complexes containing flavonoid ligands. *Inorg. Chem.* **2020**, *59*, 4424–4434. [CrossRef] [PubMed]
56. Munteanu, A.-C.; Musat, M.G.; Mihaila, M.; Badea, M.; Olar, R.; Nitulescu, G.M.; Rădulescu, F.Ș.; Brasoveanu, L.I.; Uivarosi, V. New heteroleptic lanthanide complexes as multimodal drugs: Cytotoxicity studies, apoptosis, cell cycle analysis, DNA interactions, and protein binding. *Appl. Organomet. Chem.* **2020**, e6062. [CrossRef]
57. Marques, J.; Silva, A.M.S.; Marques, M.P.M.; Braga, S.S. Ruthenium(II) trithiacyclononane complexes of 7,3′,4′-trihydroxyflavone, chrysin and tectochrysin: Synthesis, characterisation, and cytotoxic evaluation. *Inorg. Chim. Acta* **2019**, *488*, 71–79. [CrossRef]
58. Marques, J.; Fernandes, J.A.; Almeida Paz, F.A.; Marques, M.P.M.; Braga, S.S. Isolation, crystal structure, and cytotoxicity on osteosarcoma of a ruthenium(III) complex with coordinated acetonitrile. *J. Coord. Chem.* **2012**, *65*, 2489–2499. [CrossRef]
59. Naso, L.G.; Lezama, L.; Valcarcel, M.; Salado, C.; Villacé, P.; Kortazar, D.; Ferrer, E.G.; Williams, P.A.M. Bovine serum albumin binding, antioxidant and anticancer properties of an oxidovanadium(IV) complex with luteolin. *J. Inorg. Biochem.* **2016**, *157*, 80–93. [CrossRef] [PubMed]
60. Lopez-Lazaro, M. Distribution and biological activities of the flavonoid luteolin. *Mini Rev. Med. Chem.* **2009**, *9*, 31–59. [CrossRef] [PubMed]
61. Ferrer, E.G.; Salinas, M.V.; Correa, M.J.; Naso, L.; Barrio, D.A.; Etcheverry, S.B.; Lezama, L.; Rojo, T.; Williams, P.A.M. Synthesis, characterization, antitumoral and osteogenic activities of quercetin vanadyl(IV) complexes. *J. Biol. Inorg. Chem.* **2006**, *11*, 791–801. [CrossRef] [PubMed]
62. Etcheverry, S.B.; Ferrer, E.G.; Naso, L.; Rivadeneira, J.; Salinas, V.; Williams, P.A.M. Antioxidant effects of the VO(IV) hesperidin complex and its role in cancer chemoprevention. *J. Biol. Inorg. Chem.* **2007**, *13*, 435–447. [CrossRef] [PubMed]
63. Islas, M.S.; Naso, L.G.; Lezama, L.; Valcarcel, M.; Salado, C.; Roura-Ferrer, M.; Ferrer, E.G.; Williams, P.A.M. Insights into the mechanisms underlying the antitumor activity of an oxidovanadium(IV) compound with the antioxidant naringenin. Albumin binding studies. *J. Inorg. Biochem.* **2015**, *149*, 12–24. [CrossRef] [PubMed]
64. Seelinger, G.; Merfort, I.; Wölfle, U.; Schempp, C.M. Anti-carcinogenic effects of the flavonoid luteolin. *Molecules* **2008**, *13*, 2628–2651. [CrossRef] [PubMed]
65. Sharfalddin, A.A.; Al-Younis, I.M.; Mohammed, H.A.; Dhahri, M.; Mouffouk, F.; Abu Ali, H.; Anwar, M.J.; Qureshi, K.A.; Hussien, M.A.; Alghrably, M. Therapeutic Properties of Vanadium Complexes. *Inorganics* **2022**, *10*, 244. [CrossRef]
66. Deng, S.-P.; Yang, Y.-L.; Cheng, X.-X.; Li, W.-R.; Cai, J.-Y. Synthesis, spectroscopic study and radical scavenging activity of kaempferol derivatives: Enhanced water solubility and antioxidant activity. *Int. J. Mol. Sci.* **2019**, *20*, 975. [CrossRef] [PubMed]
67. Tu, L.-Y.; Pi, J.; Jin, H.; Cai, J.-Y.; Deng, S.-P. Synthesis, characterization and anticancer activity of kaempferol–zinc(II) complex. *Bioorg. Med. Chem. Lett.* **2016**, *26*, 2730–2734. [CrossRef] [PubMed]
68. Liu, L.; Xu, K.; Wang, H.; Tan, P.K.J.; Fan, W.; Venkatraman, S.S.; Li, L.; Yang, Y.-Y. Self-assembled cationic peptide nanoparticles as an efficient antimicrobial agent. *Nat. Nanotechnol.* **2009**, *4*, 457–463. [CrossRef] [PubMed]
69. Yuan, L.; Wang, Y.; Wang, J.; Xiao, H.; Liu, X. Additive effect of zinc oxide nanoparticles and isoorientin on apoptosis in human hepatoma cell line. *Toxicol. Lett.* **2014**, *225*, 294–304. [CrossRef] [PubMed]
70. Gunduz, S.; Goren, A.C.; Ozturk, T. Syntheses of 3-hydroxyflavones. *Org. Lett.* **2012**, *14*, 1576–1579. [CrossRef] [PubMed]
71. Meshram, R.J.; Bagul, K.T.; Pawnikar, S.P.; Barage, S.H.; Kolte, B.S.; Gacche, R.N. Known compounds and new lessons: Structural and electronic basis of flavonoid-based bioactivities. *J. Biomol. Struct. Dyn.* **2020**, *38*, 1168–1184. [CrossRef] [PubMed]
72. Clinical and Laboratory Standards Institute (CLSI). *Performance Standards for Antimicrobial Disk Susceptibility Tests*; Approved Standard; CLSI: Wayne, PA, USA, 2012.
73. Samsonowicz, M.; Regulska, E.; Kalinowska, M. Hydroxyflavone metal complexes—Molecular structure, antioxidant activity and biological effects. *Chem. Biol. Interact.* **2017**, *273*, 245–256. [CrossRef] [PubMed]

inorganics

MDPI

Article

Ligand Rigidity and π-Surface Modulate Biomolecular Interactions and Cytotoxicity in Ru(II) Polypyridyl Complexes

Patrícia Alves de Matos [1], Marcos Eduardo Gomes do Carmo [1], André Luis Araújo Parussulo [2], Clara Maria Faria Silva [3], Ricardo Campos Lino [3], Henrique Eisi Toma [2], Marcelo Emílio Beletti [4], Robson José de Oliveira Júnior [3], Antônio Otávio de Toledo Patrocinio [1], Tiago Araújo Matias [5,*] and Tayana Mazin Tsubone [1,*]

1 Institute of Chemistry, Federal University of Uberlândia (UFU), Uberlândia 38408-100, MG, Brazil; patricia.matos@ufu.br (P.A.d.M.); marcos.go.carmo@gmail.com (M.E.G.d.C.); otaviopatrocinio@ufu.br (A.O.d.T.P.)

2 Institute of Chemistry, University of São Paulo (USP), São Paulo 05508-900, SP, Brazil; henetoma@iq.usp.br (H.E.T.)

3 Institute of Biotechnology, Federal University of Uberlândia (UFU), Uberlândia 38402-045, MG, Brazil; clara.faria@ufu.br (C.M.F.S.); ricardo.lino@ufu.br (R.C.L.); oliveirajunior@ufu.br (R.J.d.O.J.)

4 Institute of Biomedical Sciences, Federal University of Uberlândia (UFU), Uberlândia 38408-100, MG, Brazil; mebeletti@ufu.br

5 Chemistry Department, Federal University of Espírito Santo (UFES), Vitória 29075-910, ES, Brazil

* Correspondence: tiago.matias@ufes.br (T.A.M.); tayana.tsubone@ufu.br (T.M.T.)

Abstract

The complexes *cis*-[Ru(dmbpy)$_2$Cl(bpy)](PF$_6$) (Rubpy) and *cis*-[Ru(dmbpy)$_2$Cl(bpe)](PF$_6$) (Rubpe) (dmbpy = 4,4′-Dimethyl-2,2′-dipyridyl, bpy= 4,4′-dipyridyl and bpe = 1,2-bis(4-pyridyl)ethane) were synthesized and spectroelectrochemically characterized. Both Ru(II) complexes exhibited absorption bands assigned to intraligand and metal-to-ligand charge transfer (MLCT) transitions, and their spectral stability in PBS buffer (pH 7.4) supports their suitability for biological studies involving biomolecules or living cells. Fluorescence quenching assays revealed strong interactions with bovine serum albumin (BSA), with binding constants (K_b) values were 2.89×10^5 M^{-1} for Rubpy and 1.97×10^5 M^{-1} for Rubpe, and a stoichiometry of one binding site per albumin molecule. DNA-binding studies demonstrated non-covalent interactions with ss-DNA, evidenced by a hyperchromic effect in the MLCT bands, suggesting a partial intercalation or groove-binding mechanism. Cellular uptake assays indicated moderate incorporation of both complexes in tumor cells, with uptake levels of 52% (Rubpy) and 47% (Rubpe) in HeLa cells, and 42% (Rubpy) and 32% (Rubpe) in MDA-MB-231 cells. Despite the similar uptake profiles, cytotoxicity assays showed that Rubpe is approximately 2.4 times more potent than Rubpy, with IC$_{50}$ values of 9 μM (HeLa) and 12 μM (MDA-MB-231), compared to 22 μM and 29 μM for Rubpy, respectively. These results highlight the relevance of these Ru(II) complexes as molecular platforms for exploring structure–activity relationships in anticancer agents.

Keywords: ruthenium(II) polypyridine; albumin (BSA); cellular toxicity; biomolecule interaction

1. Introduction

Although platinum(II)-based anticancer compounds have achieved significant success in clinical oncology, their prolonged use is often associated with severe limitations, including increased drug efflux, reduced cellular uptake, rapid DNA repair, and poor tumor selectivity, ultimately contributing to drug resistance [1]. As alternatives, ruthenium-based complexes have been identified as promising candidates due to their distinct chemical properties and biological behavior [2]. In particular, Ru(II) polypyridine complexes have attracted considerable attention because of their lower toxicity, slower ligand dissociation, reduced resistance development, strong interactions with biomacromolecules, and cytotoxic activity in various cancer models [2–5]. Several Ru-based agents have progressed to clinical evaluation, including the Ru(III) compounds NAMI-A and KP1019/KP1339, and the Ru(II) phototherapeutic TLD1433 [2]. While NAMI-A showed limited efficacy in phase II trials, KP1019's solubility issues were overcome in its sodium derivative BOLD-100, which has demonstrated clinical activity and is currently in global phase II trials for gastrointestinal cancers [6,7]. Meanwhile, TLD1433 advanced to phase II investigations for the photodynamic treatment of bladder cancer [8].

The antitumor properties of Ru(II) complexes are highly dependent on ligand design, coordination geometry, and the presence of labile sites within the metal center. Subtle modifications in ligand electronics or sterics can significantly alter DNA and protein interactions [9]. For example, complexes of the type $[Ru(R\text{-}tpy)(LL)Cl]^{n+}$ (R = Chloro or 4-chlorophenyl; LL = bidentate ligand) exhibit strong cytotoxic activity against multiple tumor cell lines, where the aromaticity of the terpyridine substituent has been shown to enhance biological activity [9–11]. Similar trends have been observed for mixed-ligand complexes such as $[Ru(ppy)(phen)(NCCH_3)_2]^+$ and $[Ru(bpy)(ppy)(dppz)]^+$, whose hydrophobic ligands facilitate rapid cellular uptake and nuclear accumulation [12]. Overall, Ru(II) polypyridyl complexes containing hydrophobic bidentate ligands such as bpy, phen, or derivatives often display higher cytotoxicity due to enhanced membrane permeability and intracellular accumulation [5,9,12,13].

The biological activity of ruthenium complexes is influenced not only by interactions with DNA but also by their ability to bind proteins [14,15]. Protein binding plays a central role in drug pharmacokinetics and distribution, and regulatory agencies such as the FDA require early-stage evaluation of protein interactions during drug development [16]. Serum albumin, which is highly abundant in plasma, is particularly important because it modulates drug transport, bioavailability, and accumulation in tumor tissues [2]. Consequently, BSA is commonly used as a model for studying protein binding and its implications for cytotoxicitys [14].

Inspired by the relevance of ligand-controlled structure–activity relationships in Ru(II) polypyridyl scaffolds, we examined two complexes differing only in their ancillary ligands (Figure 1) as potential antitumor candidates: *cis*-$[Ru(dmbpy)_2Cl(bpy)](PF_6)$ (or Rubpy) and *cis*-$[Ru(dmbpy)_2Cl(bpe)](PF_6)$ (or Rubpe) (dmbpy = 4,4′-Dimethyl-2,2′-bipyridyl, bpy = 4,4′-bipyridyl and bpe = 1,2-bis(4-pyridyl)ethane)). Although Ru(II) polypyridyl complexes have been widely studied, direct comparisons that isolate the influence of ligand rigidity and π-surface, such as the rigid, fully conjugated bpy versus the flexible, reduced-π bpe, remain limited. These ligands provide a minimal and well-defined structural variation, enabling a controlled assessment of how such differences affect biomolecular interactions and cellular responses. From a structure–property perspective, bpy offers a rigid π-conjugated platform that favors π-stacking with nucleic acids, whereas bpe provides greater conformational freedom and a reduced aromatic surface, features that may influence hydrophobic partition, membrane interactions, and intracellular reorganization. This framework establishes the rationale for interpreting the distinct biomolecular and cytotoxic behaviors observed for Rubpy and Rubpe.

https://doi.org/10.3390/inorganics14020063

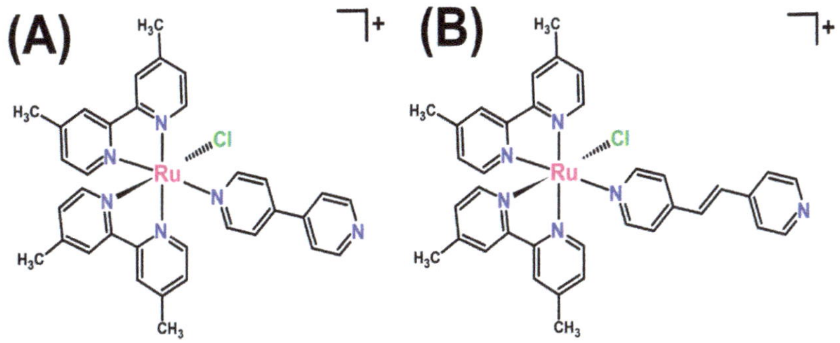

Figure 1. Chemical structures of studied Ru(II) complexes. (**A**) *cis*-[Ru(dmbpy)$_2$Cl(bpy)](PF$_6$) (or Rubpy) and (**B**) *cis*-[Ru(dmbpy)$_2$Cl(bpe)](PF$_6$) (or Rubpe). dmbpy = 4,4′-Dimethyl-2,2′-bipyridyl, bpy = 4,4′-dipyridyl and bpe = 1,2-bis(4-pyridyl)ethane.

2. Results and Discussion

2.1. Characterization and Physicochemical Properties

2.1.1. Electronic Structure and Redox Properties of the Ru(II) Complexes

To gain insight into the physicochemical behavior of Rubpy and Rubpe and to ensure the integrity of the species investigated in biomolecular and cellular assays, we first examined their electronic, redox, and spectroelectrochemical properties, as well as their stability under aqueous and physiological conditions.

Both complexes display the characteristic absorption profile of Ru(II) polypyridyl scaffolds (Figure 2A), with intense intraligand $\pi \rightarrow \pi^*$ transitions near 292 nm and broad metal-to-ligand charge transfer (MLCT) bands centered around 400–460 nm. These features arise from Ru(II)$\rightarrow\pi^*$ transitions involving the coordinated polypyridyl ligands and are fully consistent with previously reported Ru(II) complexes of similar composition [17]. The electrochemical behavior also follows the expected pattern for this family of compounds: cyclic voltammograms reveal a reversible Ru(III)/Ru(II) oxidation at $E_{1/2} \approx 0.81$–0.83 V, followed by two ligand-centered reductions at negative potentials (Figure 2B). These reductive processes, observed at -1.38 and -1.57 V for Rubpy and at -1.29 and -1.50 V for Rubpe, correspond to successive one-electron reductions in the bipyridyl-type ancillary ligands ($L^{0}/^{-}$ and $L^{-}/^{2-}$ for L = bpy or bpe), in agreement with literature data [17].

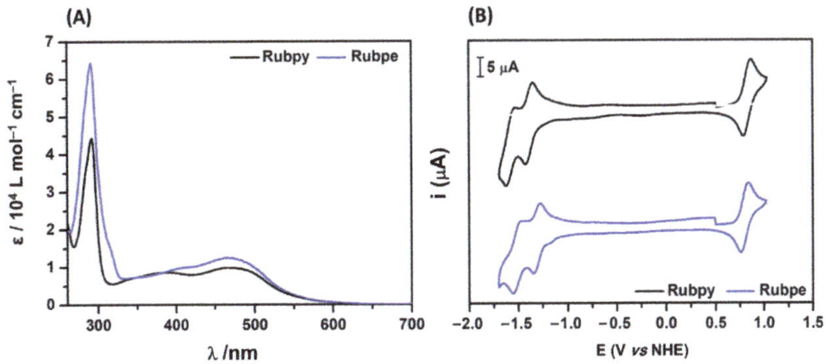

Figure 2. (**A**) Electronic absorption spectra of Rubpy and Rubpe complexes in methanol solution. (**B**) Cyclic voltammograms of Rubpy and Rubpe complexes in 1.0 mM of (Bu$_4$N)(ClO$_4$) in DMF, at scan rate of 100 mVs^{-1}.

https://doi.org/10.3390/inorganics14020063

Spectroelectrochemical measurements (Figures S1–S3) further corroborate these assignments. Oxidation of the ruthenium center results in the extinction of the MLCT bands between 383 and 530 nm and a decrease in the ligand-centered $\pi \rightarrow \pi^*$ transition near 292 nm (Figure S1A). Upon electrochemical reduction in *cis*-[Ru^{III}(dmbpy)$_2$Cl(bpy)]$^{2+}$ back to the Ru(II) form, the MLCT bands at 383 and 470 nm reappear, together with the restored $\pi \rightarrow \pi^*$ band (Figure S2A). Such reversibility is typical of inert Ru(II) complexes of the type *cis*-[Ru(bpy)L(py)]$^{n+}$, which are well known for their high stability and use in various catalytic applications [18].

Reduction of the bpy or bpe ancillary ligand induces a characteristic decrease in the intraligand absorption near 300 nm, accompanied by an increase in absorbance above 348 nm and a bathochromic shift in the MLCT bands (Figures S1B and S3B). A similar pattern is observed in the second ligand-centered reduction (Figures S1C and S3C), with further enhancement of low-energy transitions and attenuation of the $\pi \rightarrow \pi^*$ band [17]. The close structural similarity between Rubpy and Rubpe results in nearly identical spectroelectrochemical profiles, with comparable redox potentials and analogous spectral changes upon oxidation and reduction.

Notably, no redox processes were detected in the potential window between -0.24 and -0.15 V, a range associated with the intracellular redox state of proliferating (-0.24 V), apoptotic (-0.17 V), and necrotic cells (-0.15 V) [19]. The absence of redox activity in this biologically relevant region suggests that both complexes are electrochemically stable under typical cellular redox environments.

2.1.2. Stability of the Complexes in Aqueous and Physiological Media

Water constitutes approximately 70% of the cellular environment and acts as the primary solvent for physiological buffer systems such as phosphate-buffered saline (PBS). Under these conditions, water molecules (H_2O) and phosphate species ($HPO_4{}^{2-}/H_2PO_4{}^{-}$) may function as Lewis bases capable of coordinating to metal centers. Chloro-ruthenium polypyridyl complexes can undergo ligand substitution reactions in aqueous media, forming aqua species through chloride dissociation [9,20]. Therefore, evaluating the solution stability of Rubpy and Rubpe is essential before examining their interactions with biomacromolecules.

Both complexes were initially dissolved in minimal methanol and diluted to 60 μM in water, and their UV–Vis spectra were monitored for 6 h (Figure S4). No changes in the MLCT band positions were observed during this period, indicating that chloride substitution by water did not occur [21–23]. This invariance demonstrates that both complexes exhibit good stability in aqueous solution, supporting their suitability for biological studies in predominantly aqueous environments.

To further assess their solution behavior, UV–Vis spectra were recorded in PBS (pH 7.4) over 96 h. As shown in Figure 3, no shifts in the MLCT or intraligand absorption bands were detected for either complex. In similar Ru(II) polypyridyl systems, chloride substitution by water typically produces hypsochromic shifts in ca. 30 nm [21–23]; the absence of such changes confirms that Rubpy and Rubpe remain intact under physiological buffer conditions. This stability is consistent with literature reports describing slower substitution kinetics for Ru–Cl bonds trans to π-acceptor ligands, such as dmbpy, compared to those trans to strong σ-donors like primary amines [9,20].

Because proteins contain nucleophilic amino acid residues (e.g., $-NH_2$, –OH) capable of coordinating to metal centers, the stability of Rubpy and Rubpe in the presence of bovine serum albumin (BSA) was also evaluated (Figure S5). Similar to the results in PBS, no λ_{max} shifts were observed, indicating the absence of ligand exchange at Ru(II). A modest decrease in absorption intensity was detected, consistent with weak complex–

https://doi.org/10.3390/inorganics14020063

protein association rather than ligand substitution. This hypochromic effect supports the occurrence of non-covalent interactions with BSA, which were subsequently quantified using fluorescence spectroscopy, a more sensitive method for evaluating binding affinity.

Having established that both complexes remain structurally and spectroscopically stable in aqueous, physiological, and protein-containing environments, we proceeded to investigate their interactions with serum albumin and DNA, two key biomolecular targets involved in cellular transport, distribution, and molecular recognition.

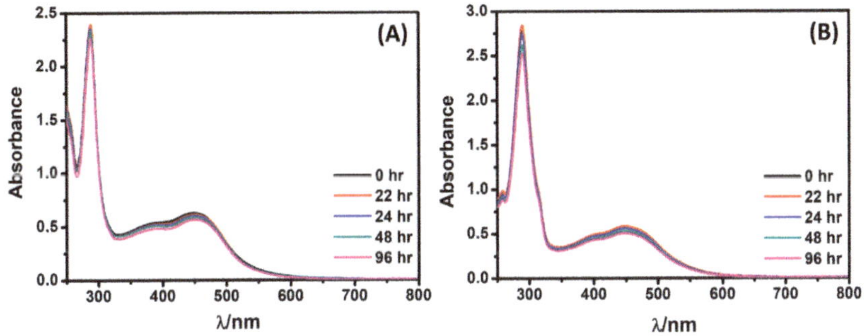

Figure 3. UV–Vis spectra of (**A**) Rubpy and (**B**) Rubpe in PBS buffer (pH 7.4) over a 96 h period. [Ru complex] = 60 μM, T = 25 °C.

2.2. *Experimental Biomolecular Interactions*

2.2.1. Interaction with Serum Albumin: Fluorescence Quenching and Binding Constants

Serum albumins represent the predominant protein fraction in blood plasma, accounting for around 60% of total protein and providing approximately 80% of blood osmotic pressure [24]. This biomolecule can bind reversibly to a variety of endogenous and exogenous molecules, such as drugs and other small bioactive compounds, playing an important role in drug transport [25]. Bovine serum albumin (BSA) is the most studied protein in biomimetic systems due to its structural homology with human serum albumin (HSA), with an identity of approximately 76% [26]. The primary structure of BSA consists of around 580 amino acids, including tryptophan and tyrosine residues, both fluorescent [27]. The fluorescence of bovine serum albumin is observed at 350 nm (λ_{ex} = 270 nm) primarily due to the emission of tryptophan, as the indole molar extinction coefficient of this residue is approximately five times greater than that of the phenyl group in the tyrosine residue [28]. Thus, the interaction between BSA and small molecules can modify the microenvironment surrounding the secondary structure of the protein chain, typically leading to a reduction in the fluorescence emitted by the intrinsic tryptophan group within this biomolecule [28].

As shown in Figure 4, an increase in the concentration of ruthenium complexes in the BSA solution systems causes a gradual decrease in the macromolecule's fluorescence intensity. This effect is associated with the quenching of the fluorophore's fluorescence. The observed quenching can be credited to changes in the tertiary structure of the protein, leading to alterations in the tryptophan environment of BSA and, therefore, indicating the binding of each complex to albumin [28]. There are two types of fluorescence quenching mechanisms: static and dynamic. In the former, a collision occurs between albumin and the ruthenium compound in the ground state, resulting in the formation of a non-fluorescent BSA–ruthenium complex. In the dynamic mechanism, the collision occurs between the fluorophore in the excited state and the ruthenium compound, forming a complex that returns to the ground state without emitting light. The efficiency of fluorescence quenching is related to the Stern–Volmer constant (K_{SV}) by the Stern-Volmer relation (Equation (1)), where F_0 is the fluorescence intensity in the absence of the quencher, F is the fluorescence

https://doi.org/10.3390/inorganics14020063

intensity in the presence of the quencher and [Q] is the concentration of the quencher, the ruthenium complexes.

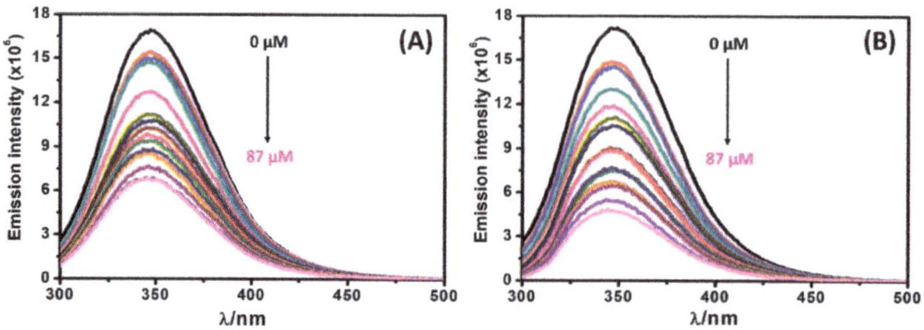

Figure 4. Steady-state fluorescence emission spectra for the interaction between BSA and (**A**) Rubpy and (**B**) Rubpe in PBS buffer (pH 7.4), at 25 °C. The complexes concentration ranged from 0 to 87 μM, and BSA presented fixed concentrations at 3 μM. λ_{exc} = 280 nm, excitation and emission slits of 5.0 nm and 1.0 nm, respectively.

Figure 5 displays a graph illustrating F_0/F as a function of the concentration of the ruthenium complexes. The data deviate from the linearity described by Equation (1), suggesting the presence of both static and dynamic quenching phenomena [9,25,28–30].

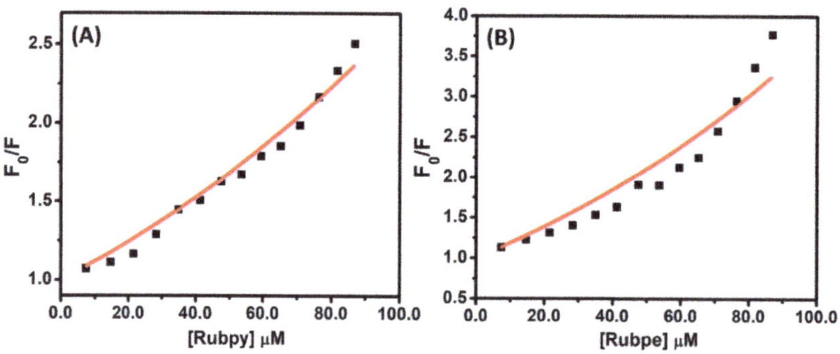

Figure 5. Stern-Volmer plots for the interaction (**A**) BSA-[Rubpy] and (**B**) BSA-[Rubpe].

In addition, the upward curve observed in Figure 5 is associated with a static component in the quenching mechanism. Therefore, a factor exp(V[Q]), where V is defined as the static quenching constant and is incorporated into the Stern–Volmer relationship (Equation (3)), to describe both quenching modes. The constant, *V* was determined from a plot of $F_0/FeV[Q]$ versus [Q] and adjusting *V* until a linear plot was obtained (Figure S6). The highest correlation coefficient was used as the criterion for linearity to obtain a precise value of *V* [9]. The static quenching constant (V) for compounds was 5000 L mol^{-1} while the dynamic collisional quenching constant, K_{sv} was 0.61 × 10^5 M^{-1} for Rubpy and 1.27 × 10^5 M^{-1} for Rubpe (Table 1). Both complexes showed high values of the quenching constant indicating their great efficiency to interact strongly with BSA.

The values of the BSA-binding constant (K_b) and the number of binding sites per albumin (*n*) were calculated (Figure 6), and the results are shown in Table 1. For both complexes, the K_b value was around 10^5, indicating a strong interaction between the complex and the BSA protein [31]. Rubpy exhibited a slightly higher K_b value (2.89 × 10^5 M^{-1}) than Rubpe (1.97 × 10^5 M^{-1}), suggesting a marginally stronger interaction of Rubpy compared to Rubpe. The analysis of binding constants is highly useful for inferring how a molecular species will be distributed in blood plasma; the binding constant considered ideal falls

https://doi.org/10.3390/inorganics14020063

within the range of 10^4–10^6 L mol^{-1} [32]. Thus, the values obtained for Rubpy and Rubpe are sufficiently high for the compounds to bind to BSA for transport, but still low enough to allow their release from albumin upon reaching the target cells [9]. In addition, the number of quencher binding sites (n) calculated for both the [Rubpy]-BSA and [Rubpe]-BSA systems was approximately 1, indicating a single binding site for BSA for both complexes.

Table 1. Stern-Volmer (K_{sv}), binding constant (K_b) number of binding sites (n) of ruthenium(II) complexes to BSA (3 µM) in PBS, pH 7.4 at 25 °C. Data are represented as mean ± standard deviation of experiments performed in duplicate.

Complexes	K_{SV} ($\times 10^5$ M^{-1})	K_b ($\times 10^5$ M^{-1})	n
Rubpy	0.61 ± 0.02	2.89 ± 0.04	1.31
Rubpe	1.27 ± 0.08	1.97 ± 0.03	1.23

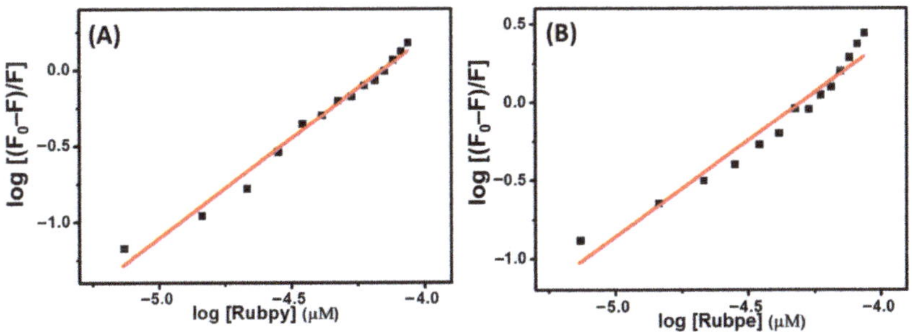

Figure 6. The plot of log [(F$_0$ − F/F] versus log [complex] for (**A**) Rubpy and (**B**) Rubpe.

2.2.2. DNA Interaction and Binding Constants (K_b) of Ruthenium Complexes

Beyond their interaction with albumin, which plays a central role in transport and pharmacokinetics, we also examined the interaction of the complexes with DNA, a key intracellular target for several metal-based therapeutics. Many clinically approved or clinically investigated anticancer agents act by binding to DNA and perturbing replication and transcription processes, making DNA–metal interactions a long-standing subject of interest in medicinal inorganic chemistry [33–35]. Such interactions may occur covalently, requiring a labile coordination site for metal–nucleobase binding, or non-covalently, through intercalation, groove binding, or electrostatic association with the negatively charged phosphate backbone [33–35].

The UV–Vis spectroscopy was employed to study [Ru-Complexe]-DNA interactions, the absorbance spectra of the ruthenium compounds in the presence of various concentrations of ss-DNA are shown in Figure S7. The ss-DNA biomolecule exhibited an absorption band (260 nm), while did not exhibit absorption in the MLCT band region [36]. As with the interaction of the complexes with BSA presented previously, no significant shift was observed in the λ_{max} of absorption values of the MLCT's bands for both complexes. The absence of MLCT band shift indicates that there is no exchange of coordinated ligands to the metal ruthenium(II) in the presence of ss-DNA. On the other hand, was observed a hyperchromic effect in MLCT's bands for both complexes (Figure S7A,B), this suggests that the electrostatic attraction or groove binding plays a dominant role, contributing to local destabilization of the DNA double helix [33–35]. In addition, the hypochromic effect (reduced absorbance) was associated with intercalating-type interactions arising from π-π-type stacking between ligands in the complex with the nitrogenous bases of DNA for many compounds [33–35]. To conclusively support that Rubpy and Rubpe complexes display an electrostatic interactions or interactions with the DNA groove, additional ex-

https://doi.org/10.3390/inorganics14020063

periments still would be needed, such as viscosity, thermal denaturation and/or circular dichroism [33–35]. Nevertheless, our research primarily focused on assessing the strength of the interaction between [Ru-Complex] and DNA by K_b rather than the mechanism of DNA interaction. Additionally, we employed ss-DNA as a model due to its reproducibility and well-defined optical signature in UV–Vis assays; however, we acknowledge that binding to double-stranded DNA may differ in magnitude or mode and warrants future biophysical analyses.

The intrinsic binding constant (Kb) was derived from the slope to the intercept obtained from the plot of $A_0/(A - A_0)$ versus 1/[DNA] (Figure S7C,D) and summarized in Table 2. Again, Rubpy complex showed the highest K_b value with DNA, corroborating previous data showing this compound has the highest interaction with BSA. The intrinsic binding constant is also an important indicator of the way of interaction of a metal complex with DNA. The K_b values for classical intercalators such as ethidium bromide (EB) and [Ru(phen)(dppz)] (phen = 1,10-phenanthroline; dppz = dipyrido[3,2-a:2′,3′-c]phenazine) are in the order of 10^6–10^7 M^{-1}, and smaller values were an indicator of electrostatic and partial intercalation type of mode of interaction, which are consistent with our previous hypothesis observed in the absorption spectrum of the complexes in the presence of DNA through the hyperchromic effect in the MLCT bands.

Table 2. Binding constant (Kb) of the interaction between Rubpy and Rubpe complexes and DNA. Values represent the mean ± standard error.

Complexes	K_b ($\times 10^3$ M^{-1})
Rubpy	1.47 ± 0.50
Rubpe	0.13 ± 0.03

Both complexes were studied as racemic mixtures, this approach is consistent with previous studies reporting similar binding constants for Δ/Λ enantiomers in Ru(II) polypyridyl systems [37–47]. Although chiral ruthenium polypyridyl complexes can interact differently with chiral biomolecules such as BSA, DNA, or RNA, numerous studies report that the two enantiomers often display binding constants within the same order of magnitude, with only minor differences between them [37–47]. Therefore, the use of racemic mixtures is not expected to significantly alter the interpretation of the binding affinities reported here.

The moderate K_b values obtained for both complexes suggest that DNA engagement is not the dominant contributor to cytotoxicity. To determine whether these interactions produce functional consequences at the DNA level, we next evaluated their capacity to induce plasmid cleavage under physiological conditions, described further below.

2.3. *Computational Modeling of Biomolecular Binding*

2.3.1. Docking to Serum Albumin: Structural Insights into Binding Sites I and II

While the experimental assays quantify binding strength in solution, docking studies were performed to visualize how Rubpy and Rubpe arrange within the BSA binding pockets and to identify the main non-covalent forces that stabilize these associations. This computational approach provides qualitative structural insight complementary to the fluorescence data.

Docking analysis confirmed that both complexes interact favorably with BSA, particularly within subdomain IIA, which contains the well-established Trp213 and Trp134 binding sites. For Rubpy, the lowest-energy pose at Trp213 (−8.18 kJ/mol; Figure 7) involved π–π T-shaped interactions with Trp213, hydrophobic contacts with Pro446, Cys447 and Lys439, and π–cation/π–anion contributions from Arg194 and Asp450. Additional stabilization

https://doi.org/10.3390/inorganics14020063

was provided by amide–π interactions with Pro446 and Cys447. At the Trp134 pocket (Figure 8), Rubpy also displayed a low-energy arrangement (−8.52 kJ/mol), driven by a hydrogen bond with Asn161, alkyl interactions with Leu283 and Pro281, π-alkyl contacts with Pro281 and Lys159, and π-anion contributions from Gln284.

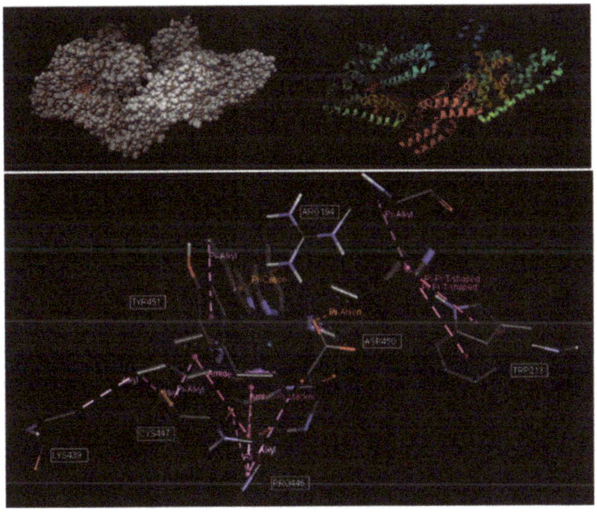

Figure 7. Molecular docking between the Rubpy complex and the tryptophan 213 site of BSA.

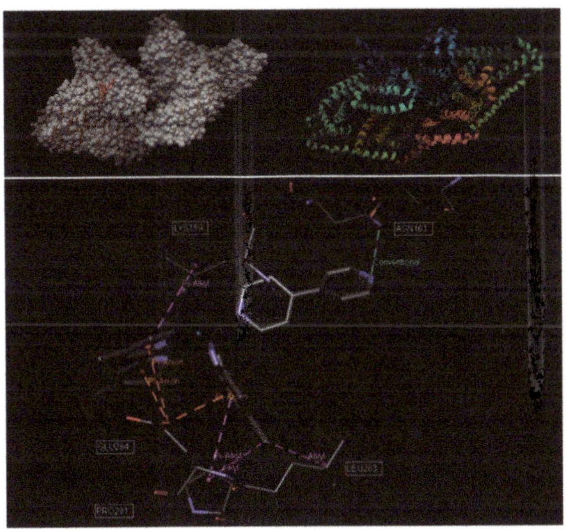

Figure 8. Molecular docking between the Rubpy complex and the tryptophan 134 site of BSA.

Rubpe exhibited a comparable interaction profile. At Trp213, its most stable configuration (−8.63 kJ/mol; Figure S8) was stabilized by alkyl, π-alkyl and π–σ interactions with Ala324, Arg208, Val215, Val234, Leu346, Lys350, Ala349 and Phe227. Amide–π contacts with Asp323 and Ala324, along with a hydrogen-bond–type interaction between a ligand nitrogen and the carbonyl oxygen of Leu346, further supported binding. A similarly favorable arrangement was obtained at the Trp134 pocket (−8.88 kJ/mol; Figure

https://doi.org/10.3390/inorganics14020063

S9), dominated by π–π stacking with Trp134 and π-alkyl interactions with Lys131, Lys159, Tyr155 and Val163.

Overall, the docking results show that both Rubpy and Rubpe preferentially occupy the hydrophobic cavities of BSA subdomain IIA, engaging residues around Trp213 and Trp134 through aromatic, hydrophobic and electrostatic interactions. These structural models are consistent with the fluorescence quenching experiments, which measured the relative binding affinities under physiological conditions, while the docking simulations provide molecular-level insight into the spatial organization and types of interactions governing complex–protein association.

2.3.2. Docking to DNA: Predicted Binding Modes and Interaction Energies

Molecular docking simulations were used to investigate how Rubpy and Rubpe arrange within the DNA double helix and to identify the main stabilizing interactions governing association. Both complexes displayed a clear preference for the DNA minor groove.

For Rubpy, the lowest-energy pose (−7.8 kJ/mol; Figure 9) positioned the complex along the minor groove, stabilized primarily by a T-shaped π–π interaction with adenine 12 (chain A) and π-alkyl contacts with guanines 13 and 14 (chain B). These interactions were complemented by additional hydrophobic contributions involving the coordinated ligands. Rubpe exhibited a similar binding pattern, with its most stable configuration (−9.87 kJ/mol; Figure S10) also occupying the minor groove. Stabilization arose from π-alkyl interactions with thymine 9 and T-shaped π–π associations with thymine 9 and cytosine 6 of chain A. These results are consistent with general structure–binding principles, as compact aromatic ligands frequently accommodate the geometry of the minor groove, where π-based interactions with nucleobases are particularly favorable. The docking outcomes for Rubpy and Rubpe therefore reinforce the minor groove as the preferred region of interaction for both complexes.

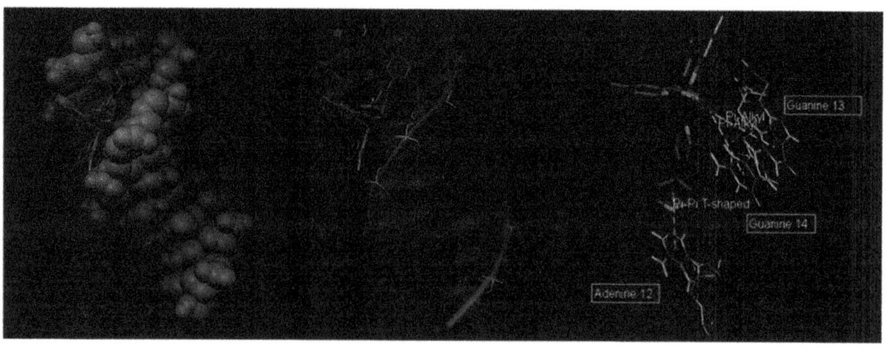

Figure 9. Molecular docking between the Rubpy complex and the macromolecule DNA.

Although docking suggested slightly stronger interactions for Rubpe, the experimental binding constants indicated the opposite trend, with Rubpy consistently showing higher affinity for DNA. This divergence is expected: docking evaluates geometric complementarity using rigid structures and simplified solvation models, whereas experimental measurements incorporate the full thermodynamic contributions of binding in solution, including solvent reorganization, ligand flexibility, hydrophobic effects and local structural rearrangements within DNA. In particular, the conformational flexibility and π-surface of the bpe ligand can artificially favor docking scores by enabling additional low-energy poses, even though these same features weaken π-stacking and reduce true binding affinity in solution. Thus, the docking preference for Rubpe reflects structural accommodation rather

https://doi.org/10.3390/inorganics14020063

than actual thermodynamic stability. These structural considerations also help explain the experimental results: the π-conjugated bpy ligand in Rubpy promotes stronger π-stacking and more stable non-covalent interactions with nucleobases, resulting in a higher intrinsic binding constant despite the slightly lower docking score. This behavior is fully consistent with previous reports showing that ligand planarity and π-surface are dominant contributors to the experimental DNA binding affinity of Ru(II) polypyridyl complexes. Highly planar, rigid ligands enhance π–π stacking and groove interactions, whereas non-planar or conformationally flexible ligands display significantly weaker binding despite favorable docking geometries [48,49]. Thus, while docking accurately identifies preferred binding geometries and interaction motifs, experimental affinities more faithfully reflect biologically relevant conditions [50]. The combined analysis therefore reconciles the apparent discrepancy and highlights how ligand rigidity and π-surface contribute differently to docking outcomes and experimental binding behaviors.

Together, the computational and spectroscopic analyses show that both complexes associate with DNA through non-covalent, minor-groove binding modes. To determine the functional consequences of these interactions, we next evaluated whether the complexes could induce structural alterations in DNA and how these molecular effects correlate with cellular uptake and cytotoxic responses.

2.4. Biological Consequences of DNA and Protein Interaction

2.4.1. DNA Cleavage Activity of Ruthenium Complexes

To further examine the functional consequences of complex–DNA association, plasmid cleavage assays were performed using pUC19 under physiological conditions. These tests complement the binding studies by assessing the ability of Rubpy and Rubpe to induce structural modifications in supercoiled DNA.

For Rubpy, concentrations of 50 and 100 µM promoted single-strand breaks (SSBs), with the extent of cleavage increasing at higher concentration (Figure 10). The electrophoretic profile revealed only Form I (supercoiled) and Form II (open circular), with no detectable linear Form III. Form II originates from SSBs that relax the helix, whereas Form III, observed only in the positive control containing restriction endonuclease, reflects double-strand breaks (DSBs) and migrates between the supercoiled and circular bands. These results indicate that Rubpy induces modest but detectable nuclease-like activity without generating DSBs.

A protective effect of DMSO was observed, consistent with partial quenching of reactive intermediates, whereas hydrogen peroxide did not significantly enhance DNA damage. This behavior suggests that cleavage proceeds mainly through structural destabilization or localized perturbation of the helix rather than through ROS-mediated oxidative pathways.

Rubpe displayed a similar concentration-dependent pattern (Figure S11), also limited to SSB formation and partially attenuated by DMSO, with no enhancement upon addition of hydrogen peroxide. These findings align with the spectroscopic Kb measurements: the moderate binding affinities observed for both complexes are sufficient to induce local unwinding of the double helix and facilitate single-strand scission. This is consistent with previous reports of Ru(II) polypyridyl complexes, where ligand electronic and steric properties modulate both DNA affinity and the extent of nuclease-like activity [51,52].

Overall, the cleavage data support a non-intercalative binding mode involving surface association accompanied by shallow groove interactions. Such interactions can destabilize the plasmid structure enough to promote SSBs but do not generate extensive or oxidative cleavage. These observations emphasize the key role of ligand environment in controlling DNA reactivity and motivate further mechanistic investigations into the specific interactions responsible for the observed nuclease-like behavior.

https://doi.org/10.3390/inorganics14020063

The absence of double-strand breaks and the modest level of single-strand scission indicate that DNA is not the primary cytotoxic target for these complexes, in agreement with the dissociation observed between DNA affinity and antiproliferative potency. These findings suggest that the biological activity of Rubpy and Rubpe cannot be explained solely by their interactions with DNA. Because DNA cleavage requires intracellular availability of the complexes, we next evaluated their ability to enter cancer cells and accumulate at biologically relevant concentrations.

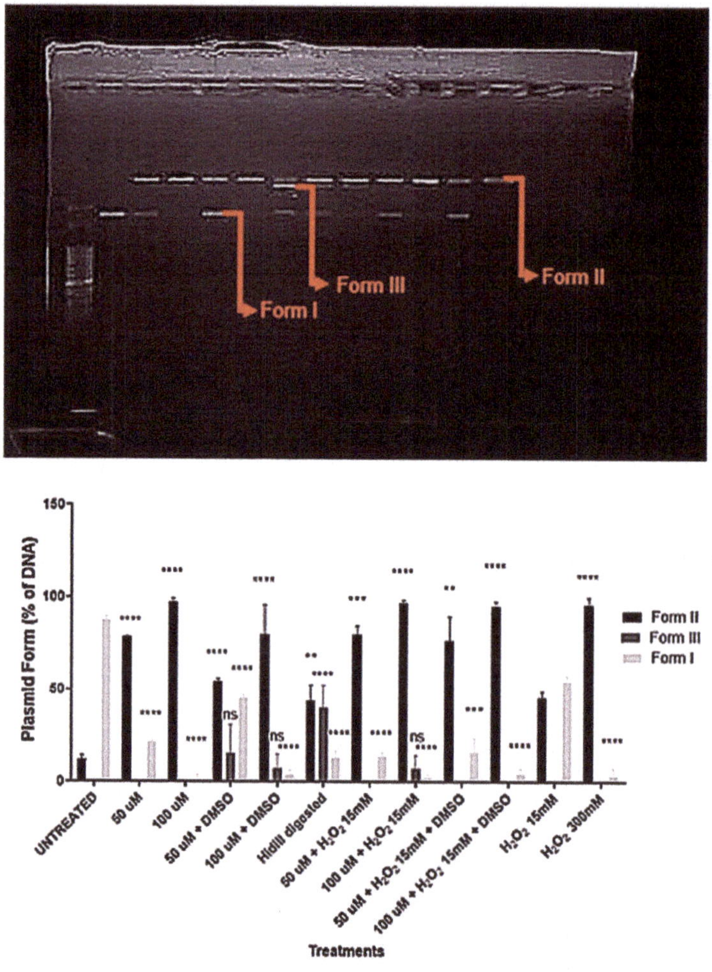

Figure 10. Image of plasmid degradation agarose gel and banding patterns found in each treatment, where 1: Marker; 2: Untreated control; 3: Rubpy (50 µM); 4: Rubpy (100 µM); 5: Rubpy (50 µM) + DMSO; 6: Rubpy (100 µM) + DMSO; 7: Hind III; 8: Rubpy (50 µM); 9: Rubpy (100 µM) + 15 mM H_2O_2; 10: Rubpy (50 µM) + 15 mM H_2O_2 + DMSO; 11: Rubpy (100 µM) + 15 mM H_2O_2 + DMSO; 12: 15 mM H_2O_2; 13: 300 mM H_2O_2. Graph representing the quantification of plasmid DNA in the bands found by agarose gel electrophoresis. Data expressed as the mean ± standard deviation of three assays. Statistical analysis was performed by 2-way ANOVA and multiple comparison by Bonferroni test * $p < 0.05$; ** $p < 0.01$; *** $p < 0.001$; **** $p < 0.0001$; ^{ns} not significant. Results were compared to the untreated control (in the absence of H_2O_2) and to the 15 mM H_2O_2 control (in the presence of H_2O_2).

https://doi.org/10.3390/inorganics14020063

2.4.2. Cellular Uptake in Cancer Cell Lines

The cellular uptake/incorporation properties of a molecule are critical for its application as a therapeutic agent. To exert cytotoxic effects, the compound must traverse the cell membrane, primarily composed of lipids and membrane proteins that regulate the passage of substances in and out of the cell [53]. Research on ruthenium complexes indicates that cellular uptake correlates directly with hydrophobicity, and higher cellular uptake leads to elevated cytotoxicity [54–58]. Thus, we exploited the cell uptake of ruthenium complexes into HeLa and MDA-MB-231 cells.

In HeLa cells, approximately 52% of Rubpy and around 47% of Rubpe were observed to incorporate (Figure 11). Meanwhile, the incorporation rates were 42% for Rubpy and 32% for Rubpe in MDA-MB-231 cells (Figure 11). Notably, the incorporation rates between Rubpy and Rubpe were very similar in both tested tumor cell lines. No significant difference in uptake percentages was observed between complexes in any of the examined tumor cell lines. The comparable uptake levels suggest that the structural modification in Rubpe, involving the insertion of a vinylene group (–CH=CH–), does not significantly alter the ability to permeate the cell membrane. This observation is consistent with the physicochemical similarity between the two complexes, including charge and size, which typically drive uptake via passive diffusion, facilitated diffusion, or endocytosis. The higher incorporation observed in HeLa cells compared to MDA-MB-231 cells may reflect differences in membrane composition or transport processes specific to each cell line.

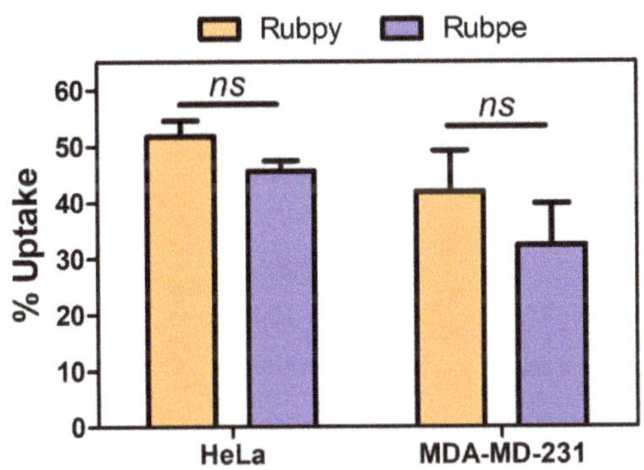

Figure 11. % Cellular uptake of Ru complexes (60 μM) incubated during 24 h in HeLa or MDA-MB-231 cells. Bars correspond to the mean ± standard deviation of three independent experiments with 4 replications in each experiment (n = 12), *ns* indicates that there is no statistical difference determined by the *Holm–Sidak* method, α= 0.05.

Ruthenium complexes can enter the cell through several mechanisms that include endocytosis and active transport (an energy-dependent process) or passive and facilitated (or energy-independent) diffusion [53]. These mechanisms generally depend on characteristics such as lipophilicity/hydrophilicity, charge, and size of the molecular structures. Given that the compounds share very similar structures, including their charges, it was expected that their cellular incorporation behaviors would yield comparable outcomes.

2.4.3. In Vitro Cytotoxicity Activity in HeLa and MDA-MB-231 Cells (MTT Assay)

To determine whether the molecular interactions observed in solution translate into biological activity, the cytotoxicity of Rubpy and Rubpe was evaluated in HeLa and MDA-MB-231 cells using the MTT assay after 24 h of incubation. This time was selected to minimize secondary metabolic effects often observed at longer exposures (48–72 h), enabling a clearer correlation with immediate molecular interactions. Both complexes reduced

https://doi.org/10.3390/inorganics14020063

cell viability in a dose-dependent manner (Figure 12), with Rubpe exhibiting consistently higher cytotoxicity than Rubpy in both cell lines.

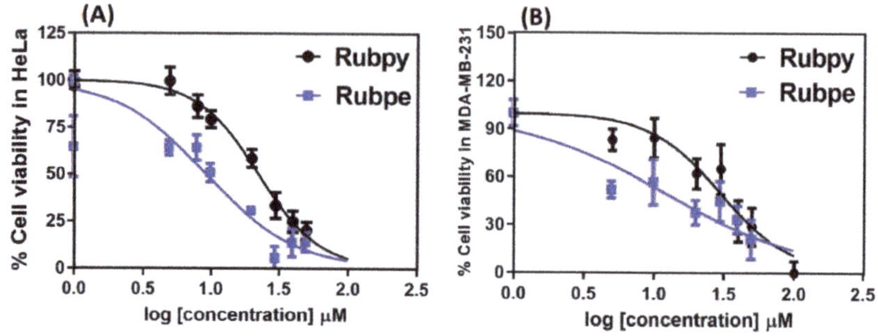

Figure 12. Dose–response curve on (**A**) HeLa cells and (**B**) MDA-MB-231 cells treated with Rubpy (black curves) and Rubpe (blue curves), during 24 h incubation under 5% CO_2 at 37 °C. Error bars represent the mean ± standard deviation of three independent experiments.

Rubpy displayed IC_{50} values of 22.37 μM (HeLa) and 29.69 μM (MDA-MB-231), whereas Rubpe showed IC_{50} values of 9.41 μM and 12.08 μM, respectively (Table 3). Thus, Rubpe was approximately 2.4-fold more potent than Rubpy in both models. These values fall well within the ranges reported for other ruthenium complexes acting on HeLa (13–200 μM) and MDA-MB-231 cells (5–72 μM) [59], confirming that both compounds exhibit comparatively strong cytotoxic effects [59].

Table 3. Cytotoxic effects expressed as IC_{50} values (μM) of the complexes $[Ru(dmbpy)_2Cl(bpy)](PF_6)$ and $[Ru(dmbpy)_2Cl(bpe)](PF_6)$, after 24 h treatment, toward two cancer cell lines, HeLa and MDA-MB-321.

Complexes	IC_{50} (μM)	
	HeLa	MDA-MB-231
Rubpy	22.37 ± 0.03	29.69 ± 0.06
Rubpe	9.41 ± 0.04	12.08 ± 0.10

To contextualize the cytotoxicity of Rubpy and Rubpe, their IC_{50} values against HeLa cells were compared with representative Ru(II) polypyridyl complexes reported in the literature under dark conditions (Table 4). HeLa cells were selected as the reference model because they are among the most frequently used human cancer cell lines in studies of Ru(II) polypyridyl complexes, allowing a more meaningful qualitative comparison across different reports. As summarized in Table 4, most Ru(II) polypyridyl complexes evaluated in HeLa cells exhibit IC_{50} values in the range of approximately 10–40 μM. Complexes containing extended planar ligands, such as phenanthroline- or dppz-based systems, typically display moderate cytotoxicity, with reported IC_{50} values between ~18 and 36 μM [60–62]. Similarly, Ru(II) complexes designed to target specific DNA motifs or transcriptional regulators show IC_{50} values in the low-to-mid micromolar range in HeLa cells [63,64].

Within this context, the IC_{50} values obtained for Rubpy (22.37 μM) and Rubpe (9.41 μM) fall well within, and in the case of Rubpe at the lower end of, the range reported for Ru(II) polypyridyl complexes in HeLa cells. Notably, Rubpe displays cytotoxicity comparable to or higher than several literature-reported Ru(II) complexes that incorporate more extended aromatic ligands or additional DNA-binding motifs, despite its comparatively simpler coordination environment. This observation highlights that subtle variations in ligand rigidity and π-surface can be sufficient to modulate cytotoxic responses in HeLa

https://doi.org/10.3390/inorganics14020063

cells, without the need for highly elaborate ligand frameworks. For comparison with a clinically established reference drug, cisplatin exhibits an IC_{50} value of 12.0 ± 0.52 μM in HeLa cells under comparable conditions [62]. Under the experimental conditions employed in this study, Rubpe shows cytotoxicity approaching that of cisplatin, whereas Rubpy is less potent but remains within the same order of magnitude. These results indicate that Rubpe, in particular, achieves cytotoxic efficacy in HeLa cells that is competitive with cisplatin.

Table 4. Cytotoxicity (IC_{50}) of some Ru(II) polypyridyl complexes and cisplatin in HeLa cells.

Complexes	IC_{50} (μM)	Ref.
Rubpy	22.37 ± 0.03	This work
Rubpe	9.41 ± 0.04	This work
$[Ru(phen)_2(tip)]^{2+}$	18.56 ± 0.67	[63]
$[Ru(phen)_2(bppp)]^{2+}$	32.0 ± 1.2	[61]
$[Ru(phen)_2(5\text{-}idip)]^{2+}$	11.7 ± 1.2	[64]
$[Ru(phen)_2BrIPC]^{2+}$	18.15	[60]
$[Ru(phen)_2(dppz\text{-}7,8\text{-}(OMe)_2)]^{2+}$	36.5	[60]
Cisplatin	12.0 ± 0.52	[61]

phen: 1,10-phenantroline, tip: 2-thiophene-(1H-imidazo-4,5-f][1,10]phenanthroline, bppp = 11-bromo-pyrido[2′,3′:5,6]pyrazino[2,3-f][1,10]phenanthroline], 5-idip: 2-indole-[4,5-f][1,10]phenanthroline, BrIPC: 6-bromo-3-(1H-imidazo[4,5-f][1,10]-phenanthroline, dppz-7,8-$(OMe)_2$: 7,8-Dimethoxydipyrido[3,2-a:2′,3′-c]phenazine.

Notably, the IC_{50} values do not follow the same trend as the binding constants measured for BSA and DNA, where Rubpy exhibited slightly stronger affinity. This lack of correlation has been reported for other Ru complexes, where rapid aquation or strong protein binding does not necessarily predict cytotoxicity [59]. For instance, 4-amino-1,2,4-triazolium [trans-$RuCl_4$(4-amino-1,2,4-triazole)(dmso-S)] displays the fastest aquation and highest BSA affinity within its series but is also the least cytotoxic (IC_{50} = 621 μM in HT-29 cells) [59].

Cellular uptake is often associated with increased cytotoxicity [65–68]. However, Rubpy and Rubpe showed comparable internalization levels, approximately 52% and 47% in HeLa, and 42% and 32% in MDA-MB-231, respectively, suggesting that uptake alone does not account for the difference in antiproliferative activity. Furthermore, DNA binding and cleavage studies indicated that Rubpy interacts more strongly with DNA, yet Rubpe is the more cytotoxic compound, reinforcing the notion that DNA engagement is not the primary driver of activity for these complexes.

Although logP values were not experimentally determined, the increased flexibility and reduced π-surface of bpe relative to bpy are structural features that are known to modulate lipophilicity and membrane partition in Ru(II) polypyridyl systems. These characteristics can reduce π-stacking with DNA while favoring intracellular redistribution, which may contribute to the higher cytotoxicity observed for Rubpe [69,70]. We note that this interpretation is qualitative and is intended to provide a mechanistic context consistent with established structure–property trends in the literature [69,70]. The greater cytotoxic efficacy of Rubpe, despite its lower affinity for DNA and BSA, suggests contributions from additional biological pathways. In Ru(II) polypyridyl systems, mitochondrial accumulation, disruption of redox homeostasis and modulation of apoptotic signaling have frequently been reported as non-DNA mechanisms [5,13,65,71,72] and these processes may also be relevant here.

Beyond membrane partition, the conformational flexibility of the bpe ligand may facilitate dynamic ligand reorganization within the intracellular environment, enabling alternative interactions with proteins or organelles that are not accessible to the more rigid bpy analogue. Such behavior has been increasingly recognized in Ru(II) systems and may

https://doi.org/10.3390/inorganics14020063

contribute to the superior cytotoxicity observed for Rubpe. Further mechanistic studies will be required to clarify the intracellular pathways underlying these differences

Although the present study provides a controlled comparison between bpy- and bpe-containing Ru(II) complexes, some constraints should be acknowledged. ss-DNA was used as a spectroscopic model and may underestimate affinity relative to dsDNA; enantiomers were not separated; and mechanistic cellular pathways beyond DNA binding were not resolved. These limitations do not compromise the comparative trends but highlight opportunities for future refinement.

3. Materials and Methods

3.1. Materials

Analytical-grade reagents were used without purification in all experiments. The compounds *cis*-[Ru(dmbpy)$_2$Cl(bpy)](PF$_6$) and *cis*-[Ru(dmbpy)$_2$Cl(bpe)](PF$_6$) were prepared following the report by Toma and collaborators [73]. Product purity was confirmed by elemental analysis, mass spectrometry, and ^{1}H Nuclear Magnetic Resonance (NMR). Methanol, N,N′-dimethylformamide (DMF), dimethyl sulfoxide (DMSO), Bovine Serum Albumin (BSA), phosphate-buffered solution (PBS—10.0 mM phosphate buffer, 2.70 mM potassium chloride and 137 mM sodium chloride at pH 7.4), Dulbecco's Modified Eagle medium (DMEM) and 3-(4,5-dimethylthiazol-2-yl)-2,5-diphenyltetrazolium bromide (MTT) were purchased from Sigma-Aldrich (St. Louis, MO, USA).

3.2. Spectroscopic and Electrochemical Measurements

UV/Vis absorption spectra were achieved by using a Hewlett Packard 8453A diode array spectrophotometer (Palo Alto, CA, USA). Electrochemical measurements were recorded on an Autolab PGSTAT30 potentiostat (Eco-Chemie, Utrecht, The Netherlands) employing a standard three-electrode setup, with a platinum working electrode, a platinum wire auxiliary electrode, and a reference electrode consisting of Ag/AgNO$_3$ (10 mM; +0.503 V vs. NHE). Tetrabutylammonium perchlorate (Bu$_4$NClO$_4$; 0.10 M) was used as the electrolytes. UV/Vis absorption spectroelectrochemistry measurements were carried out using a custom-designed electrochemical cell equipped with a gold minigrid working electrode, an Ag/AgNO$_3$ (10 mM in acetonitrile) reference electrode, and a platinum wire auxiliary electrode mounted inside a quartz cuvette with a path length of 25 μm. The application of potentials was controlled by an EG&G PAR 173 potentiostat (Oak Ridge, TN, USA), and the spectra were recorded with an HP 8453A spectrophotometer. Steady-state emission measurements were carried out in a Horiba FluoroMax 4 spectrofluorimeter (Edison, NJ, USA). Samples were measured at 298 K in a 1.000 cm quartz cuvette with a closed cap. The solutions were derated with Ar before the measurements. Mass spectra were obtained on an Esquire 3000 Plus Bruker Daltonics Mass Spectrometer (Bremen, Germany), 4000 V capillary and 180 μL h^{-1} flow injection. Elemental Analyses were executed by using a Perkin Elmer 2400 series II analyzer (Waltham, MA, USA). The ^{1}H Nuclear Magnetic Resonance spectra were performed on a Bruker DRX 300 Spectrometer (Rheinstetten, Germany) using acetone-d_6 as solvent.

3.3. Studies of Complexes in Aqueous Solution

The stability of the complexes solutions was analyzed by UV–Vis spectroscopy recording the electronic spectra in water and in PBS buffer (pH 7.4). The final concentration of each complex was achieved by diluting small aliquots of freshly prepared concentrated methanolic stock solutions, 6×10^{-5} M. The resulting solutions were examined over 96 h at 25 °C.

https://doi.org/10.3390/inorganics14020063

It also evaluated the stability of the Rubpy and Rubpe complexes in the presence of bovine serum albumin (BSA). For that, the stock solution of the complex in methanol was evaporated using nitrogen gas to produce a film. Then, a solution of BSA in the PBS buffer (pH = 7.4) was added to the complex film. The final concentration of the ruthenium complex was 50 μM, and absorbance readings were taken as a function of time. The spectrum of BSA (3 μM) without complex was used to subtract from the spectra of the complex and BSA mixture solutions, better to visualize the overlap and similarity of the MLCT bands. The absorption spectra were recorded in a Shimadzu spectrophotometer model UV-2501 BC (Kyoto, Japan).

3.4. Molecular Docking Studies and Ligand Preparation

The ligands were developed employing Avogadro software version 1.1.1 and saved in mol2 format following structural optimization. Using the same software and its extensions tool, the structures were optimized based on the Universal Force Field (UFF). In AutoDock 1.5.7, ligand preparation involved adding polar hydrogen atoms, assigning Gasteiger charges, and removing potential water molecules.

Molecular docking simulation of ruthenium complexes were conducted using AutoDock 1.5.7 with DNA and BSA biomolecules. The crystal structures of DNA (126D) with the sequence CATGGCCATG (resolution: 2 Å) and BSA (3V03) (resolution: 2.7 Å) were retrieved from the Protein Data Bank (www.rcsb.org, accessed on 6 October 2025). Polar hydrogen atoms and Kollman charges were added to these structures, and water molecules surrounding the DNA/BSA were removed.

For blind docking, the DNA binding analysis used a grid box with dimensions 62 × 58 × 96 for the three complexes. BSA binding was assessed at two possible sites near tryptophan residues: tryptophan 134 (coordinates x = 45.989286, y = 36.667143, z = 25.500143) and tryptophan 213 (coordinates x = 101.169786, y = 28.225500, z = 19.778429), with a scan box size of 70 × 76 × 78.

Ligand-receptor binding calculations utilized the Lamarkian genetic algorithm, with 100 runs and 2.5 million evaluations. Visualization of DNA/BSA interactions was performed with Biovia Discovery Studio Visualizer version 2021.

3.5. Albumin Binding Studies

The study of protein binding was conducted through tryptophan fluorescence quenching experiments, employing bovine serum albumin (BSA) [28]. Quenching of the emission intensity of tryptophan residues of BSA at 350 nm was evaluated using the complexes Rubpy and Rubpe as quenchers. For that, the ruthenium complexes in different concentrations (0–87 μM) were added to a BSA (3 μM) solution in PBS (pH 7.4). Fluorescence spectra were collected in the range of 300–550 nm at an excitation wavelength of 280 nm, at 25 °C. All intrinsic quenching fluorescence experiments were executed in duplicate. The absorption spectra of the compounds in buffered solutions were recorded under the same experimental conditions, and to compensate for the inner filter effect [28], fluorescence intensity values were corrected for the absorbance of the individual complex using the Parker equation (Equation (1)) [74,75]:

$$F_{cor} = F_{obs} \times \frac{2,3dA_{exc}}{1-10^{-dA_{exc}}} \times 10^{gA_{em}} \times \frac{2,3sA_{em}}{1-10^{-sA_{em}}} \quad (1)$$

where F_{Cor} and F_{Obs} correspond to the corrected and observed fluorescence intensities, × respectively. A_{exc} and A_{em} denote the absorbances at the excitation λ_{exc} = 280 nm and λ_{em} = 350 nm, respectively. The optical path length of the cuvette (d) was 1.0 cm; "g" represents the distance between the edge of the excitation beam and "d" the edge of the cuvette (0.40 cm); and "s" is the excitation beam thickness (0.10 cm) [74–76].

https://doi.org/10.3390/inorganics14020063

The Stern-Vomer constant (K_{SV}) was calculated for ruthenium complexes from Stern-Volmer Equation (2) [28]:

$$\frac{F_0}{F} = 1 + K_{SV} \cdot [Q] \quad (2)$$

where F_0 and F represent the fluorescence intensities recorded in the absence and presence of the quencher, respectively. K_{sv} denotes the Stern–Volmer quenching constant, and [Q] corresponds to the concentration of the quencher.

The regular Stern–Volmer plots were not applicable for complexes Rubpy and Rubpe as the plot between F_0/F and [Q] displayed an upward curvature. An exponential term, $\exp^{(V[Q])}$, where V represents the static quenching constant, can be introduced into the Stern–Volmer equation (Equation (3)) to account for both quenching modes (static and dynamic) [9,29]:

$$F_0/F = (1 + K_{sv}[Q])e^{V[Q]} \quad (3)$$

The static quenching constant, V was determined by applying the equation and plotting $\{F_0/Fe^{V[Q]}\} - 1$ vs. [Q], systematically varying V until a linear relationship was observed. The optimal value of V was determined by identifying the plot that yielded the highest correlation coefficient. The dynamic (collisional) quenching constant, Ksv, was subsequently determined from the slope of the corresponding linear plots [9,29].

The strength of interaction between the compounds and BS) was determined by binding constant (K_b) obtained employing the modified Stern-Volmer (Equation (4)) below:

$$log\frac{(F_0 - F)}{F} = log\, K_b + n\, log\, [Q] \quad (4)$$

where F_0 and F are, respectively, the fluorescence intensity in the absence and presence of the quencher, K_b is the apparent binding constant of the quencher to the BSA, "n" is the number of binding sites of the quencher to the BSA and [Q] is the concentration of the quencher [28].

3.6. Cell Culture

The HeLa (human cervical cancer) cell line was obtained from Laboratory of Photoinduced Processes and Interfaces coordinated by Prof. Dr. Maurício Baptista (Institute of Chemistry, University of São Paulo, São Paulo—SP, Brazil). The MDA-MB-231 (human breast cancer) cell line was provided by the Laboratory of Nanobiotechnology Dr. Luiz Ricardo Goulart coordinated by Dr. Luciana Machado Bastos (Institute of Biotechnology, Federal University of Uberlandia, Uberlandia—Minas Gerais, Brazil).

HeLa (human cervical cancer) and MDA-MB-231 (human breast cancer) cell lines were cultivated as monolayer culture in Dulbecco's Modified Eagle Medium (DMEM) high glucose (4500 mg/L glucose, sodium pyruvate, L-glutamine, sodium bicarbonate and phenol red) supplemented with 10% fetal bovine serum (FBS), 100 UI/mL penicillin and 100 μg/mL streptomycin. The cells were kept at 37 °C, in an incubator with a 5% CO_2 atmosphere.

3.7. Cellular Uptake

HeLa or MDA-MB-231 cells (1.0×10^5 cell/well) were seeded in 1 mL of culture medium in 12-well plates and incubated at 37 °C and 5% CO_2 atmosphere during 24 h, for attachment of cells to the bottom of the well. After that, cells were treated with ruthenium complexes (60 μM) in 1 mL supplemented medium (DMEM with 10% FBS, 100 UI/mL penicillin and 100 μg/mL streptomycin) for 24 hours in the dark, 5% CO_2 at 37 °C. After this period, 500 μL of the supernatant solutions were removed from the well. The remaining cells in the well plate were washed with PBS, after which 500 μL of methanol was added to each well to extract the intracellularly incorporated complex. The percentage of cellular uptake was calculated as described previously [77,78]. Incubations

https://doi.org/10.3390/inorganics14020063

were performed in the dark to avoid uncontrolled photochemical processes that could induce ROS formation or alter Ru(II) electronic states, ensuring that all cytotoxic effects arose solely from ground-state interactions.

3.8. Cytotoxicity Activity (MTT Assay)

The cytotoxic activity of ruthenium complexes was analyzed by MTT assay [79–81]. For that, the cells (2.8×10^4 cells/well) were seeded in 300 μL of the medium in 48-well plates and incubated at 37 °C, in a 5% CO_2 atmosphere for 24 h. Next, the complexes were dissolved in methanol, which evaporated under argon flux, producing a film. Thus, the complex solutions were prepared by hydration of its film with supplemented culture medium (DMEM with 10% fetal bovine serum (FBS), 100 UI/mL penicillin and 100 μg/mL streptomycin). After 24 h of growth, cells were exposed to 300 μL of each concentration of complexes Rubpy or Rubpe (0, 1, 5, 8, 10, 20, 30, 40, 50 and 100 μM) for 24 h. Then, 100 μL of MTT (0.75 mg/mL in PBS, pH 7.4) solution was added to each well. After a further period of MTT incubation (3 h at 37 °C in 5% CO_2), the medium was removed, and the formazan crystals were solubilized by DMSO. The absorbance was registered using an automated microplate reader at 570 nm [79–81]. The percent of cell viability was calculated as Equation (5):

$$\%\ Cell\ viability = \frac{A_{treated(570\mathrm{nm}-800\mathrm{nm})}}{A_{control(570\mathrm{nm}-800\mathrm{nm})}} \times 100\% \tag{5}$$

where $A_{treated}$ is the Absorbance (in 570 nm–800 nm) of cells treated with ruthenium complexes and $A_{control}$ is the Absorbance (in 570 nm–800 nm) of control cells.

The IC_{50} values were determined by plotting the percentage viability vs. concentration on a logarithmic graph and fitting the dose–response curve with GraphPad Prism 5.0 Software.

3.9. DNA Interaction Assay by UV–Vis Spectroscopy

A stock solution of salmon sperm DNA (ss-DNA) was prepared by dissolving 5 mg of ss-DNA in 5 mL of PBS buffer (50 mM) containing 5 mM NaCl at pH 7.4. Prior to the experiments, DNA purity was verified by confirming an $Abs_{260nm}/Abs_{280nm} \geq 1.8$ The ss-DNA concentration was determined using the molar extinction coefficient at 260 nm ε_{260} = 6600 M^{-1} cm^{-1} [36]. The ss-DNA stock solution was maintained at 2–10 °C in an ice bath throughout the experiments. Solutions of Rubpy or Rubpe were prepared at a fixed concentration of 50 μM in PBS buffer (50 mM) containing NaCl (5 mM). Sequential additions of 10 μL aliquots of ss-DNA were then made, and after each addition the mixture was incubated for 5 min prior to recording the absorption spectra over an ss-DNA concentration range of 0–50 μM. The intrinsic binding constant (K_b) was derived from the slope/intercept relationship of th $A_0/(A - A_0)$ vs. 1/[DNA] according to the following Equation [36]

$$\frac{A_0}{A - A_0} = \frac{e_0}{e_{HG} - e_G} \times \frac{1}{K_b[DNA]} \tag{6}$$

3.10. DNA Cleavage Studies

To investigate the interaction of the complexes with DNA, we combined *cis*-$[Ru(dmbpy)_2Cl(bpy)](PF_6)$ and *cis*-$[Ru(dmbpy)_2Cl(bpe)](PF_6)$ with plasmid DNA from the pUC-19 vector (Lot. 1212425A, Clontech Laboratories, Inc., Takara Bio USA, Mountain View, CA, USA). The isolated plasmid was added to a series of 11 solutions, each with a total volume of 20 μL. Each solution contained 30 ng/μL of plasmid DNA treated with either 50 μM or 100 μM of *cis*-$[Ru(dmbpy)_2Cl(bpy)](PF_6)$ or *cis*-$[Ru(dmbpy)_2Cl(bpe)](PF_6)$ in PBS 1×, with or without dimethyl sulfoxide (DMSO, 0.05%), hydrogen peroxide (H_2O_2,

https://doi.org/10.3390/inorganics14020063

15 mM), or a combination of both, as well as with H_2O_2 alone (300 mM). Controls included untreated plasmid DNA, plasmid DNA digested with the site-specific restriction enzyme Hind III, and plasmid DNA treated solely with H_2O_2. The reaction mixtures were incubated at 37 °C for 12 h, followed by the addition of 3 µL of loading buffer (0.25% bromophenol blue, 0.25% xylene cyanol, 30% glycerol, and 10 mM EDTA). Electrophoresis was then performed on a 0.8% agarose gel containing 0.05% ethidium bromide (10 µg/mL) in 0.5× TBE buffer (90 mM Tris-borate, pH 8.0, and 20 mM EDTA). The gels were run at 80 V for 3 h, and the resulting DNA bands were visualized under ultraviolet light and quantified using ImageJ software, version 1.53k (Java 1.8.0_172).

4. Conclusions

The two Ru(II) polypyridyl complexes, Rubpy and Rubpe, displayed closely related spectroscopic and redox properties, consistent with their structural similarity. Both complexes remained stable in aqueous solution and in the presence of BSA, ensuring that the species probed in biomolecular and cellular assays correspond to intact Ru(II) complexes. Fluorescence quenching experiments revealed binding constants on the order of 10^5 M^{-1}, indicating appreciable affinity toward serum albumin. UV–Vis titrations with ss-DNA showed hyperchromic MLCT responses and moderate intrinsic binding constants, with Rubpy exhibiting slightly higher affinity, consistent with a non-covalent interaction mode involving electrostatic and/or shallow groove association.

Cellular uptake levels were comparable for the two complexes in both HeLa and MDA-MB-231 cells. Nevertheless, Rubpe exhibited approximately 2.4-fold greater cytotoxicity than Rubpy, underscoring that neither uptake nor DNA binding alone accounts for the observed antiproliferative effects. This distinction suggests that additional intracellular pathways, beyond direct DNA involvement, likely contribute to the biological activity of Rubpe.

Overall, the integrated spectroscopic, computational, and cellular data highlight how subtle modifications in ligand rigidity and π-surface can modulate biomolecular interactions and cytotoxic responses in Ru(II) polypyridyl complexes, offering mechanistic insight that can guide future ligand design.

Supplementary Materials: The following supporting information can be downloaded at: https://www.mdpi.com/article/10.3390/inorganics14020063/s1, Figure S1. Spectroelectrochemical changes associated with the oxidation (A), the first (B) e second (C) reduction of Rubpy complex; Figure S2. Spectroelectrochemical changes associated with the oxidation following by reduction of Rubpy (A) and Rubpe (B) complexes; Figure S3. Spectroelectrochemical changes associated with the oxidation (A), the first (B) e second (C) reduction of Rubpe complex; Figure S4. UV–Vis spectra of (A) *cis*-[RuCl(dmbpy)(bpy)]$^+$ and (B) *cis*-[RuCl(dmbpy)(bpe)]$^+$ in 80% H_2O:20% MeOH over a period of 377 min. [Ru complex] = 60 µM, T = 25 °C; Figure S5. UV–Vis spectra of (A) *cis*-[RuCl(dmbpy)(bpy)]$^+$ and (B) *cis*-[RuCl(dmbpy)(bpe)]$^+$ in presence of 3 µM BSA. [Ru complex] = 50 µM, room temperature. Insert: MLCT band of complexes between 383 nm and 470 nm, highlighting hypochromic effect; Figure S6. Determination of the static suppression constant (V) using the Stern-Volmer equation; Figure S7. Absorption spectra of the *cis*-[RuCl(dmbpy)(bpy)]$^+$ (A) and *cis*-[RuCl(dmbpy)(bpe)]$^+$ (B) both at 50 µM, with successive additions of ss-DNA aliquots (ranging from 0 to 50 µM, in 50 mM PBS buffer containing 5 mM NaCl, pH 7.4) with 5 min of incubation at room temperature. Plots of A_0/A-A_0 versus 1/[DNA] derived to calculate K_b values of *cis*-[RuCl(dmbpy)(bpy)]$^+$ (C) and *cis*-[RuCl(dmbpy)(bpe)]$^+$ (D); Figure S8. Molecular docking between the Rubpe complex and the tryptophan 213 site of BSA; Figure S9. Molecular docking between the Rubpe complex and the tryptophan 134 site of BSA; Figure S10. Molecular docking between the Rubpe complex and the macromolecule DNA; Figure S11. Image of plasmid degradation agarose gel and banding patterns found in each treatment, where: 1: Marker;

https://doi.org/10.3390/inorganics14020063

2: Untreated control; 3: Rubpe (50 μM); 4: Rubpe (100 μM); 5: Rubpe (50 μM) + DMSO; 6: Rubpe (100 μM) + DMSO; 7: Hind III; 8: Rubpe (50 μM); 9: Rubpe (100 μM) + 15 mM H_2O_2; 10: Rubpe (50 μM) + 15 mM H_2O_2 + DMSO; 11: Rubpe (100 μM) + 15 mM H_2O_2 + DMSO; 12: 15 mM H_2O_2; 13: 300 mM H_2O_2. Graph representing the quantification of plasmid DNA in the bands found by agarose gel electrophoresis. Data expressed as the mean ± standard deviation of three assays. Statistical analysis was performed by 2-way ANOVA and multiple comparison by Bonferroni test * $p < 0.05$; ** $p < 0.01$; *** $p < 0.001$; **** $p < 0.0001$; ^{ns} not significant. Results were compared to the untreated control (in the absence of H_2O_2) and to the 15 mM H_2O_2 control (in the presence of H_2O_2).

Author Contributions: Investigation, formal analysis and validation, P.A.d.M., M.E.G.d.C., A.L.A.P., C.M.F.S. and R.C.L.; methodology, H.E.T., M.E.B., R.J.d.O.J., A.O.d.T.P., T.A.M. and T.M.T.; writing—original draft preparation, P.A.d.M., H.E.T., M.E.B., R.J.d.O.J., A.O.d.T.P., T.A.M. and T.M.T.; writing—review and editing, T.A.M. and T.M.T.; supervision, T.A.M. and T.M.T.; project administration, T.M.T.; funding acquisition, T.M.T. All authors have read and agreed to the published version of the manuscript.

Funding: This research was funded by *Fundação de Amparo à Pesquisa do Estado de Minas Gerais* (FAPEMIG, APQ-00704-21; APQ-02393-24; APQ-01044-21) and *Conselho Nacional de Desenvolvimento Científico e Tecnológico* (CNPq, 407282/2023-8; 308094/2025-5). This study was financed in part by *Coordenação de Aperfeiçoamento de Pessoal de Nível Superior* (CAPES—Finance Code 001 and PROEX).

Data Availability Statement: All data are contained within the manuscript. Raw data used for analysis will be shared upon request.

Acknowledgments: The authors acknowledge the financial support provided by FAPEMIG, CNPq and CAPES. Patrícia Alves de Matos thanks PhD scholarship grant from CAPES. Tayana M. Tsubone thanks the grant received by L'Oréal Brazil-UNESCO-ABC through "*For Women in Science*" Award 2023, in chemistry field.

Conflicts of Interest: The authors declare that they have no known competing financial interests or personal relationships that could have appeared to influence the work reported in this paper.

Abbreviations

The following abbreviations are used in this manuscript:

Rubpy	*cis*-[Ru(dmbpy)$_2$Cl(bpy)](PF$_6$)
Rubpe	*cis*-[Ru(dmbpy)$_2$Cl(bpe)](PF$_6$)
MLCT	metal-to-ligand charge transfer
BSA	bovine serum albumin

References

1. Kelland, L. The resurgence of platinum-based cancer chemotherapy. *Nat. Rev. Cancer* **2007**, *7*, 573–584. [CrossRef] [PubMed]
2. Thota, S.; Rodrigues, D.A.; Crans, D.C.; Barreiro, E.J. Ru(II) Compounds: Next-Generation Anticancer Metallotherapeutics? *J. Med. Chem.* **2018**, *61*, 5805–5821. [CrossRef] [PubMed]
3. Alessio, E. Thirty Years of the Drug Candidate NAMI-A and the Myths in the Field of Ruthenium Anticancer Compounds: A Personal Perspective. *Eur. J. Inorg. Chem.* **2016**, *2017*, 1549–1560. [CrossRef]
4. Bashir, M.; Mantoo, I.A.; Arjmand, F.; Tabassum, S.; Yousuf, I. An overview of advancement of organoruthenium (II) complexes as prospective anticancer agents. *Coord. Chem. Rev.* **2023**, *487*, 215169. [CrossRef]
5. Rilak Simović, A.; Masnikosa, R.; Bratsos, I.; Alessio, E. Chemistry and reactivity of ruthenium(II) complexes: DNA/protein binding mode and anticancer activity are related to the complex structure. *Coord. Chem. Rev.* **2019**, *398*, 113011. [CrossRef]
6. Baier, D.; Mendrina, T.; Schoenhacker-Alte, B.; Pirker, C.; Mohr, T.; Rusz, M.; Regner, B.; Schaier, M.; Sgarioto, N.; Raynal, N.J.; et al. The Lipid Metabolism as Target and Modulator of BOLD-100 Anticancer Activity: Crosstalk with Histone Acetylation. *Adv. Sci.* **2023**, *10*, e2301939. [CrossRef]
7. Skoczynska, A.; Lewinski, A.; Pokora, M.; Paneth, P.; Budzisz, E. An Overview of the Potential Medicinal and Pharmaceutical Properties of Ru(II)/(III) Complexes. *Int. J. Mol. Sci.* **2023**, *24*, 9512. [CrossRef]

https://doi.org/10.3390/inorganics14020063

8. Monro, S.; Colon, K.L.; Yin, H.; Roque, J., III; Konda, P.; Gujar, S.; Thummel, R.P.; Lilge, L.; Cameron, C.G.; McFarland, S.A. Transition Metal Complexes and Photodynamic Therapy from a Tumor-Centered Approach: Challenges, Opportunities, and Highlights from the Development of TLD1433. *Chem. Rev.* **2019**, *119*, 797–828. [CrossRef]
9. Milutinovic, M.M.; Rilak, A.; Bratsos, I.; Klisuric, O.; Vranes, M.; Gligorijevic, N.; Radulovic, S.; Bugarcic, Z.D. New 4′-(4-chlorophenyl)-2,2′:6′,2″-terpyridine ruthenium(II) complexes: Synthesis, characterization, interaction with DNA/BSA and cytotoxicity studies. *J. Inorg. Biochem.* **2017**, *169*, 1–12. [CrossRef]
10. Canovic, P.; Simovic, A.R.; Radisavljevic, S.; Bratsos, I.; Demitri, N.; Mitrovic, M.; Zelen, I.; Bugarcic, Z.D. Impact of aromaticity on anticancer activity of polypyridyl ruthenium(II) complexes: Synthesis, structure, DNA/protein binding, lipophilicity and anticancer activity. *JBIC J. Biol. Inorg. Chem.* **2017**, *22*, 1007–1028. [CrossRef] [PubMed]
11. Lazic, D.; Arsenijevic, A.; Puchta, R.; Bugarcic, Z.D.; Rilak, A. DNA binding properties, histidine interaction and cytotoxicity studies of water soluble ruthenium(ii) terpyridine complexes. *Dalton Trans.* **2016**, *45*, 4633–4646. [CrossRef]
12. Meng, X.; Leyva, M.L.; Jenny, M.; Gross, I.; Benosman, S.; Fricker, B.; Harlepp, S.; Hebraud, P.; Boos, A.; Wlosik, P.; et al. A ruthenium-containing organometallic compound reduces tumor growth through induction of the endoplasmic reticulum stress gene CHOP. *Cancer Res.* **2009**, *69*, 5458–5466. [CrossRef]
13. Huang, H.; Zhang, P.; Yu, B.; Chen, Y.; Wang, J.; Ji, L.; Chao, H. Targeting nucleus DNA with a cyclometalated dipyridophenazineruthenium(II) complex. *J. Med. Chem.* **2014**, *57*, 8971–8983. [CrossRef]
14. Khanvilkar, P.; Pulipaka, R.; Shirsath, K.; Devkar, R.; Chakraborty, D. Organometallic binuclear Ru(II) complexes: Design, synthesis, DNA/BSA binding interactions and in-vitro cytotoxicity against HeLa cell line. *Inorg. Chem. Commun.* **2019**, *102*, 134–140. [CrossRef]
15. Matos, C.P.; Valente, A.; Marques, F.; Adão, P.; Paula Robalo, M.; de Almeida, R.F.M.; Pessoa, J.C.; Santos, I.; Helena Garcia, M.; Tomaz, A.I. New polydentate Ru(III)-Salan complexes: Synthesis, characterization, anti-tumour activity and interaction with human serum proteins. *Inorganica Chim. Acta* **2013**, *394*, 616–626. [CrossRef]
16. Nehru, S.; Veeralakshmi, S.; Kalaiselvam, S.; Subin David, S.P.; Sandhya, J.; Arunachalam, S. Protein binding and antioxidant studies of diimine based emissive surfactant-ruthenium(II) complexes. *J. Biomol. Struct. Dyn.* **2021**, *39*, 1535–1546. [CrossRef]
17. Toma, S.H.; Uemi, M.; Nikolaou, S.; Tomazela, D.M.; Eberlin, M.N.; Toma, H.E. {trans-1, 4-Bis [(4-pyridyl) ethenyl] benzene}(2, 2′-bipyridine) ruthenium (II) Complexes and Their Supramolecular Assemblies with β-Cyclodextrin. *Inorg. Chem.* **2004**, *43*, 3521–3527. [CrossRef] [PubMed]
18. Meyer, T.J.; Huynh, M.H.V. The remarkable reactivity of high oxidation state ruthenium and osmium polypyridyl complexes. *Inorg. Chem.* **2003**, *42*, 8140–8160. [CrossRef] [PubMed]
19. Schafer, F.Q.; Buettner, G.R. Redox environment of the cell as viewed through the redox state of the glutathione disulfide/glutathione couple. *Free Radic. Biol. Med.* **2001**, *30*, 1191–1212. [CrossRef]
20. Rilak, A.; Bratsos, I.; Zangrando, E.; Kljun, J.; Turel, I.; Bugarcic, Z.D.; Alessio, E. New water-soluble ruthenium(II) terpyridine complexes for anticancer activity: Synthesis, characterization, activation kinetics, and interaction with guanine derivatives. *Inorg. Chem.* **2014**, *53*, 6113–6126. [CrossRef]
21. Matias, T.A.; Mangoni, A.P.; Toma, S.H.; Rein, F.N.; Rocha, R.C.; Toma, H.E.; Araki, K. Catalytic Water-Oxidation Activity of a Weakly Coupled Binuclear Ruthenium Polypyridyl Complex. *Eur. J. Inorg. Chem.* **2016**, *2016*, 5547–5556. [CrossRef]
22. Matias, T.A.; Rein, F.N.; Rocha, R.C.; Formiga, A.L.B.; Toma, H.E.; Araki, K. Effects of a strong pi-accepting ancillary ligand on the water oxidation activity of weakly coupled binuclear ruthenium catalysts. *Dalton Trans.* **2019**, *48*, 3009–3017. [CrossRef]
23. Matias, T.A.; Parussulo, A.L.A.; Benavides, P.A.; Guimarães, R.R.; Dourado, A.H.B.; Nakamura, M.; de Torresi, S.I.C.; Bertotti, M.; Araki, K. Polymeric binuclear ruthenium complex as efficient electrocatalyst for oxygen evolution reaction. *Electrochim. Acta* **2018**, *283*, 18–26. [CrossRef]
24. Sun, J.; Huang, Y.; Zheng, C.; Zhou, Y.; Liu, Y.; Liu, J. Ruthenium (II) complexes interact with human serum albumin and induce apoptosis of tumor cells. *Biol. Trace Elem. Res.* **2015**, *163*, 266–274. [CrossRef] [PubMed]
25. Htun, T. A negative deviation from Stern–Volmer equation in fluorescence quenching. *J. Fluoresc.* **2004**, *14*, 217–222. [CrossRef] [PubMed]
26. Kandagal, P.; Ashoka, S.; Seetharamappa, J.; Shaikh, S.; Jadegoud, Y.; Ijare, O.B. Study of the interaction of an anticancer drug with human and bovine serum albumin: Spectroscopic approach. *J. Pharm. Biomed. Anal.* **2006**, *41*, 393–399. [CrossRef]
27. Straight, R.; Spikes, J. Photosensitized oxidation of biomolecules. *Singlet O2* **1985**, *4*, 143.
28. Lakowicz, J.R. *Principles of Fluorescence Spectroscopy*; Springer: Boston, MA, US, 2006.
29. Seedher, N.; Agarwal, P. Complexation of fluoroquinolone antibiotics with human serum albumin: A fluorescence quenching study. *J. Lumin.* **2010**, *130*, 1841–1848. [CrossRef]
30. Eftink, M.R.; Ghiron, C.A. Fluorescence quenching studies with proteins. *Anal. Biochem.* **1981**, *114*, 199–227. [CrossRef]
31. B Neto, G.L.; Baptista, E.A.M.; Becca, G.H.S.; Nakatani, H.S.; Souza, V.R.D. Interações competitivas de complexos de rutênio contendo dimetilsulfóxido e ligantes N-heterocíclicos com albumina de soro humao. *Química Nova* **2020**, *43*, 261–270.

https://doi.org/10.3390/inorganics14020063

32. Lazniewska, J.; Agostino, M.; Hickey, S.M.; Parkinson-Lawrence, E.; Stagni, S.; Massi, M.; Brooks, D.A.; Plush, S.E. Spectroscopic and Molecular Docking Study of the Interaction between Neutral Re(I) Tetrazolate Complexes and Bovine Serum Albumin. *Chem. –A Eur. J.* **2021**, *27*, 11406–11417. [CrossRef]
33. Sirajuddin, M.; Ali, S.; Badshah, A. Drug-DNA interactions and their study by UV-Visible, fluorescence spectroscopies and cyclic voltametry. *J. Photochem. Photobiol. B Biol.* **2013**, *124*, 1–19. [CrossRef]
34. Rehman, S.U.; Sarwar, T.; Husain, M.A.; Ishqi, H.M.; Tabish, M. Studying non-covalent drug-DNA interactions. *Arch. Biochem. Biophys.* **2015**, *576*, 49–60. [CrossRef] [PubMed]
35. Kumar, N.; Kaushal, R.; Awasthi, P. Non-covalent binding studies of transition metal complexes with DNA: A review. *J. Mol. Struct.* **2023**, *1288*, 135751. [CrossRef]
36. de Lavor, T.S.; Teixeira, M.H.S.; de Matos, P.A.; Lino, R.C.; Silva, C.M.F.; do Carmo, M.E.G.; Beletti, M.E.; Patrocinio, A.O.T.; de Oliveira Junior, R.J.; Tsubone, T.M. The impact of biomolecule interactions on the cytotoxic effects of rhenium(I) tricarbonyl complexes. *J. Inorg. Biochem.* **2024**, *257*, 112600. [CrossRef] [PubMed]
37. Zhang, P.; Wang, J.; Huang, H.; Qiao, L.; Ji, L.; Chao, H. Chiral ruthenium(II) complexes with phenolic hydroxyl groups as dual poisons of topoisomerases I and IIalpha. *Dalton Trans.* **2013**, *42*, 8907–8917. [CrossRef]
38. Sun, D.; Liu, Y.; Liu, D.; Zhang, R.; Yang, X.; Liu, J. Stabilization of G-quadruplex DNA, inhibition of telomerase activity and live cell imaging studies of chiral ruthenium(II) complexes. *Chem. –A Eur. J.* **2012**, *18*, 4285–4295. [CrossRef]
39. Nie, Y.; Dai, Z.; Fozia; Zhao, G.; Jiang, J.; Xu, X.; Ying, M.; Wang, Y.; Hu, Z.; Xu, H. Comparative studies on DNA-binding mechanisms between enantiomers of a polypyridyl ruthenium (II) complex. *J. Phys. Chem. B* **2022**, *126*, 4787–4798. [CrossRef]
40. Zou, S.; Li, G.; Rees, T.W.; Jin, C.; Huang, J.; Chen, Y.; Ji, L.; Chao, H. Interfering with DNA High-Order Structures using Chiral Ruthenium(II) Complexes. *Chem. –A Eur. J.* **2018**, *24*, 690–698. [CrossRef]
41. Wang, Y.C.; Qian, C.; Peng, Z.L.; Hou, X.J.; Wang, L.L.; Chao, H.; Ji, L.N. Dual topoisomerase I and II poisoning by chiral Ru(II) complexes containing 2-thiophenylimidazo[4,5-f][1,10]phenanthroline derivatives. *J. Inorg. Biochem.* **2014**, *130*, 15–27. [CrossRef]
42. Dong, Z.; Liu, X.; Tan, L. Biophysical insights into the interaction of two enantiomers of Ru (II) complex [Ru (bpy) 2 (7-CH 3-dppz)] 2+ with the RNA poly (UA∗ U) triplex. *JBIC J. Biol. Inorg. Chem.* **2020**, *25*, 1085–1095. [CrossRef]
43. Gao, F.; Chao, H.; Wang, J.Q.; Yuan, Y.X.; Sun, B.; Wei, Y.F.; Peng, B.; Ji, L.N. Targeting topoisomerase II with the chiral DNA-intercalating ruthenium(II) polypyridyl complexes. *JBIC J. Biol. Inorg. Chem.* **2007**, *12*, 1015–1027. [CrossRef] [PubMed]
44. Jiang, L.; Liu, X.; Tan, L. Synthesis and characterization of chiral Ru(II) polypyridyl complexes and their binding and stabilizing effects toward triple-helical RNA. *J. Inorg. Biochem.* **2020**, *213*, 111263. [CrossRef]
45. Wang, J.Q.; Zhang, P.Y.; Qian, C.; Hou, X.J.; Ji, L.N.; Chao, H. Mitochondria are the primary target in the induction of apoptosis by chiral ruthenium(II) polypyridyl complexes in cancer cells. *JBIC J. Biol. Inorg. Chem.* **2014**, *19*, 335–348. [CrossRef]
46. Yuan, F.; Chen, X.; Liu, Y.; Zhang, T.; Sun, D.; Liu, J. Chiral ruthenium complexes induce apoptosis of tumor cell and interact with bovine serum albumin. *Chirality* **2012**, *24*, 174–180. [CrossRef]
47. Wen, B.; Liu, X.; Tan, L. Binding and stabilizating effect of RNA triplex poly(U)·poly(A)*poly(U) by enantiomers of ruthenium(II) polypyridyl complex $[Ru(bpy)_2(dppx)]^{2+}$. *J. Biol. Inorg. Chem.* **2023**, *28*, 509–517. [CrossRef]
48. Xu, H.; Zheng, K.-C.; Chen, Y.; Li, Y.-Z.; Lin, L.-J.; Li, H.; Zhang, P.-X.; Ji, L.-N. Effects of ligand planarity on the interaction of polypyridyl Ru (II) complexes with DNA. *Dalton Trans.* **2003**, *11*, 2260–2268. [CrossRef]
49. Zhen, Q.-X.; Ye, B.-H.; Zhang, Q.-L.; Liu, J.-G.; Li, H.; Ji, L.-N.; Wang, L. Synthesis, characterization and the effect of ligand planarity of $[Ru(bpy)_2L]^{2+}$ on DNA binding affinity. *J. Inorg. Biochem.* **1999**, *76*, 47–53. [CrossRef]
50. Kitchen, D.B.; Decornez, H.; Furr, J.R.; Bajorath, J. Docking and scoring in virtual screening for drug discovery: Methods and applications. *Nat. Rev. Drug Discov.* **2004**, *3*, 935–949. [CrossRef] [PubMed]
51. Hu, X.; Liu, N.Y.; Deng, Y.Q.; Wang, S.; Liu, T.; Liu, X.W. Photoinduced DNA Cleavage and Photocytotoxic of Phenanthroline-Based Ligand Ruthenium Compounds. *Molecules* **2021**, *26*, 3471. [CrossRef] [PubMed]
52. Sangeetha, S.; Murali, M. Cytotoxic Ruthenium(II) Complexes Containing a Dangling Pyridine: Selectivity for Diseased Cells Mediated by pH-Dependent DNA Binding. *Inorg. Chem.* **2022**, *61*, 2864–2882. [CrossRef]
53. Sudhindra, P.; Ajay Sharma, S.; Roy, N.; Moharana, P.; Paira, P. Recent advances in cytotoxicity, cellular uptake and mechanism of action of ruthenium metallodrugs: A review. *Polyhedron* **2020**, *192*, 114827. [CrossRef]
54. Chen, L.M.; Peng, F.; Li, G.D.; Jie, X.M.; Cai, K.R.; Cai, C.; Zhong, Y.; Zeng, H.; Li, W.; Zhang, Z.; et al. he studies on the cytotoxicity in vitro, cellular uptake, cell cycle arrest and apoptosis-inducing properties of ruthenium methylimidazole complex $[Ru(MeIm)_4(p\text{-}cpip)]^{2+}$. *J. Inorg. Biochem.* **2016**, *156*, 64–74. [CrossRef] [PubMed]
55. Mehanna, S.; Mansour, N.; Audi, H.; Bodman-Smith, K.; Mroueh, M.A.; Taleb, R.I.; Daher, C.F.; Khnayzer, R.S. Enhanced cellular uptake and photochemotherapeutic potential of a lipophilic strained Ru(ii) polypyridyl complex. *RSC Adv.* **2019**, *9*, 17254–17265. [CrossRef] [PubMed]
56. Piccolo, M.; Misso, G.; Ferraro, M.G.; Riccardi, C.; Capuozzo, A.; Zarone, M.R.; Maione, F.; Trifuoggi, M.; Stiuso, P.; D'Errico, G.; et al. Exploring cellular uptake, accumulation and mechanism of action of a cationic Ru-based nanosystem in human preclinical models of breast cancer. *Sci. Rep.* **2019**, *9*, 7006. [CrossRef]

https://doi.org/10.3390/inorganics14020063

57. Tan, C.; Lai, S.; Wu, S.; Hu, S.; Zhou, L.; Chen, Y.; Wang, M.; Zhu, Y.; Lian, W.; Peng, W.; et al. Nuclear permeable ruthenium(II) beta-carboline complexes induce autophagy to antagonize mitochondrial-mediated apoptosis. *J. Med. Chem.* **2010**, *53*, 7613–7624 [CrossRef] [PubMed]
58. Zhang, S.Q.; Meng, T.T.; Li, J.; Hong, F.; Liu, J.; Wang, Y.; Gao, L.H.; Zhao, H.; Wang, K.Z. Near-IR/Visible-Emitting Thiophenyl-Based Ru(II) Complexes: Efficient Photodynamic Therapy, Cellular Uptake, and DNA Binding. *Inorg. Chem.* **2019**, *58*, 14244–14259. [CrossRef]
59. Omondi, R.O.; Ojwach, S.O.; Jaganyi, D. Review of comparative studies of cytotoxic activities of Pt(II), Pd(II), Ru(II)/(III) and Au(III) complexes, their kinetics of ligand substitution reactions and DNA/BSA interactions. *Inorganica Chim. Acta* **2020**, *512*, 119883. [CrossRef]
60. Hess, J.; Huang, H.; Kaiser, A.; Pierroz, V.; Blacque, O.; Chao, H.; Gasser, G. Evaluation of the medicinal potential of two ruthenium (II) polypyridine complexes as one-and two-photon photodynamic therapy photosensitizers. *Chem. –A Eur. J.* **2017**, *23*, 9888–9896. [CrossRef]
61. Srishailam, A.; Gabra, N.M.; Kumar, Y.P.; Reddy, K.L.; Devi, C.S.; Kumar, D.A.; Singh, S.S.; Satyanarayana, S. Synthesis, characterization; DNA binding and antitumor activity of ruthenium (II) polypyridyl complexes. *J. Photochem. Photobiol. B Biol.* **2014**, *141*, 47–58. [CrossRef]
62. Deepika, N.; Devi, C.S.; Kumar, Y.P.; Reddy, K.L.; Reddy, P.V.; Kumar, D.A.; Singh, S.S.; Satyanarayana, S. DNA-binding, cytotoxicity, cellular uptake, apoptosis and photocleavage studies of Ru (II) complexes. *J. Photochem. Photobiol. B Biol.* **2016**, *160*, 142–153. [CrossRef]
63. Wang, C.; Yu, Q.; Yang, L.; Liu, Y.; Sun, D.; Huang, Y.; Zhou, Y.; Zhang, Q.; Liu, J. Ruthenium (II) polypyridyl complexes stabilize the bcl-2 promoter quadruplex and induce apoptosis of Hela tumor cells. *Biometals* **2013**, *26*, 387–402. [CrossRef]
64. Yu, Q.; Liu, Y.; Xu, L.; Zheng, C.; Le, F.; Qin, X.; Liu, Y.; Liu, J. Ruthenium (II) polypyridyl complexes: Cellular uptake, cell image and apoptosis of HeLa cancer cells induced by double targets. *Eur. J. Med. Chem.* **2014**, *82*, 82–95. [CrossRef]
65. Zeng, L.; Chen, Y.; Huang, H.; Wang, J.; Zhao, D.; Ji, L.; Chao, H. Cyclometalated Ruthenium(II) Anthraquinone Complexes Exhibit Strong Anticancer Activity in Hypoxic Tumor Cells. *Chem. –A Eur. J.* **2015**, *21*, 15308–15319. [CrossRef]
66. Erenburg, S.B.; Trubina, S.V.; Yukhin, Y.M.; Sharafutdinov, M.R. Structural characteristics of amorphous K-Bi citrate (De-Nol) and its aqueous solutions from EXAFS spectra. *J. Inorg. Biochem.* **2017**, *166*, 94–99. [CrossRef]
67. Mendoza-Ferri, M.G.; Hartinger, C.G.; Mendoza, M.A.; Groessl, M.; Egger, A.E.; Eichinger, R.E.; Mangrum, J.B.; Farrell, N.P.; Maruszak, M.; Bednarski, P.J.; et al. Transferring the concept of multinuclearity to ruthenium complexes for improvement of anticancer activity. *J. Med. Chem.* **2009**, *52*, 916–925. [CrossRef]
68. Giannini, F.; Paul, L.E.H.; Furrer, J.; Therrien, B.; Süss-Fink, G. Highly cytotoxic diruthenium trithiolato complexes of the type $[(\eta^6\text{-p-MeC}_6H_4Pri)_2Ru_2(\mu_2\text{-SR})_3]+$: Synthesis, characterization, molecular structure and in vitro anticancer activity. *New J. Chem.* **2013**, *37*, 3503–3511. [CrossRef]
69. Conti, L.; Macedi, E.; Giorgi, C.; Valtancoli, B.; Fusi, V. Combination of light and Ru (II) polypyridyl complexes: Recent advances in the development of new anticancer drugs. *Coord. Chem. Rev.* **2022**, *469*, 214656. [CrossRef]
70. Zeng, L.; Gupta, P.; Chen, Y.; Wang, E.; Ji, L.; Chao, H.; Chen, Z.-S. The development of anticancer ruthenium (II) complexes: From single molecule compounds to nanomaterials. *Chem. Soc. Rev.* **2017**, *46*, 5771–5804. [CrossRef] [PubMed]
71. Wang, J.-Q.; Zhao, Z.-Z.; Bo, H.-B.; Chen, Q.-Z. Synthesis, characterization, and antitumor properties of ruthenium(II) anthraquinone complexes. *J. Coord. Chem.* **2016**, *69*, 177–189. [CrossRef]
72. Li, G.; Sun, L.; Ji, L.; Chao, H. Ruthenium(ii) complexes with dppz: From molecular photoswitch to biological applications. *Dalton Trans.* **2016**, *45*, 13261–13276. [CrossRef]
73. Parussulo, A.L.A.; Matias, T.A.; Guimaraes, R.R.; Toma, S.H.; Araki, K.; Toma, H.E. Accessing the charge separation effects in dye-sensitized solar cells based on a vectorial planning of supramolecular ruthenium dyes. *Inorganica Chim. Acta* **2016**, *453*, 764–770. [CrossRef]
74. Parker, C.; Barnes, W. Some experiments with spectrofluorimeters and filter fluorimeters. *Analyst* **1957**, *82*, 606–618. [CrossRef]
75. Kumar Panigrahi, S.; Kumar Mishra, A. Inner filter effect in fluorescence spectroscopy: As a problem and as a solution. *J. Photochem. Photobiol. C Photochem. Rev.* **2019**, *41*, 100318. [CrossRef]
76. MacDonald, B.C.; Lvin, S.J.; Patterson, H. Correction of fluorescence inner filter effects and the partitioning of pyrene to dissolved organic carbon. *Anal. Chim. Acta* **1997**, *338*, 155–162. [CrossRef]
77. do Carmo, M.E.G.; de Matos, P.A.; Maia, P.I.S.; Machado, A.E.H.; Beletti, M.E.; Tsubone, T.M.; Patrocinio, A.O.T. The photophysics of Ir(III) cyclometalated complexes containing the 2-(2-pyridyl)benzimidazole ancillary ligand: Protonation effect and their potential as specific lysosome probes in cells. *J. Photochem. Photobiol. A Chem.* **2024**, *448*, 115339. [CrossRef]
78. Tsubone, T.M.; Martins, W.K.; Pavani, C.; Junqueira, H.C.; Itri, R.; Baptista, M.S. Enhanced efficiency of cell death by lysosome-specific photodamage. *Sci. Rep.* **2017**, *7*, 6734. [CrossRef]
79. Mosmann, T. Rapid colorimetric assay for cellular growth and survival: Application to proliferation and cytotoxicity assays. *J. Immunol. Methods* **1983**, *65*, 55–63. [CrossRef]

https://doi.org/10.3390/inorganics14020063

80. Riss, T.; Niles, A.; Minor, L. Cell viability assays assay guidance manual. In *Assay Guide Manual*; Eli Lilly & Company: Bethesda, MD, USA, 2004.
81. *ISO 10993-5:2009*; Biological Evaluation of Medical Devices—Part 5: Tests for In Vitro Cytotoxicity Standardization. International Organization for Standardization: Geneva, Switzerland, 2009.

https://doi.org/10.3390/inorganics14020063

inorganics

MDPI

Article

DNA Binding with Dipyrromethene Ruthenium(II) Complexes

Maria Isabel Murillo [1], Carlos Felipe Mejia [1], Andrés Restrepo-Acevedo [1], Benjamin Barraud [1], Adrian L. Orjuela [2], Marcos Flores-Alamo [3], Rubén A. Toscano [1], Jorge Alí-Torres [4], Alexander D. Ryabov [5] and Ronan Le Lagadec [1,*]

[1] Instituto de Química UNAM, Circuito Exterior s/n, Ciudad Universitaria, Ciudad de México 04510, Mexico; isabel.murillo@iquimica.unam.mx (M.I.M.); cfmejiag@comunidad.unam.mx (C.F.M.); acrestrepoa@comunidad.unam.mx (A.R.-A.); bbarraudgutierrez@gmail.com (B.B.); toscano@unam.mx (R.A.T.)
[2] Instituto de Investigaciones Científicas y Servicios de Alta Tecnología (INDICASAT AIP), Panama City 0843-01103, Panama; alorjuelar@unal.edu.co
[3] Facultad de Química, Circuito Exterior s/n, Coyoacán, Ciudad Universitaria, Ciudad de México 04510, Mexico; mfa@unam.mx
[4] Departamento de Química, Universidad Nacional de Colombia, Bogotá 111321, Colombia; jialit@unal.edu.co
[5] Department of Chemistry, Carnegie Mellon University, 4400 Fifth Avenue, Pittsburgh, PA 15213, USA; ryabov@andrew.cmu.edu
* Correspondence: ronan@unam.mx

Abstract: Four new arene–ruthenium(II) complexes [(η^6-*p*-cymene)RuCl(dpm)], where dpm are hexa-(**L3–L5**) and *meso*-substituted (**L6**) dipyrromethene ligands, were synthesized. These ligands and the corresponding complexes were thoroughly characterized by elemental analysis and spectroscopic techniques (MS, IR, ^{1}H, ^{13}C NMR, and UV–vis), and the structures of one ligand and three ruthenium complexes were determined by X-ray single-crystal analysis. The DNA-binding ability of the **Ru-3–Ru-6** complexes was evaluated by UV–vis DNA titration. Compound **Ru-3** exhibited the highest binding energy, outperforming the complexes containing a dipyrrin ligand substituted by chlorides (**Ru-4** and **Ru-5**) or a *meso*-substituted dipyrrin (**Ru-6**). Molecular docking revealed that the hypothetical **Ru-1** and **Ru-2** complexes, which contain iodide ligands in the dipyrrin structures, showed higher DNA-binding affinities than **Ru-3**. Computational calculations supported the experimental results, confirming that **Ru-3** has a higher affinity for DNA than the other complexes.

Keywords: ruthenium complexes; dipyrrin ligands; DNA interactions; molecular docking calculations

1. Introduction

Nitrogen mustards, the first DNA-alkylating agents used in cancer chemotherapy, represent probably the most widely studied DNA crosslinking agents [1]. Nitrogen mustards, like cyclophosphamide (2-(bis(2-chloroethyl)amino)tetrahydro-2*H*-1,3,2-oxazaphosphorine 2-oxide) and chlorambucil (4-[4-[bis(2-chloroethyl)amino]phenyl]butanoic acid), are among the most widely used clinical anticancer agents [2,3]. Platinum derivatives are also associated with alkylating agents, as they can form coordination bonds with DNA nucleobases [4]. As such, platinum(II) compounds have been implemented to treat many types of tumors in chemotherapy [5,6]. However, these agents come with various disadvantages, like the generation of resistance mechanisms and severe side effects entailing the development of metal complexes different than platinum [7,8]. Ruthenium complexes have emerged as an alternative to overcome such drawbacks of platinum-based drugs in cancer treatment. The ruthenium derivatives KP1019 ([IndH][*trans*-$RuCl_4(Ind)_2$], Ind = indazole) and NAMI-A ([ImH][*trans*-$RuCl_4$(dmso-S)(Im)], Im = imidazole) have shown great promise as new drugs due to their possible therapeutic applications, low toxicity, and activity against primary tumors and metastases [9]. Ruthenium complexes that can act as DNA intercalators have been extensively studied as cytotoxic agents because of their unique redox and photophysical properties [10–12]. In addition, the antitumor activity of various ruthenium(II)–arene complexes has been associated with their high binding affinity to DNA through covalent and/or non-covalent interactions [13,14]. RAPTA (ruthenium arene PTA, PTA = 1,3,5-triaza-7-phosphaadamantane) organometallic complexes have shown great potential for preclinical cancer research. While their in vitro cytotoxicity is low, they have shown elevated antitumor activity in vivo [15,16]. The RAPTA structure features a ruthenium center; an η^6-coordinated arene, like *p*-cymene in RAPTA-C; a 1,3,5-triaza-7-phosphoadamantane ligand (PTA); and two chloride-leaving groups. Meggers et al. demonstrated the relevance of three-dimensional structures for the interactions between ruthenium–arene complexes and biomacromolecules [17]. Additionally, numerous ruthenium complexes have been investigated as inhibitors of enzymes. For instance, staurosporine derivatives coordinated to ruthenium–arene moieties have shown remarkable inhibitory activity on various enzymes in cancer cells [18]. However, the metal center of these ruthenium complexes is not involved in direct interactions with the enzymes. Instead, the metal controls the orientation of the staurosporine in the receptor space of the target enzyme, generating three-dimensional structures that are complementary in shape and functional-group transport to the active site [19]. Dipyrromethene derivatives, also called dipyrrins, feature π-conjugated systems and consist of two pyrroles linked by a *meso* carbon. They are precursors for the synthesis of porphyrins with potential applications as dyes [20] and as ligands in anticancer metal agents [21,22]. Hexasubstituted dipyrrins, structurally related to verteporfin, can interact with cellular proteins, altering the transcription of genetic material [23].

An alternative approach to multi-target ruthenium–arene compounds is their binding to bioactive ligands, such as 3-hydroxyflavones [24–26], lapachol [27], paullone [28], and lonidamine derivatives [29]. This intramolecular combination therapy may also allow for a reduction in the disadvantages associated to antineoplastic alkylating agents [30]. To explore possible modes of DNA intercalation or alkylation, Pandey's group performed diverse modifications at the *meso* carbon of dipyrrins. Thioether, fluorides, and 2-methoxypyridyl groups were introduced (Figure 1). Ruthenium–arene complexes with *meso*-substituted dipyrrin-type ligands showed strong interactions with DNA [31–35]. Modern drugs with a halide-leaving group bind covalently to biological targets and modulate their electronic, lipophilic, and steric properties [30,36].

Here, we describe the synthesis and properties of half-sandwich ruthenium complexes with hexasubstituted dipyrrin-like ligands related to verteporfin and their impact on the DNA structure. These compounds represent a promising avenue for developing biomolecular agents capable of interacting with genetic materials, which may open doors to new therapies against various diseases, including cancer.

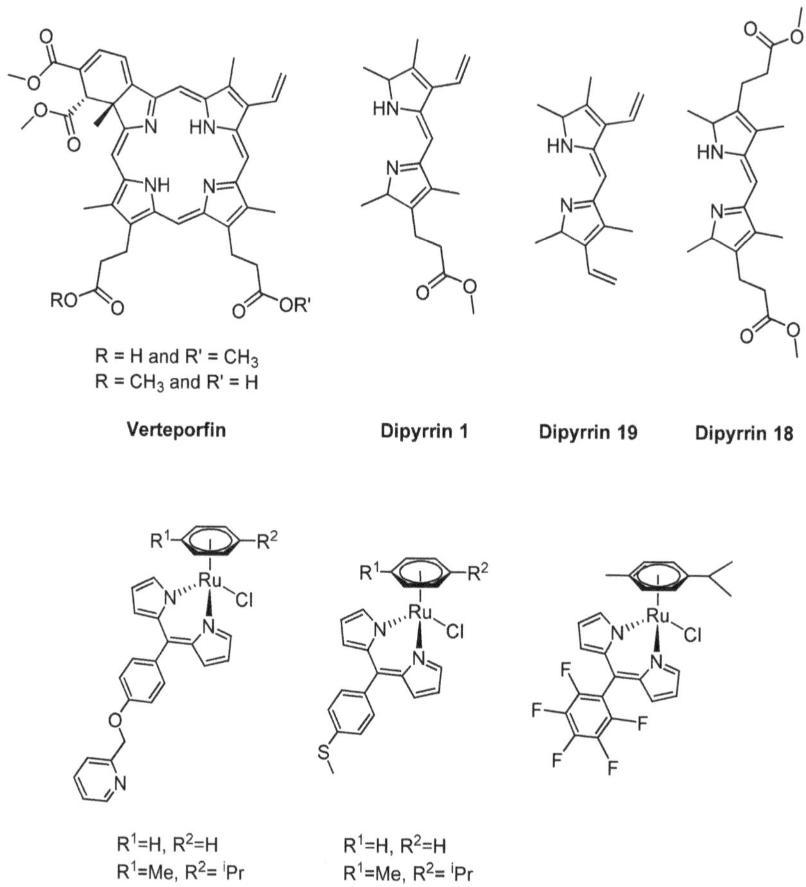

Figure 1. Structures of biologically active verteporfin and dipyrrins [23] (**top**) and ruthenium–arene complexes with *meso*-substituted dipyrrin ligands, which have shown interactions with DNA [31–35] (**bottom**).

2. Results

2.1. Synthesis of Ligands L1–L5

An optimized synthesis of new ligands, **L1**, **L2**, and **L3**, was developed based on the published methods for dipyrrin 1 [23] and dipyrrin 18 [37]. In dipyrrin 1 and dipyrrin 18, the ethyl acetate fragment occupies the 8 or the 2 and 8 positions. In contrast, in compounds **L1** and **L3** (Figure 2), those positions are occupied by methyl acetate. This slight modification allowed for a standardized synthesis in higher yields.

Condensation between the corresponding β-diketone and 2-oxopropanal oxime was carried out to give rise to **pyrrole A**, a precursor of the other pyrrole derivatives (Scheme 1) [38]. **Pyrrole A** underwent formylation after decarboxylation of the *tert*-butyl ester group, giving **pyrrole B**. A selective reduction of the methyl acetate moiety of **pyrrole A** to alcohol gave **pyrrole C**, which then underwent iodination of the hydroxyl group to yield **pyrrole D**. Due to the importance of halogen atoms for interactions with DNA, the synthesis of **pyrroles F** and **G**, homologous to **pyrroles D** and **E** with chlorine instead of iodine, was also carried out [39–41].

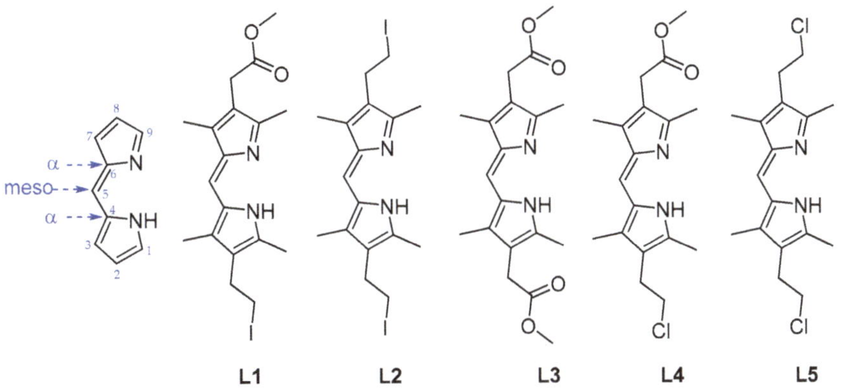

Figure 2. Chemical numbering used to describe dipyrrins and structures of new ligands, **L1–L5**.

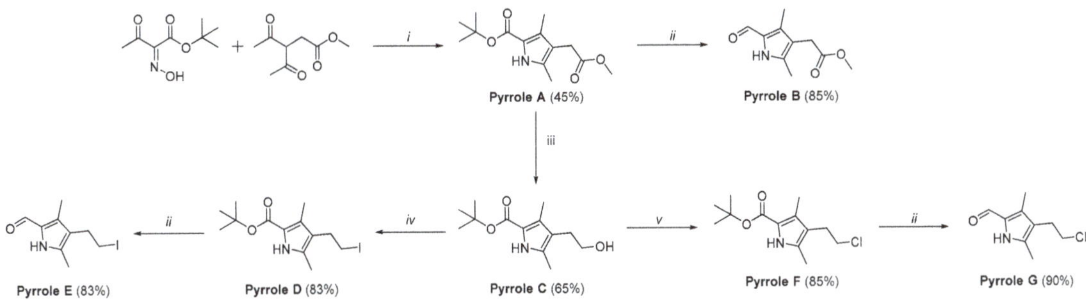

Scheme 1. Synthesis of pyrrole precursors. Reagents and conditions: (*i*) Zn, AcONa, and AcOH; 80 °C; 2 h; (*ii*) CF_3COOH, $CH(OCH_3)_3$, and CH_2Cl_2; 0 °C; 4 h; (*iii*) DIBAL-H and THF; −40 °C; 12h; (*iv*) imidazole, PPh_3, I_2, and CH_2Cl_2; 0 °C; 4 h; (*v*) PPh_3, CCl_4, and CH_2Cl_2; 0 °C; 14 h.

Dipyrrins **L1–L5** were synthesized from **pyrroles A–G** following McDonald's methodology via an acid-catalyzed condensation of 2-formylpyrrole with a non-substituted pyrrole at position 2 [42]. The condensation of 2-formyl pyrroles (**pyrrole B**, **pyrrole E**, and **pyrrole G**) and 2-*tert*-butylpyrroles (**pyrrole A**, **pyrrole D**, and **pyrrole F**) led to the formation of the desired **L1–L5** derivatives in 94, 93, 87, 75, and 85% yields, respectively (Scheme 2). It should be noted that iododipyrrins **L1** and **L2** decomposed readily after 24 h, either in solution in various solvents, such as acetonitrile, DMSO, or dichloromethane, or in the solid state.

Analytical techniques (NMR, mass, UV–vis, and IR spectroscopies) corroborated the formation of the ligands. The formation of **L1** was confirmed by ^{1}H NMR, where all the characteristic signals corresponding to the proposed structure are observed (Figure S1). The signal for the methine group (*meso*) appears at 6.70 ppm. At high field, the signals for the CH_3 substituents at positions 1, 3, 7, and 9 of the dipyrrin core are observed at 2.15 and 2.32 ppm. The substituents at positions 2 and 8 also exhibited characteristic signals. Triplet signals were observed at 2.95 and 3.15 ppm for the ethyl iodide groups, and singlet signals were observed at 3.38 and 3.66 ppm for the methyl acetate moieties. The IR spectra (Figure S6) show a characteristic NH stretching band at 3209 cm^{-1}. A band at 1604 cm^{-1} corresponds to C=N stretching, and a band at 1730 cm^{-1} is associated with a C=O stretching vibration.

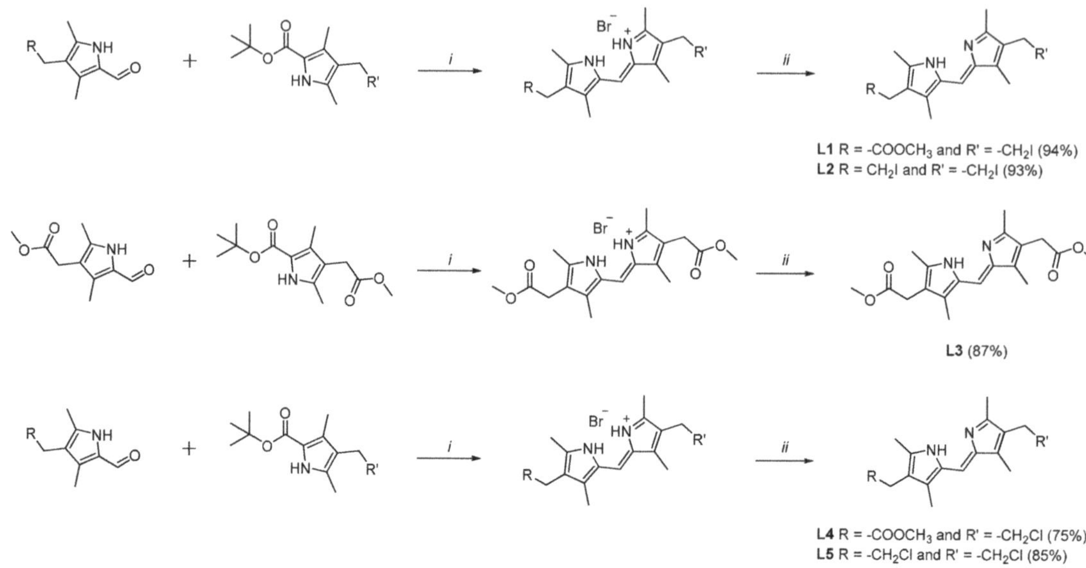

Scheme 2. Synthesis of ligands **1–5**. Reagents and conditions: (*i*) CF_3COOH, CH_2Cl_2, and HBr/AcOH; 25 °C; 2 h; (*ii*) NEt_3 and H_2O; 25 °C; 30 min.

The condensation of **pyrrole A** and **pyrrole B** to obtain **L3** was evidenced by the formation of the methine group, which was observed at 6.70 ppm in ^{1}H NMR and at 116.61 ppm in ^{13}C NMR spectra. In the IR spectra, the NH stretching band is observed at 3211 cm^{-1} and the stretching bands for C=O and C=N are observed at 1727 and 1607 cm^{-1}, respectively.

Dipyrrins **L2** and **L5** have similar structures; **L5** differs from **L2** by having chlorides instead of iodides. Both dipyrrins possess a symmetry axis through the *meso* carbon, so their ^{1}H NMR spectra show similar signals, except for the chemical shifts corresponding to the $–CH_2CH_2I$ and $–CH_2CH_2Cl$ substituents. In the IR spectra, the stretching bands for NH are observed at 3208 and 3187 cm^{-1} for **L2** and **L5**, respectively.

In the UV–vis spectroscopy, the absorption bands corresponding to a π–π^* transition of the dipyrrins are observed at around 440 nm, showing a slight red shift (445 nm) for **L5**, which contains chlorides, compared to **L3** (441 nm) without chloride.

In addition, dipyrrin **L3** was characterized by single-crystal X-ray diffraction crystallography. The pyrrole rings are in the α,α' position with respect to the nitrogen atoms, providing a typical stable conformation [22,43]. Both the pyrrole rings and the *meso* carbon in the dipyrrins are coplanar (Figure S20).

2.2. *Synthesis of Complexes* **Ru-1–Ru-6**

Dipyrrins **L1** and **L2** reacted with 0.5 equivalents of the [Ru(η^6-*p*-cymene)Cl_2]$_2$ dimer (Scheme 3) in the presence of NEt_3. However, the formed complexes decomposed upon purification using alumina column chromatography and could not be isolated and satisfactorily characterized. Therefore, the proposed **Ru-1** and **Ru-2** complexes, which incorporate iodide in the structure of the dipyrrin ligands (Scheme 3), were only evaluated for molecular docking calculations.

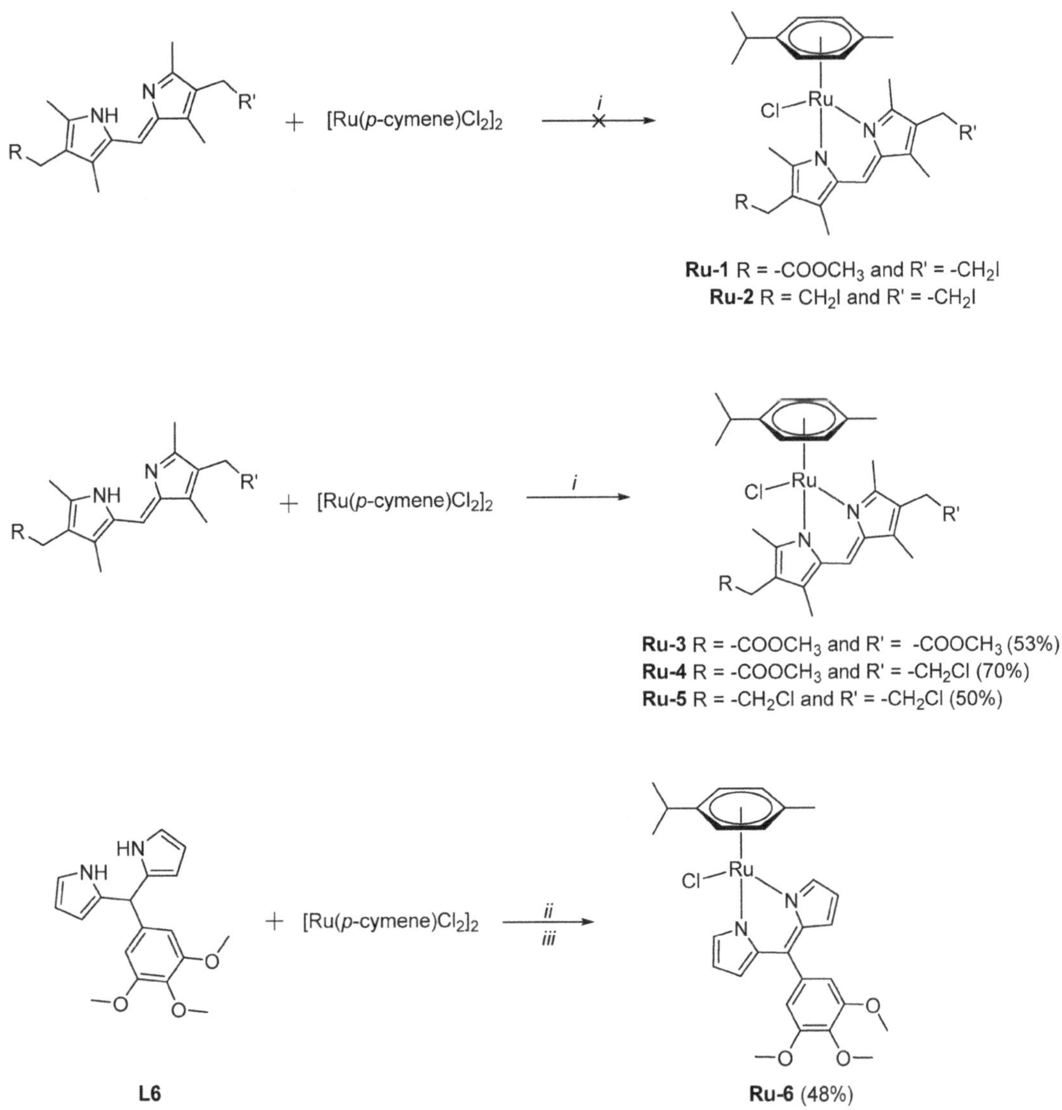

Scheme 3. Synthesis of **Ru-3–Ru-6** complexes. Reagents and conditions: (*i*) NEt_3 and CH_2Cl_2; 25 °C; 12 h; (*ii*) DDQ (2,3-dichloro-5,6-dicyano-1,4-benzoquinone) and benzene/CH_2Cl_2; 0 °C; 1 h; (*iii*) NEt_3 and CH_2Cl_2; 25 °C; 24 h.

Under similar conditions, dipyrrin **L3** was successfully attached to the ruthenium–*p*-cymene moiety to form the complex **Ru-3** in a 53% yield (Scheme 3). In the ^{1}H-NMR spectrum, the dipyrrin CH_3 groups, α to the nitrogen atom, resonate at low field from 2.32 ppm in the dipyrrin **L3** to 2.70 ppm in **Ru-3**. The C–H (*meso* carbon) signal is also affected and shifts from 6.70 ppm in the free ligand to 6.83 ppm in **Ru-3**. Although the dipyrrin retains the symmetry observed in its free form (**L3**), the ^{1}H NMR spectrum (Figure S33) of the **Ru-3** complex shows geminal coupling (^2J = 15.7 Hz) in the CH_2 group of the methyl acetate substituents, showing a signal in the 3.29 to 3.42 ppm region. In the ^{13}C NMR spectra, the signals corresponding to the quaternary carbons (C–CH_3) in the α-position to the dipyrrin nitrogen atoms shifted from 151 ppm in the free ligand (**L3**) to

161 ppm in the **Ru-3** complex. The coordination of the ligand was also confirmed by IR spectroscopy. The NH stretching band observed in **L3** at 3211 cm^{-1} disappeared in the complex. However, instead of a single C=O stretching band at 1727 cm^{-1} observed in **L3**, two bands are seen for the complex, one of medium intensity at 1742 cm^{-1} and another of strong intensity at 1728 cm^{-1}. Additionally, the **Ru-3** complex was characterized by ESI spectrometry and elemental analysis.

The **Ru-4** and **Ru-5** complexes were obtained under the same conditions as for **Ru-3** in 70 and 50% yields, respectively (Scheme 3). Their structures were corroborated by NMR, IR, mass spectrometry, elemental analysis, and single-crystal X-ray crystallography. In the ^{1}H NMR spectra of **Ru-5**, a change in the multiplicity of the signals corresponding to the -CH_2CH_2Cl fragment can be observed. Instead of two triplets at 2.87 and 3.54 ppm in **L5**, two multiplets at 2.92 and 3.50 ppm can be seen, confirming the alteration of the chemical environment of the ethyl chloride of the dipyrrin when coordinated to ruthenium. In **Ru-4**, a similar change in the multiplicity is observed in the aliphatic region of ^{1}H NMR.

The UV–vis absorption spectra of complexes **Ru-3**–**Ru-6** were recorded in DMSO at a concentration of 10 µM (Figures S40, S48, S56 and S65). The spectra exhibited two main bands. Intense low-energy bands were observed at around 500 nm (513 for **Ru-3**, 494 for **Ru-4**, 516 for **Ru-5**, and 493 nm for **Ru-6**), which were assigned to metal-to-ligand charge transfer (MLCT) transitions, and weaker bands at around 440 nm (436 for **Ru-3**, 443 for **Ru-4**, 435 for **Ru-5**, and 430 nm for **Ru-6**), attributed to π–π^* charge transfers from the conjugated dipyrrin core [34,44]. These assignations were confirmed by TD-DFT (Table S2).

Complex **Ru-6**, with a 3,4,5-trimethoxyphenyl-substituted *meso* carbon, was also prepared [45]. The synthesis of **Ru-6** was carried out in a single reactor (one-pot), starting from dipyrromethane **L6** and generating the dipyrrin in situ by reacting with DDQ (2,3-dichloro-5,6-dicyano-1,4-benzoquinone), as shown in Scheme 3 [45]. A dark, reddish-brown solid was obtained in a 48% yield. Suitable crystals for X-ray diffraction were grown in a dichloromethane/benzene/hexane diffusion system (Figure 3).

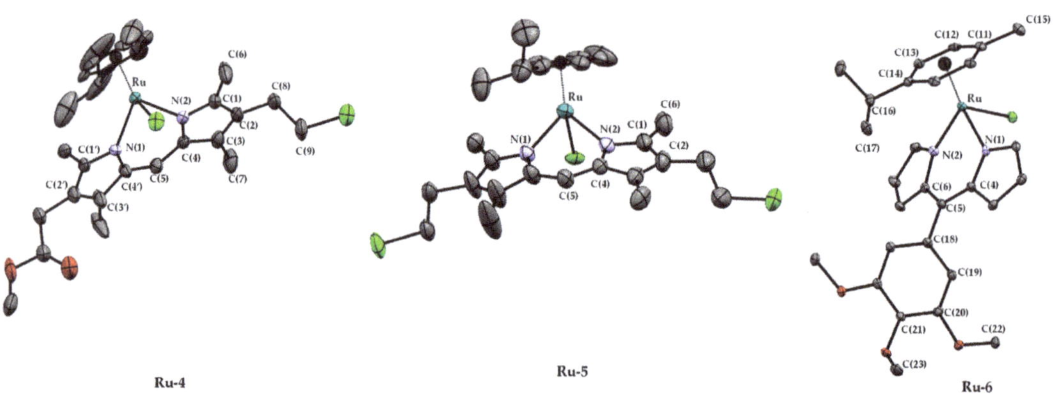

Figure 3. ORTEP diagrams for **Ru-4**, **Ru-5**, and **Ru-6**. Thermal ellipsoids are shown with a 50% probability level. Hydrogen atoms are omitted for clarity.

In the ^{1}H NMR spectrum of **Ru-6** at room temperature, broadening was observed for the signals corresponding to the *para*- and *meta*-methoxy groups of the 3,4,5-trimethoxyphenyl moiety. To restrict the rotation of the structure, the ^{1}H NMR spectrum was recorded at −30 °C in CD_2Cl_2, where well-defined singlets were detected at 3.73 and 3.80 ppm (Figure S58).

2.3. Crystal Structures

The structures of **L3**, **Ru-4**, **Ru-5**, and **Ru-6** were determined by single-crystal X-ray diffraction crystallography. Details on data collection, solutions, and refinements are summarized in Table 1. Geometrical parameters are given in Table 2, and the molecular structures are shown in Figures4 and S53. **L3** and **Ru-6** crystallized in the triclinic *P*-1 space group. Complexes **Ru-4** and **Ru-5** crystallized in the monoclinic *C*2/*c* space group for **Ru-4** and $P2_1/c$ for **Ru-5**. Crystal structures revealed a typical "piano-stool" geometry at the metal center in **Ru-4**, **Ru-5**, and **Ru-6**. The coordination sites around the metal center in these complexes are occupied by two pyrrolic nitrogen atoms from the dipyrrin ligand, a chloride ligand, and an η^6-bound arene ring.

Table 1. Crystallographic data for L3, Ru-4, Ru-5, and Ru-6.

Compound	L3	Ru-6	Ru-4	Ru-5
Emp. Formula	$C_{19}H_{24}N_2O_4$	$C_{28}H_{31}Cl_1N_2O_3Ru$	$C_{28}H_{36}Cl_2N_2O_2Ru$	$C_{27}H_{35}Cl_3N_2Ru$
FW (g/mol)	344	580	604	594
Temperature (K)	150(2)	130(2)	173(2)	173(2)
λ (Å)	0.71073	0.71073	1.54178	0.71073
Crystal system	Triclinic	Triclinic	Monoclinic	Monoclinic
Space group	*P*-1	*P*-1	*C*2/*c*	*P*21/*c*
a (Å)	7.346(4)	11.586(14)	30.047(5)	14.276(5)
b (Å)	7.866(4)	12.987(2)	11.541(2)	11.337(3)
c (Å)	8.466(4)	14.649(3)	17.2579(3)	17.579(6)
α (°)	102.167(2)	111.498(16)	90	90
β (°)	104.778(2)	108.775(13)	113.8080(10)	98.990(14)
γ (°)	99.512(2)	100.096(12)	90	90
Volume ($Å^3$)	449.82(4)	1831.00(6)	5475.35(17)	2810.30(15)
Z	1	2	8	4
ρ calcd (mg × m^3)	1.271	1.502	1.504	1.477
Abs. Coeff. (mm^{-1})	0.090	0.831	7.965	1.190
F(000)	184	848	2544	1271
θ range (°)	2.95 to 32.02	3.46 to 29.44	3.22 to 77.46	2.31 to 26.69
Reflections collected/unique	6215/3127	8506	52,372/5798	29,361/5668
[R(int)]	[0.0199]	[0.0291]	[0.0494]	[0.0363]
Completeness (%)	99.6	99.7	99.5	99.2
Data/restraints/parameters	3127/930/469	8506/12/395	5798/1/334	5668/59/343
GoF on F^2	1.136	1.083	1.317	1.049
R1 [I > 2σ(I)]	0.0514	0.0442	0.0922	0.0540
wR2 [I > 2σ(I)]	0.0615	0.0538	0.1052	0.0749
Final R-index	0.05	0.04	0.09	0.05

The piano-stool arrangement at the ruthenium centroid is further reflected in the small dipyrrin ligand bite angles [N(1)–Ru(1)–N(2)] 84.86° (**Ru-6**), [N(1)–Ru(1)–N(2)] 86.03° (**Ru-4**), and [N(1)–Ru(1)–N(2)] 86.39° (**Ru-5**) [46]. The bond distances from ruthenium to the arene centroid are 1.442 (**Ru-6**), 1.456 (**Ru-4**), and 1.453 Å (**Ru-5**), which turn out to be shorter than the distances reported in similar Ru–arene complexes [35]. The pyrrolic rings in **Ru-4** and **Ru-5** complexes are coplanar. In contrast, the **Ru-6** complex does not show the coplanarity between the pyrroles and the ring attached to the *meso* carbon, which is twisted out of the methene plane by 71.79°. The Ru(1)–N(1) bond lengths of 2.114 (**Ru-4**), 2.104 (**Ru-5**), and 2.078 (**Ru-6**) Å are similar to those of related complexes [47]. The Ru–Cl distances are 2.509 (**Ru-4**), 2.582 (**Ru-5**), and 2.413 (**Ru-6**) Å and agree with the values reported for similar complexes [35,48].

Table 2. Experimental (Exp.) and calculated (DFT) bond distances for **Ru-4**, **Ru-5**, and **Ru-6**.

Compound	Exp. Distance (Å)	DFT Distance (Å)	Bond
Ru-4	2.509	2.42	
Ru-5	2.582	2.42	Ru–Cl
Ru-6	2.413	2.37	
Ru-4	1.456	1.82	
Ru-5	1.453	1.82	Ru–arene centroid
Ru-6	1.442	1.98	
Ru-4	2.114	2.15	
Ru-5	2.104	2.15	Ru–N1
Ru-6	2.078	2.03	
Ru-4	2.084	2.14	
Ru-5	2.079	2.14	Ru–N2
Ru-6	2.068	1.97	

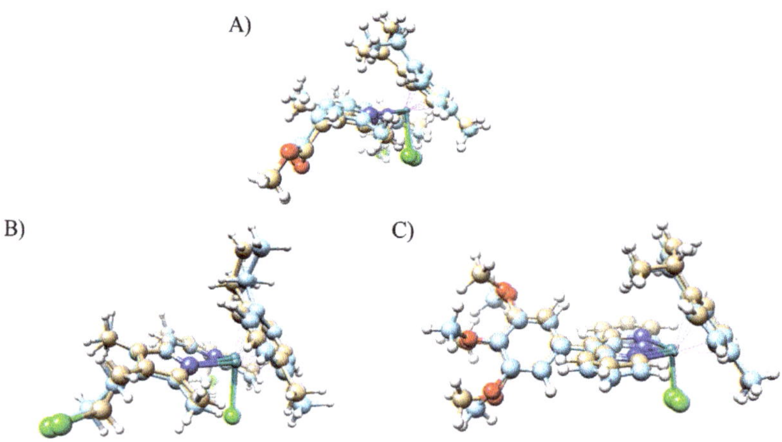

Figure 4. Superposition of the crystal structure and the DFT-optimized geometry for complex (**A**) **Ru-4**, (**B**) **Ru-5**, and (**C**) **Ru-6** (blue). Carbon atoms are shown in brown for the crystal structure and in cyan for the DFT-optimized geometry.

The structures of **Ru-4**, **Ru-5**, and **Ru-6** were optimized using DFT calculations. A comparison with their corresponding crystal structures revealed RMSD values of 1.10 Å, 1.27 Å, and 1.45 Å, respectively (Figure 4). These deviations are primarily attributed to the conformational flexibility of the methoxy and methyl groups. Notably, when the methyl moieties of the methoxy substituents were excluded from the comparison, the RMSD significantly decreased to 0.87 Å for **Ru-6**, highlighting the influence of peripheral group dynamics. These results confirm the accuracy and reliability of the DFT methodology in reproducing experimental geometries and validate its use as a robust approach for modeling related ruthenium complexes.

2.4. Frontier Molecular Orbitals (FMOs)

FMOs have been widely used to analyze the reactivity and regioselectivity of various structures with biological properties [49]. The frontier molecular orbitals of a molecule are called HOMO (corresponding to the highest occupied molecular orbital in energy) and LUMO (lowest unoccupied molecular orbital in energy), and the energy gap between these orbitals helps to characterize the chemical reactivity and selectivity of the molecule in terms of global parameters, such as electronegativity, hardness, and softness [50]. The FMOs of the **Ru-1–Ru-6** complexes are shown in Figure 5.

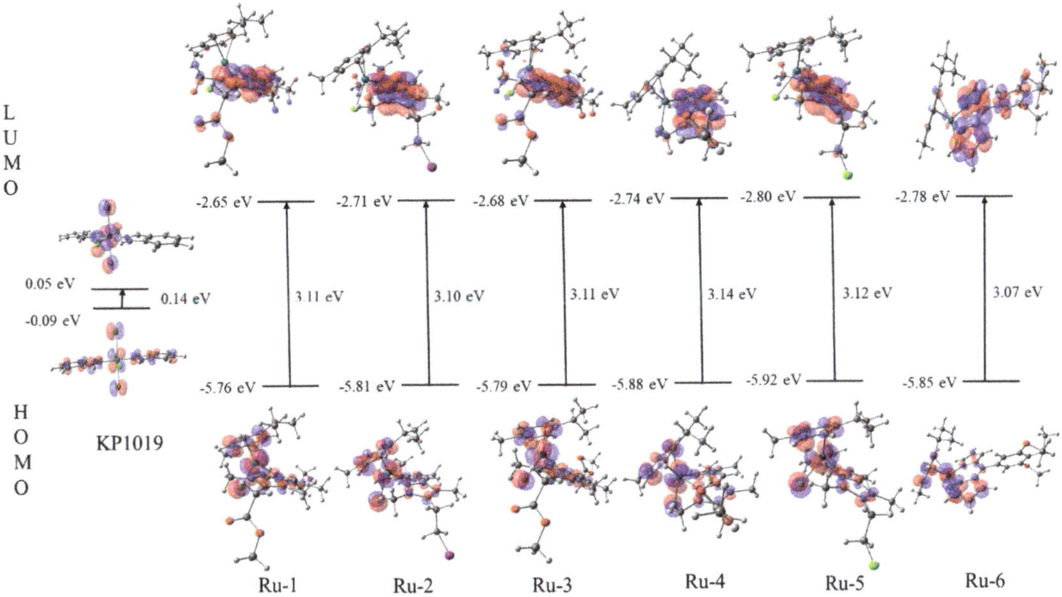

Figure 5. HOMO and LUMO distributions for **Ru-1–Ru-6** and **KP1019**, showing σ-donation in HOMOs and π-backdonation in LUMOs.

The conceptual DFT descriptors presented in Table 3 reveal consistent electronic trends across the **Ru-1** to **Ru-6** complexes. The chemical potential (μ) decreases from −4.21 eV in **Ru-1** to −4.26 eV in **Ru-2** and from −4.31 eV in **Ru-4** to −4.36 eV in **Ru-5**, indicating a lower tendency to lose electrons for complexes with **L2** and **L5** ligands. This is accompanied by a slight rise in electronegativity (χ), from 4.21 eV in **Ru-1** to 4.26 eV in **Ru-2** and 4.31 eV in **Ru-4** and 4.36 eV in **Ru-5**, suggesting a stronger electron-attracting character as the ligand environment becomes more electron-withdrawing. The hardness (η) remains relatively stable across the series (3.10–3.14 eV), with similar softness (S) values (~0.32 eV^{-1}), reflecting comparable polarizability among the complexes.

Table 3. Conceptual DFT descriptors for **Ru-1–Ru-6** complexes, calculated from HOMO and LUMO energies (in eV). Chemical potential (μ), electronegativity (χ), hardness (η), softness (S), and electrophilicity index (ω) are reported with values rounded to three significant figures.

Complex	μ (eV)	χ (eV)	η (eV)	S (eV^{-1})	ω (eV)
Ru-1	−4.21	4.21	3.11	0.322	2.84
Ru-2	−4.26	4.26	3.1	0.323	2.93
Ru-3	−4.24	4.24	3.11	0.322	2.88
Ru-4	−4.31	4.31	3.14	0.318	2.96
Ru-5	−4.36	4.36	3.12	0.321	3.05
Ru-6	−4.31	4.31	3.07	0.326	3.03

The electrophilicity index (ω) shows a gradual increase from 2.84 eV in **Ru-1** to 3.05 eV in **Ru-5**, indicating a growing ability to accept electron density. This trend is especially pronounced in **Ru-5** and **Ru-6**, which exhibit higher ω and χ values, suggesting that they may act as more effective electrophilic species. These results confirm that subtle changes in the ligand structure influence the electronic properties of the complexes, providing a predictive framework for adjusting the reactivity and coordination behavior of Ru(II)-based systems.

2.5. Stability Studies

The stability of the ruthenium complexes was evaluated in DMSO and H_2O/DMSO solutions, since these aqueous media are similar to the one used in UV–vis titration studies. The complexes were dissolved in water containing 0.1% DMSO, and the concentration was 10 µM. The solutions were monitored by UV–vis, keeping the temperature at 25 °C for 24 h. The spectra were recorded at 0, 2, 4, 8, 16, and 24 h.

Ligands **L1** and **L3** were stable in both DMSO and aqueous solutions, while **L5** showed lower stability in DMSO. Over time, a slight decrease in intensity of λ_{max} can be observed without a significant shift or appearance of new bands, suggesting that the structure of the primary chromophore remains intact (Figures S8, S19 and S32 for **L1**, **L3**, and **L5**, respectively). Complexes **Ru-3–Ru-6** exhibited high stability in DMSO and aqueous solutions. Specifically, all four complexes were stable during the first 2 h in the aqueous solutions, slightly decomposing after 2 h (between 1–5%), as seen in Figure 6 for **Ru-4** and **Ru-5**. Figures S41, S49, S57 and S66 show the absorption spectra for the ruthenium complexes. These results confirm that the ruthenium derivatives are stable enough to be used in DNA titration studies, as each study took less than 1 h.

A) B)

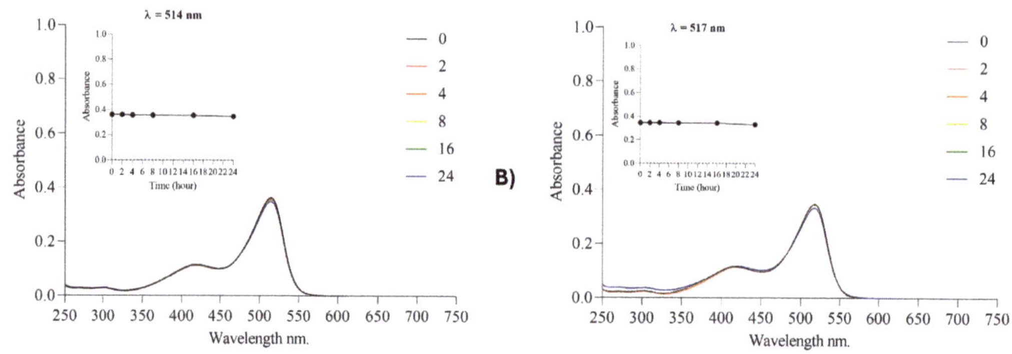

Ru-4-H_2O/DMSO **Ru-5**-H_2O/DMSO

Figure 6. UV–vis spectra of **Ru-4** (left, (**A**)) and **Ru-5** (right, (**B**)) at 10 µM in a H_2O/DMSO (0.1%) solution for 24 h at 25 °C.

2.6. UV–vis Titration Studies

An analysis of changes in UV–vis spectra of metal compounds during the progressive addition of DNA is widely used to determine the binding mode and kinetics of interaction between complexes and DNA [51]. Salmon sperm DNA was used to simulate the interaction of **L1–L5** and **Ru-3–Ru-6** complexes with the double helix. After the continuous addition of DNA, the absorbance bands of compounds **L1–L4** and **Ru-3–Ru-6** decreased (Figures7, S67–S69). The intensity of the absorbance bands at around 500 nm for **Ru-4** and **Ru-5** showed a significant decrease after DNA addition, and an isosbestic point appeared, indicating the generation of a new species [52]. Compound **L5** did not interact with DNA, as the band at 489 nm remained unchanged upon adding DNA (Figure S68). Previous reports described a decrease in the absorbance of the compound upon interaction with DNA and the appearance of a new band at 260 nm, attributed to the DNA bound by the interaction of other structures. However, in our study, only the dipyrrin **L3** presents a significant increase in the absorption peak at 260 nm after interaction with DNA. On the other hand, according to the performed binding energy calculations, the ruthenium complexes have a high binding energy to DNA [53].

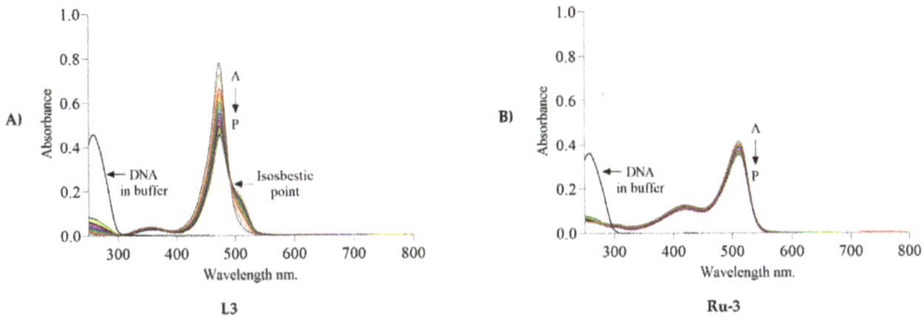

Figure 7. Effects of DNA on the absorption spectra of **L3** (**A**) (10 μM) and **Ru-3** (**B**) (10 μM) in buffered aqueous solutions (5 mM of trizma buffer at pH 7.4 and 25 °C). Final DNA concentration is 90 μM.

To obtain information on the binding affinity of compounds **L1–L4** and **Ru-3–Ru-6** to DNA, the intrinsic binding constants (K_b) were estimated using Equation (1) and are listed in Table 3. The value of K_b indicates how strong the interactions between compounds and DNA are. According to the data in Table 3, the **Ru-3** complex displays a higher binding affinity than **Ru-6**, with a K_b of 6.5×10^{11} M^{-1} for **Ru-3** and 9.6×10^{10} M^{-1} for **Ru-6**. The K_b values of known intercalators, such as ethidium bromide, epirubicin, and proflavine, are 1.50×10^5 [54], 3.40×10^4 [55], and 2.32×10^4 M^{-1} [55], respectively, demonstrating the strong affinity of compounds **L1–L4** and **Ru-3–Ru-6** for binding to DNA with K_b values higher than for known intercalators.

The thermal stability of nucleic acid complexes was evaluated using the Gibbs free energy changes. The $\Delta G°$ for ligands **L1–L4** and complexes **Ru-3–Ru-6** was calculated using the binding constant (K_b) in Equation (2) and are listed in Table 4. Negative values of the Gibbs free energy changes indicate the spontaneity of these interactions [53,56].

Table 4. DNA-binding constants (K_b) and binding energies, $\Delta G°$ (kJ/mol), for interactions between salmon sperm DNA and **L1–L4** and **Ru-3–Ru-6** compounds. Three independent experiments were performed.

Compounds	K_b (M^{-1})	Binding Energy $\Delta G°$ (kJ/Mol)
L1	$1.1 \pm 0.1 \times 10^7$	-36.8 ± 0.7
L2	$1.2 \pm 0.5 \times 10^{10}$	-52.7 ± 0.5
L3	$5.7 \pm 1.8 \times 10^{10}$	-56.2 ± 0.5
L4	$3.4 \pm 0.8 \times 10^{11}$	-60.3 ± 1.1
Ru-3	$6.5 \pm 0.5 \times 10^{11}$	-61.8 ± 1.0
Ru-4	$2.0 \pm 0.6 \times 10^{10}$	-53.9 ± 0.6
Ru-5	$3.7 \pm 3.3 \times 10^{10}$	-55.0 ± 0.2
Ru-6	$9.6 \pm 1.3 \times 10^{10}$	-57.4 ± 0.6

2.7. Molecular Modeling

The molecular docking analysis performed on **Ru-1–Ru-6** and KP1019 revealed substantial differences in their binding affinities to DNA (Table 5). Among the series, **Ru-1** and **Ru-2** exhibited the highest binding affinities, surpassing the reference compound cisplatin (CP) and KP1019, suggesting significant potential for anticancer activity [9]. The calculated binding energies of −29.41 kJ/mol for **Ru-1** and −28.70 kJ/mol for **Ru-2** indicate strong interactions with the DNA backbone and nucleobases, thereby supporting their biological relevance. Experimental UV–vis titration measurements further confirmed these interactions, displaying even stronger binding affinities than those predicted by docking, suggesting that these ruthenium complexes exhibit enhanced stabilization with DNA in solution.

Table 5. Binding energies (kJ/mol) of ruthenium complexes (**Ru-1–Ru-6**) docked onto DNA, calculated using molecular docking. Cisplatin (CP) and KP1019 were included as reference compounds.

Complex	Ru-1	Ru-2	Ru-3	Ru-4	Ru-5	Ru-6	CP	KP1019
Binding energy (kJ/mol)	−29.41	−28.70	−27.69	−27.57	−26.36	−26.61	−25.73	−14.72

The molecular docking results of **Ru-1**, **Ru-2**, and **Ru-3** revealed high spatial alignment within the DNA major groove (Figure S71), with all three complexes binding in similar regions. However, their distinct substituents—chlorine, iodine, and acetate—strongly influenced the mode and strength of interaction. The electrostatic potential surfaces shown in Figure S72 provide insights into these differences. For instance, **Ru-1** displays a well-defined negative potential around the acetate group, facilitating a hydrogen bond with guanine 7 (Figure 8A). This interaction is absent in **Ru-2** (Figure 8B), whose iodine substituent remains external to the DNA and lacks a localized negative potential for effective bonding, as also seen in its ESP surface.

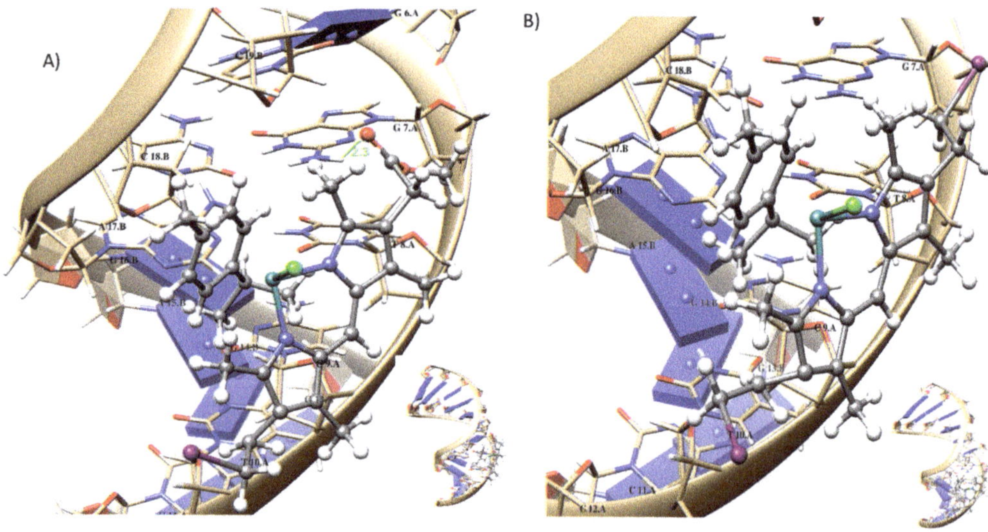

Figure 8. Binding interactions of **Ru-1** (**A**) and **Ru-2** (**B**) with DNA. Insets in the bottom left corners show the overall binding sites along the DNA strand, while the main panels display close-up views of the interaction regions. Hydrogen bonds are represented by green dashed lines, with distances indicated in angstroms (Å).

Although structurally similar, **Ru-3** adopts an inverted orientation upon docking, which prevents hydrogen bonding and instead results in electrostatic and hydrophobic interactions with adenine 17 and cytosine 18 of chain B (Figure 9A). Its ESP map (Figure S72) confirms a more diffuse charge distribution, lacking the concentrated potential necessary for directed hydrogen bonding.

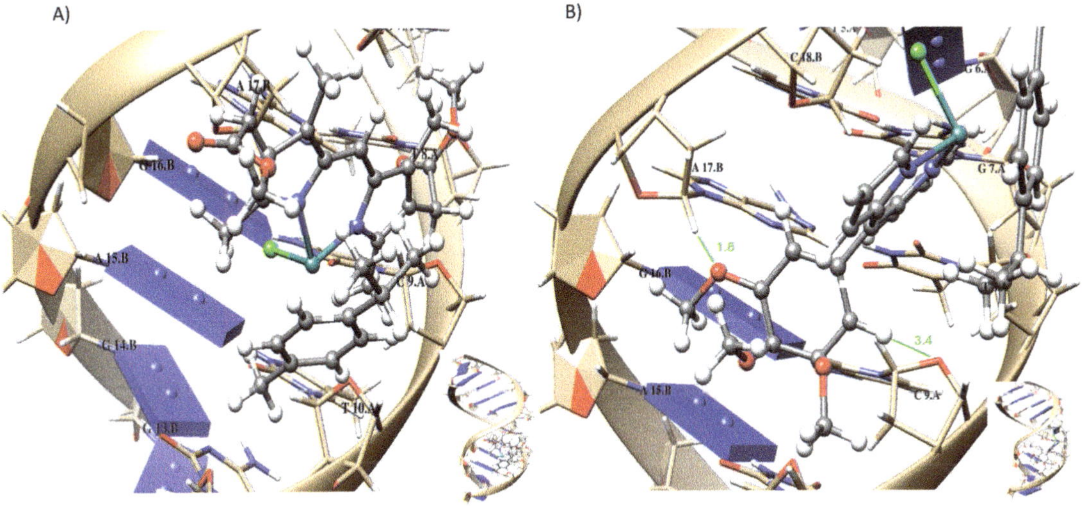

Figure 9. Binding interactions of **Ru-3** (**A**) and **Ru-6** (**B**) with DNA. Insets in the bottom left corners show the overall binding sites along the DNA strand, while the main panels display close-up views of the interaction regions. Hydrogen bonds are represented by green dashed lines, with distances indicated in angstroms (Å).

In contrast, **Ru-6** exhibits a distinct binding pattern due to the presence of methoxy substituents, which support the formation of two hydrogen bonds with DNA (Figure 9B). The ESP surface of **Ru-6** (Figure S72) exhibits intensely negative regions near the methoxy groups and the arene ring, which align with the DNA hydrogen bond acceptors. This localized potential optimizes the docking conformation, promoting strong, site-specific interactions. However, these interactions are more localized, and the shielding of the Ru center leads to reduced long-range electrostatic engagement with the DNA backbone.

KP1019 exhibited a distinct interaction with DNA, characterized by a lower binding energy of −14.72 kJ/mol, compared to the other ruthenium complexes. Despite the weaker binding energy, KP1019 formed a hydrogen bond with DNA at a distinct site, suggesting a different mode of action. The HOMO-LUMO gap analysis (0.136 eV) indicated a highly reactive electronic configuration, which favors metal-to-ligand charge transfer (MLCT) mechanisms (Table S2). This reactivity is crucial for anticancer properties involving oxidative DNA damage and apoptosis. This electronic configuration highlights the role of KP1019 in selectively targeting nucleobases and inducing cytotoxic effects, supporting its established therapeutic relevance.

3. Materials and Methods

The compounds were synthesized under a dinitrogen-inert atmosphere using a double vacuum/inert gas line. Reagents were purchased from Sigma-Aldrich (St. Louis, MO, USA) and were used without prior purification. Dichloromethane and methanol were distilled using P_2O_5 and iodine with magnesium, respectively. Chromatography columns were carried out on aluminum oxide-90 from Macherey-Nagel GmbH (Dueren, Germany). Ruthenium trichloride was purchased from Pressure Chemicals (Pittsburgh, PA, USA) and converted into $[Ru(\eta^6\text{-}p\text{-cymene})Cl_2]_2$ following literature procedures [57]. NMR spectra were recorded on a Bruker (Billerica, MA, USA) Advance 300 MHz spectrometer at 300 MHz for ^1H and 75 MHz for ^{13}C. Chemical shifts (δ) are in ppm, with TMS as a reference and using the solvent as an internal standard. The solvents used were $CDCl_3$ and CD_2Cl_2 from

Sigma-Aldrich (St. Louis, MO, USA). Two-dimensional correlated experiments (HSQC and HMBC) were used to assign chemical shifts. Coupling constants (*J*) are expressed in Hz. Multiplicity of the signals is s: singlet, d: doublet, t: triplet, q: quadruplet, and m: multiplet. The infrared spectra were recorded on an Alpha ATR spectrometer from Bruker Optics (Billerica, MA, USA) and analyzed with the OPUS software version 8.0. UV–vis spectra were measured on a Shimadzu UV-2600 spectrophotometer (Shimadzu Corporation, Kyoto, Japan) at 25 °C. Electrospray ionization mass spectra (ESI-MS) were obtained using a Bruker Esquire spectrometer. Combustion analysis was carried out on a Thermo Scientific Flash 2000 Elemental Analyzer (Thermo Fisher Scientific Inc., Waltham, MA, USA). DART (Direct Analysis in Real Time) mass spectra were obtained using a Jeol AccuTOF JMS-T100LC instrument (JEOL Ltd., Tokyo, Japan). MALDI-TOF spectra were recorded on a Bruker Microflex Mass Spectrometer (Bruker Daltonics, Bremen, Germany). Elemental analyses were carried out with an Exeter Analytical CE-440 instrument (Exeter Analytical Inc., North Chelmsford, MA, USA).

3.1. X-Ray Diffraction Crystallography

Data for compounds **L3**, **Ru-4**, and **Ru-5** were collected at 298 K on a Bruker Apex-II CCD diffractometer (Bruker AXS Inc., Madison, WI, USA) using MoKα radiation (0.71073 Å). Cell parameters were determined using the Bruker SAINT software V8. 34A. Data integration, scaling, and corrections for Lorentz-polarization effects and absorption were performed with CrysAlisPro [58]. Structures were solved by direct methods using Olex2 [59]. Hydrogen atoms were placed at calculated positions and refined with fixed displacement parameters (Uiso(H) = 1.2 Ueq or 1.5 Ueq). Molecular representations were generated with Diamond [60] and MERCURY 3.9 [61]. A suitable single crystal of compound **Ru-6** was mounted on a glass fiber, and data were collected at 130 K on an Oxford Diffraction Gemini diffractometer (Agilent Technologies, Yarnton, UK) using MoKα radiation (0.71073 Å) with a CCD Atlas area detector. The CrysAlisPro and CrysAlis RED software packages were used for data collection and integration [62]. The double-pass method of scanning was used to exclude any noise. The collected frames were integrated using an orientation matrix determined from narrow-frame scans. Final cell constants were determined by global refinement; data were collected and corrected for absorbance using analytical numerical absorption correction with a multifaceted crystal model based on expressions upon the Laue symmetry with equivalent reflections [63]. Structure solutions and refinements were carried out with the SHELXS-2018 and SHELXL-2018 packages [64,65]. The WinGX v2023 software was used to prepare the material for publication [66].

The CIF files were deposited in the Cambridge Structural Database under the codes CCDC 2441642 for **3**, 2441644 for **Ru-4**, 2441643 for **Ru-5**, and 2440997 for **Ru-6**. Copies of the data can be obtained, free of charge, at www.ccdc.cam.ac.uk.

3.2. Computational Details

Molecular complexes were modeled based on the crystal structure of complex **Ru-6**. All the complexes were optimized on their geometries through electronic structure calculations employing the B3LYP functional [67]. The DGDZVP basis set was used for the ruthenium atom and the 6-31+G(d,p) basis set for the remaining atoms [68]. Solvent effects were included using the Solvation Model Based on Density (SMD) during geometry optimization [69]. Charge model 5 (CM5), an Extension of the Hirshfeld Population Analysis, was determined to calculate the electronic charges, essential for the molecular docking protocol, while frequency calculations confirmed structural stability [70]. All quantum mechanical calculations were performed using the Gaussian 16 software [71]. These optimized structures served as critical parameters for subsequent docking studies.

Molecular docking studies used the DNA structure (PDB code: 1AIO) [72] prepared using the following protocol generated with the MetalDock software [73]: the AutoDock 4 software [74], defining the entire DNA as the binding site due to its therapeutic relevance in cisplatin treatments; docking simulations using AutoDock 4 with the Lamarckian genetic algorithm [75]; and interaction analysis with AutoDock Tools. CM5 charges were added to the PDBQT files of the ruthenium complexes to ensure accurate charge distributions, highlighting the complexes' potential to inhibit DNA interactions and enhance predictive accuracy comparable to cisplatin-based therapies. Docking results were analyzed using the Chimera [76] and Maestro software [77].

3.3. Titration Experiments

The interaction between compounds **L1–L5** and **Ru-3–Ru-6** and salmon sperm DNA (obtained from Sigma-Aldrich) was performed by electronic absorption experiments following standard methodologies and procedures reported in the literature [78,79]. A stock solution of DNA was prepared in a trizma buffer (5 mM) at pH 7.4. The concentration was determined spectrophotometrically at 25 °C. The final DNA concentration was calculated using the molar extinction coefficient (6600 M^{-1} cm^{-1}). The ratio of the absorbances at 260 and 280 nm was used to determine the purity of the DNA. The A260/A280 ratio observed between 1.8 and 1.9 indicates that the DNA was free of RNA and proteins. All compounds were dissolved in dimethyl sulfoxide at a concentration of 10 mM, and these solutions were used as stock solutions. The intrinsic binding constant (K_b) was calculated using the Wolfe–Shimmer equation (Equation (1)) [80]:

$$\frac{[DNA]}{(\varepsilon_a - \varepsilon_f)} = \frac{[DNA]}{(\varepsilon_b - \varepsilon_f)} + \frac{1}{K_b(\varepsilon_b - \varepsilon_f)} \quad (1)$$

where [DNA] is the concentration of DNA; ε_a, ε_f, and ε_b are apparent molar absorption coefficients for the compound with DNA, without DNA, and binding DNA, respectively; and K_b is the intrinsic binding constant.

Gibbs free energy changes (ΔG) for the complex–DNA interactions were calculated according to Equation (2).

$$\Delta G = -RT\ln K_b \quad (2)$$

where R is the gas constant, and T is the absolute temperature.

3.4. Synthesis of Tert-butyl 4-(2-chloroethyl)-3,5-dimethyl-1H-pyrrole-2-carboxylate (pyrrole F)

The reported procedure was modified to prepare **pyrrole F** [38]. To a solution of 250 mg (1.04 mmol) of **pyrrole C** in 10 mL of dichloromethane at 0 °C, 411 mg (1.57 mmol) of triphenylphosphine was added. Carbon tetrachloride (201.6 μL, 2.09 mmol) was added dropwise, and the mixture was stirred for 15 min at 0 °C, followed by 14 h at room temperature (25 °C). The solvent was removed under vacuum, and the crude product was purified on a silica gel chromatography column with hexane/ethyl acetate (8:2) as an eluent. A yellow solid was obtained in an 81% yield (220 mg). The analytical data agree with those reported in the literature.

3.5. Synthesis of Dipyrrin Ligands

3.5.1. Methyl (Z)-2-(2-((4-(2-methoxy-2-oxoethyl)-3,5-dimethyl-1H-pyrrol-2-yl)methylene)-3,5-dimethyl-2H-pyrrol-4-yl)acetate (L1)

Trifluoroacetic acid (2.52 mL, 32.92 mmol) was added dropwise to a solution of *tert*-butyl 4-(2-iodoethyl)-3,5-dimethyl-1*H*-pyrrole-2-carboxylate (200 mg, 0.75 mmol) in 2 mL of dichloromethane at room temperature. The solution was stirred for 30 min. A solution of methyl 3-(5-formyl-2,4-dimethyl-1*H*-pyrrol-3-yl)acetate (207 mg, 0.75 mmol) in methanol

(3 mL) was added, followed by the dropwise addition of a 33% HBr solution in acetic acid (0.86 mL, 5.15 mmol). The mixture was stirred at room temperature for 1 h, and the product was collected by filtration. The orange residue was dissolved in 40 mL of water. Triethylamine (0.27 mL, 1.97 mmol) was slowly added, and the mixture was stirred at room temperature for 30 min. The resulting precipitate was collected by filtration and washed with water. A yellow solid was obtained in a 94% yield (300 mg). ^{1}H NMR (300 MHz, $CDCl_3$): 2.14 (s, 3 H, CH_3), 2.16 (s, 3 H, CH_3), 2.32 (s, 6 H, 2 × CH_3), 2.95 (t, 2 H, J = 9.0, CH_2), 3.15 (t, 2 H, J = 9.0, CH_2), 3.38 (s, 2 H, CH_2), 3.66 (s, 3 H, CH_3), 6.70 (s, 1 H, CH); ^{13}C NMR (75 MHz, $CDCl_3$): 4.77 (CH_3), 9.78 (CH_3), 14.48 (CH_3), 14.58 (CH_3), 29.63 (CH_2), 30.61 (CH_2), 51.94 (CH_3), 116.53 (CH), 120.79, 127.13, 136.06, 136.61, 137.07, 151.00, 152.57, 171.82 (CO); Elemental analysis for [$C_{18}H_{23}IN_2O_2$]: exp. (calc.) (%) C 50.82 (50.71); H 5.47 (5.44); N 6.73 (6.57); IR (ATR, cm^{-1}) υ–NH 2907, $υ_{as}$–CH_3 2907, $υ_s$–CH_3 2853, υ–C=O (ester) 1730, υ–1604 (C=N). UV–vis (DMSO) λ_{max} (nm) 443 (s, ε = 5.32), 269 (w); LC-MS (ESI) m/z calculated: 426.08, found: 429.11 [M + 3H]$^+$, t_R = 0.25 min.

3.5.2. (Z)-3-(2-Iodoethyl)-5-((4-(2-iodoethyl)-3,5-dimethyl-2*H*-pyrrol-2-ylidene)methyl)-2,4-dimethyl-1*H*-pyrrole (**L2**)

L2 was prepared according to the reported procedure for (Z)-4-(2-iodoethyl)-2-((4-(2-iodoethyl)-3,5-dimethyl-1*H*-pyrrol-2-yl)methylene)-3,5-dimethyl-2*H*-pyrrol-1-ium [23]. Trifluoroacetic acid (1.95 mL, 25.43 mmol) was added dropwise to a solution of *tert*-butyl 4-(2-iodoethyl)-3,5-dimethyl-1*H*-pyrrole-2-carboxylate (200 mg, 0.57 mmol) in dichloromethane (2 mL) at room temperature. The solution was stirred for 30 min. A solution of 4-(2-iodoethyl)-3,5-dimethyl-1*H*-pyrrole-2-carbaldehyde (158 mg, 0.57 mmol) in methanol (2 mL) was added, followed by the dropwise addition of a 33% HBr solution in acetic acid (0.86 mL, 5.15 mmol). The mixture was stirred at room temperature for 1 h, and the precipitate was collected by filtration. The solid was dissolved in 30 mL of water. Triethylamine (0.23 mL, 1.69 mmol) was slowly added, and the mixture was stirred at room temperature for 30 min. The resulting solid was collected by filtration and washed with water, yielding 93% (271 mg) of a yellow solid. The analytical data agree with those reported in the literature. Elemental analysis for [$C_{18}H_{23}IN_2O_2$]: exp. (calc.) (%) C 40.53 (40.18); H 4.47 (4.36); N 5.83 (5.51); IR (ATR, cm^{-1}) υ–NH 3208, $υ_{as}$–CH_3 2903, $υ_s$–CH_3 2851, υ–1601 (C=N). LC-MS (ESI) m/z calculated: 507.99 found: 511.087 [M + 3H]$^+$, t_R = 0.25 min.

3.5.3. Methyl (Z)-2-(2-((4-(2-methoxy-2-oxoethyl)-3,5-dimethyl-1*H*-pyrrol-2-yl)methylene)-3,5-dimethyl-2*H*-pyrrol-4-yl)acetate (**L3**)

Trifluoroacetic acid (2.52 mL, 32.92 mmol) was added dropwise to a solution of *tert*-butyl 4-(2-methoxy-2-oxoethyl)-3,5-dimethyl-1*H*-pyrrole-2-carboxylate (200 mg, 0.75 mmol) in dichloromethane (2 mL) at room temperature. The solution was stirred for 30 min. A solution of methyl 2-(5-formyl-2,4-dimethyl-1*H*-pyrrol-3-yl)acetate (146 mg, 0.75 mmol) in methanol (3 mL) was then added, followed by the dropwise addition of a 33% HBr solution in acetic acid (0.86 mL, 6.73 mmol). The mixture was stirred at room temperature for 1 h, and the precipitate was collected by filtration. The orange solid was dissolved in 30 mL of water. Triethylamine (0.23 mL, 1.69 mmol) was slowly added, and the mixture was stirred at room temperature for 30 min. The precipitate was collected by filtration and washed with water, yielding 87% (271 mg) of a yellow solid. ^{1}H NMR (300 MHz, $CDCl_3$): 2.16 (s, 6 H, 2 × CH_3), 2.32 (s, 6 H, 2 × CH_3), 3.38 (s, 4 H, 2 × CH_2), 3.67 (s, 6 H, 2 × CH_3), 6.70 ppm (s, 1 H, CH); ^{13}C NMR (75 MHz, $CDCl_3$): 9.78 (CH_3), 14.52 (CH_3), 30.62 (CH_2), 51.93 (CH_3), 116.61 (CH), 120.55, 135.84, 136.75, 151.99, 171.88 (CO); elemental analysis for [$C_{19}H_{24}N_2O_4$]: exp. (calc.) (%) C 65.23 (66.26); H 6.86 (7.02); N 8.11 (8.13); IR (ATR, cm^{-1}) υ–NH 3211, $υ_{as}$–CH_3 2913, $υ_s$–CH_3 2856, υ–C=O (ester) 1727, υ–1607 (C=N);

UV–vis (DMSO) λ_{max} (nm) 441 (s, ε = 5.27); LC-MS (ESI) m/z calculated: 346.19, found: 347.00 $[M + H]^+$, t_R = 0.25 min.

3.5.4. Methyl (Z)-2-(2-((4-(2-chloroethyl)-3,5-dimethyl-1*H*-pyrrol-2-yl)methylene)-3,5-dimethyl-2*H*-pyrrol-4-yl)acetate (**L4**)

Trifluoroacetic acid (2.52 mL, 32.92 mmol) was added dropwise to a solution of **pyrrole A** (200 mg, 0.75 mmol) in dichloromethane (2 mL) at room temperature. The solution was stirred for 30 min. A solution of **pyrrole G** (138 mg, 0.75 mmol) in 3 mL of methanol was added, followed by the dropwise addition of a 33% HBr solution in acetic acid (0.79 mL, 3.25 mmol). The mixture was stirred at room temperature for 1 h, and the product was collected by filtration. The orange solid was dissolved in 30 mL of water, and triethylamine (0.23 mL, 1.69 mmol) was slowly added. The mixture was stirred for 30 min. The precipitate was collected by filtration and washed with water, yielding 75% (190 mg) of a yellow solid. ^{1}H NMR (300 MHz, $CDCl_3$): 2.10 (s, 6 H, 2 × CH_3), 2.26 (s, 6 H, 2 × CH_3), 2.78 (t, J = 7.7, 2 H, CH_2), 3.33 (s, 2 H, CH_2), 3.45 (t, J = 7.5, 2 H, CH_2), 3.60 (s, 3 H, CH_3), 6.63 (s, 1 H, CH), 8.13 ppm (s, 1 H, NH); elemental analysis for $[C_{18}H_{23}ClN_2O_2]$: exp. (calc.) (%) C 64.70 (64.57); H 6.93 (6.92); N 8.40 (8.37); IR (ATR, cm^{-1}) υ–NH 3204, $υ_{as}$–CH_3 2912, $υ_s$–CH_3 2856, υ–C=O (ester) 1727, υ–1604 (C=N); UV–vis (DMSO) λ_{max} (nm) 442 (s, ε = 5.31); LC-MS (MALDI-TOF) m/z calculated: 334.14, found: $[M + 2H]^+$.

3.5.5. (Z)-3-(2-Chloroethyl)-5-((4-(2-chloroethyl)-3,5-dimethyl-2*H*-pyrrol-2-ylidene)methyl)-2,4-dimethyl-1*H*-pyrrole (**L5**)

Trifluoroacetic acid (2.61 mL, 34.14 mmol) was added dropwise to a solution of **pyrrole F** (200 mg, 0.77 mmol) in dichloromethane (2 mL) at room temperature under a nitrogen atmosphere. The solution was stirred for 30 min. A solution of **pyrrole G** (144 mg, 0.77 mmol) in 3 mL of methanol was added, followed by the dropwise addition of a 33% HBr solution in acetic acid (0.83 mL, 3.38 mmol). The mixture was stirred at room temperature for 1 h, and the product was collected by filtration. The orange solid was dissolved in 40 mL of water, and triethylamine (0.23 mL, 1.69 mmol) was slowly added. The mixture was stirred for 30 min. The resulting solid was collected by filtration and washed with water, yielding 85% (214 mg) of a yellow solid. ^{1}H NMR (300 MHz, $CDCl_3$): 2.19 (s, 6 H, 2 × CH_3), 2.37 (s, 6 H, 2 × CH_3), 2.87 (t, J = 7.5, 2 H, CH_2), 3.54 (t, J = 7.5, 2 H, CH_2), 6.73 (s, 1 H, CH); ^{13}C NMR (75 MHz, $CDCl_3$): 9.80 (CH_3), 14.44 (CH_3), 28.38 (CH_2), 43.87 (CH_2), 116.54 (CH), 124.25, 152.05; elemental analysis for $[C_{17}H_{22}Cl_2N_2]$: exp. (calc.) (%) C 62.50 (62.77); H 6.79 (6.82); N8.58 (8.61); IR (ATR, cm^{-1}) υ–NH 3187, $υ_{as}$–CH_3 2909, $υ_s$–CH_3 2863, , υ–C=N 1604; UV–vis (DMSO) λ_{max} (nm) 445 (s, ε = 5.30), 495 (s); LC-MS (MALDI-TOF) m/z calculated: 324.12, found: $[M + 2H]^+$.

3.6. Synthesis of Ruthenium Complexes

The NMR assignment numbering is shown in Figure 10.

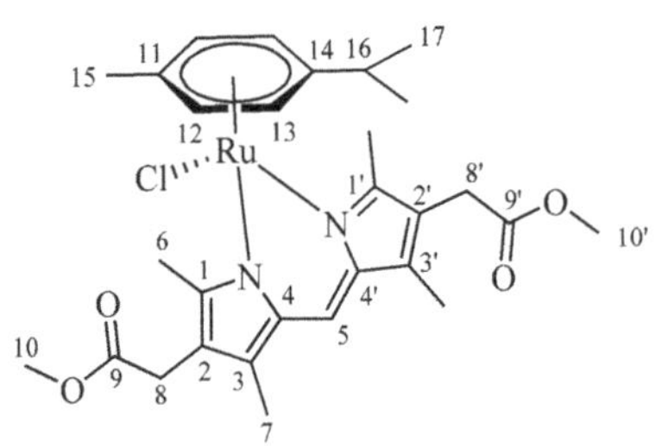

Figure 10. Atom numbering of complex **Ru-3** for NMR assignment.

3.6.1. Synthesis of Compound **Ru-3**

To a solution of **L3** (100 mg, 0.29 mmol) in dichloromethane (5 mL), triethylamine (120 μL, 0.87 mmol) was added, and the reaction mixture was stirred at room temperature for 20 min. $[Ru(\eta^6\text{-}p\text{-cymene})Cl_2]_2$ (91 mg, 0.15 mmol) was added, and the mixture was stirred for an additional 16 h at room temperature. The progress of the reaction was monitored by thin-layer chromatography (TLC) using CH_2Cl_2:MeOH (20:1, v/v). Upon completion, the solvent was removed under reduced pressure, and the resulting solid was washed with water (3 × 2 mL) to afford the crude product. The complex was purified by column chromatography on alumina, using a dichloromethane/methanol (40/1) mixture as an eluent. A red crystalline solid was obtained in a 61% yield (85 mg). ^{1}H NMR (300 MHz, $CDCl_3$): 1.02 (s, 3 H, CH_3), 1.04 (s, 3 H, CH_3), 2.23 (m, 10 H, 3 × CH_3 and CH), 2.70 (s, 6H, 2 × CH_3), 3.45 (m, 4 H, 2 × CH_2), 3.67 (s, 6 H, 2 × CH_3), 5.15 (d, J = 6, 2 H, 2 × CH), 5.20 (d, J = 6, 2 H, 2 × CH); 6.83 (s, 1 H, CH); ^{13}C NMR (75 MHz, $CDCl_3$): 9.95 (C-7 and C-7′), 17.49 (C-6 and C-6′), 18.29 (C-15), 22.37 (C-17), 30.25 (C-16), 31.3 (C-8), 51.88 (C-10), 83.33 (C-13), 84.21 (C-12), 98.97 (C-14), 102.64 (C-11), 121.15 (C-2 and C-2′), 122 (C-5), 133.45 (C-4 and C-4′), 138.65 (C-3 and C-3′), 161.87 (C-1 and C-1′), 172 (C-9 and C-9′); elemental analysis for $[C_{29}H_{37}ClN_2O_4Ru]$: exp. (calc.) (%) C 56.72 (56.71); H 6.11 (6.07); N 4.72 (4.56); IR (ATR, cm^{-1}) υ_{as}–CH_3 2906, υ–C=O (ester) 1728. UV–vis (DMSO) λ_{max} (nm): 436 (w), 513 (s). LC-MS (MALDI-TOF) m/z calculated: 614.14, found: 581.16 $[M - Cl + 1H]^+$.

3.6.2. Synthesis of Compound **Ru-4**

To a solution of **L4** (100 mg, 0.29 mmol) in dichloromethane (5 mL), triethylamine (120 μL, 0.87 mmol) was added, and the reaction mixture was stirred at room temperature for 20 min. $[Ru(\eta^6\text{-}p\text{-cymene})Cl_2]_2$ (94 mg, 0.15 mmol) was added, and the mixture was stirred for an additional 16 h at room temperature. The progress of the reaction was monitored by thin-layer chromatography (TLC) using CH_2Cl_2:MeOH (20:1, v/v). Upon completion, the solvent was removed under reduced pressure, and the resulting solid was washed with water (3 × 2 mL) to afford the crude product. The complex was purified by column chromatography on alumina, using a dichloromethane/methanol (50/1) mixture as an eluent. A red crystalline solid was obtained in a 70% yield (125 mg). ^{1}H NMR (300 MHz, $CDCl_3$): 0.94 (s, 3 H, CH_3), 0.96 (s, 3 H, CH_3), 2.15 (m, 10 H, 3 × CH_3 and CH), 2.62 (s, 6H, 2 × CH_3), 2.83 (m, 2H, CH_2), 3.38 (m, 4 H, 2 × CH_2), 3.59 (s, 6 H, 2 × CH_3), 5.06 (d, J = 6, 2 H, 2 × CH), 5.14 (d, J = 6, 2 H, 2 × CH); 6.73 (s, 1 H, CH). ^{13}C NMR (75 MHz, $CDCl_3$): 9.90 (CH_3), 9.96 (CH_3), 17.46 (CH_3), 17.53 (CH_3), 18.26 (C-15), 22.35 (C-17), 29.27 (C-16), 30.22 (CH_2), 31.30, 43.98 (CH_2), 51.84 (OCH_3), 83.32, 84.57 (C-12), 98.95 (C-14), 102.59, 122.01, 125.54, 133.13, 137.98, 138.62, 160.93, 161.62, 172.17 (CO); elemental analysis for $[C_{28}H_{36}Cl_2N_2Ru]$: exp. (calc.) C 55.69 (55.63), H 6.15 (6.00), N 4.67 (4.63); IR (ATR, cm^{-1}): υ_{as}–CH_3 2921, υ_s– CH_3 2984, υ–1732 (C=O), υ–1604 (C=N); UV–vis (DMSO) λ_{max} (nm): 437 (w, ε =4.81), 515 (s, ε = 5.17); MALDI-TOF MS m/z calculated: 604.12, found: 571.32 $[M - Cl + 2H]^+$.

3.6.3. Synthesis of Compound **Ru-5**

To a solution of **L5** (100 mg, 0.30 mmol) in dichloromethane (5 mL), triethylamine (120 μL, 0.87 mmol) was added, and the reaction mixture was stirred at room temperature for 20 min. $[Ru(\eta^6\text{-}p\text{-cymene})Cl_2]_2$ (96 mg, 0.15 mmol) was added, and the mixture was stirred for an additional 16 h at room temperature. The progress of the reaction was monitored by thin-layer chromatography (TLC) using CH_2Cl_2:MeOH (30:1, v/v). Upon completion, the solvent was removed under reduced pressure, and the resulting solid was washed with water (3 × 2 mL) to afford the crude product. The complex was purified by column chromatography on alumina, using a dichloromethane/methanol (40/1) mixture as an eluent. A red crystalline solid was obtained in a 55% yield (100 mg). ^{1}H NMR (300 MHz, $CDCl_3$): 1.02 (s, 3 H, CH_3), 1.05 (s, 3 H, CH_3), 2.23 (m, 10 H, 3 × CH_3 and CH), 2.70 (s, 6 H, 2 × CH_3), 2.91 (m, 4 H, 2 × CH_2), 3.49 (m, 4 H, 2 × CH_2), 5.14 (d, *J* = 6.0, 2 H, 2 × CH), 5.20 (d, *J* = 6.0, 2 H, 2 × CH), 6.81 (s, 1 H, CH); ^{13}C NMR (75 MHz, $CDCl_3$): 9.90 (CH_3), 17.45 (CH_3), 18.28 (C-15), 22.38 (C-17), 29.25 (C-16), 30.22 (CH_2), 43.97 (CH_2), 83.32, 84.56 (C-12), 98.97 (C-14), 102.57, 121.91, 125.61, 133.20, 138.08, 161.14; elemental analysis for [$C_{27}H_{35}Cl_3N_2Ru$]: exp. (calc.) C 54.55 (54.50), H 5.95 (5.93), N 4.74 (4.71); IR (ATR, cm^{-1}): υ_{as}–CH_3 2924, υ_s–CH_3 2863, υ–C=N 1607; UV–vis (DMSO) λ_{max} (nm) 437 (s, ε: 4.89), 517 (s, ε = 5.44). MALDI-TOF MS (*m*/*z*) calculated: 594.04, found: 561.38 $[M - Cl + 2H]^+$.

3.6.4. Synthesis of Compound **Ru-6**

A solution of 160 mg (0.48 mmol) of DDQ (2,3-dichloro-5,6-dicyano-1,4-benzoquinone) in 5 mL of benzene was added to a solution of 150 mg (56 mmol) of 2,2′-[(3,4,5-trimethoxyphenyl)methylene)bis(1*H*-pyrrole)] [45] in 2 mL of dichloromethane at 0 °C. The reaction mixture was stirred until it reached room temperature and stirred for another hour. The solvent was evaporated under vacuum, and 10 mL of CH_2Cl_2 was added, followed by 0.75 mL (2.4 mmol) of triethylamine. The mixture was stirred for fifteen minutes at room temperature. A solution of 150 mg (0.24 mmol, 0.5 eq.) of [Ru(η^6-*p*-cymene)Cl_2]$_2$ in 5 mL of CH_2Cl_2 was added, and the mixture was stirred at room temperature for 24 h. The solvent was evaporated to dryness under vacuum, and the residue was purified by column chromatography on alumina using a CH_2Cl_2:CH_3CN (70:30) mixture as an eluent. The solvent was evaporated under vacuum to give 137 mg (48%) of a dark orange solid. ^{1}H NMR (300 MHz, $CDCl_3$): 1.02 (m, 6 H, 2 × CH_3), 2.17 (s, 3 H, CH_3), 2.38 (m, 1 H, CH), 3.76 (m, 6 H, 2 × CH_3), 3.85 (s, 3 H, OCH_3), 5.24 (m, 4 H, 2 × CH_2), 6.48 (dd, *J* = 4.4, *J* = 1.4, 2 × CH), 6.56 (s, 2 H, 2 × CH), 6.62 (dd, *J* = 4.4, *J* = 1.4, 2 × CH), 7.95 (t, *J* = 1.4, 2 × CH); ^{13}C NMR (75 MHz, $CDCl_3$): 17.61 (C-15), 21.09, 29.61, 55.20, 59.99 (OCH_3), 83.70 (C-12), 99.19 (C-14), 101.06, 106.52, 107.63, 117.33, 129.96, 132.47, 133.87, 136.98, 145.14, 153.76; elemental analysis for [$C_{30}H_{36}Cl_1N_2O_3Ru$]·CH_2Cl_2: exp. (calc.) (%) C 53.54 (53.64), H 5.47 (5.52), N 4.34 (4.04); IR (ATR, cm^{-1}) υ_{as}–CH_3 2955, υ_s–CH_3 2853, υ–C-O 1540; UV–vis (CH_2Cl_2) λ_{max} (nm): 493 (s, ε = 4.43); DART-MS. *m*/*z* calculated: 580.45, found: 545.00 $[M - Cl]^+$.

4. Conclusions

A new set of dipyrrin ligands and organometallic complexes was synthesized and characterized using various spectroscopic techniques. The complexes containing iodide-substituted ligands were highly reactive, hindering further studies. Their chloride analogs remained stable under ambient conditions. The DNA-binding affinity of these compounds was evaluated through UV–vis titration, indicating their ability to interact with DNA via intercalation. A comparison of the crystal structures with those optimized by DFT confirmed the accuracy of the computational methods, particularly regarding the flexibility of the methoxy groups. Although the synthesis of the **Ru-1** and **Ru-2** complexes, which incorporate iodides into the dipyrrin ligand structures, was unsuccessful, we proceeded to investigate their molecular coupling with the DNA structure using computational methods. This analysis was crucial for determining the impact of the halogen on DNA interactions. Molecular docking revealed that the hypothetical iodide-bearing **Ru-1** and **Ru-2** complexes exhibited the highest DNA-binding affinities, featuring substantial hydrophobic and hydrogen-bonding interactions in their binding modes. The computational calculations supported the experimental results observed through UV–vis spectroscopy and confirmed that **Ru-3** has a higher affinity for DNA than **Ru-4**, **Ru-5**, and **Ru-6**. Notably, these ruthenium complexes demonstrated higher binding constants with DNA compared to cisplatin, suggesting their potential as anticancer agents.

Supplementary Materials: The following supporting information can be downloaded at: https://www.mdpi.com/article/10.3390/inorganics13060198/s1, ^{1}H-NMR, ^{13}C-NMR, HSQC-NMR, HMBC-NMR, IR, MS, and UV–vis spectra data of **L1–L5** and **Ru-3–Ru-6** (Figures S1–S66); ORTEP diagram of **L3** (Figure S20); UV–vis absorption spectra for the DNA interactions with **L1–L5** and **Ru-3–Ru-6** compounds (Figures S67–S69); binding saturation plots done(Figure S70); binding site of all complexes with DNA receptor (Figure S71); mapped electrostatic potential (ESP) surfaces (Figure S72); TD-DFT calculated UV–vis and HOMO-LUMO energies (Tables S1 and S2).

Author Contributions: Conceptualization, M.I.M., C.F.M., A.R.-A., B.B., and R.L.L.; methodology, M.I.M., C.F.M., A.R.-A., M.F.-A., R.A.T. and R.L.L.; investigation, M.I.M. and R.L.L.; computational calculations, A.L.O. and J.A.-T.; writing—original draft preparation, M.I.M., C.F.M., R.L.L., and A.L.O.; writing—review and editing, M.I.M., A.D.R., and R.L.L.; funding acquisition, R.L.L. All authors have read and agreed to the published version of the manuscript.

Funding: This research was funded by DGAPA-UNAM (PAPIIT projects IN-207725 and IN-211522).

Institutional Review Board Statement: Not applicable.

Informed Consent Statement: Not applicable.

Data Availability Statement: The data presented in this study are available in the article and Supplementary Materials.

Acknowledgments: We are grateful for support from Consejo Nacional de Humanidades, Ciencias y Tecnologías CONAHCYT (grants to M.I.M., C.F.M., and A.R.-A.). We thank M. P. Orta Pérez, E. Huerta Salazar, and E. García Ríos for obtaining the analytical data.

Conflicts of Interest: The authors declare no conflicts of interest.

References

1. Highley, M.S.; Landuyt, B.; Prenen, H.; Harper, P.G.; De Bruijn, E.A. The Nitrogen Mustards. *Pharmacol. Rev.* **2022**, *74*, 552–599. [CrossRef] [PubMed]
2. Delahousse, J.; Molina, L.; Paci, A. Cyclophosphamide and analogues; a matter of dose and schedule for dual anticancer activities. *Cancer Lett.* **2024**, *598*, 217119. [CrossRef] [PubMed]
3. Patra, S.; Dey, J.; Kar, S.; Chakraborty, A. Delivery of Chlorambucil to the Brain Using Surface Modified Solid Lipid Nanoparticles. *ACS Appl. Bio Mater.* **2024**, *7*, 3403–3413. [CrossRef]
4. Raber, J.; Zhu, C.; Eriksson, L.A. Theoretical Study of Cisplatin Binding to DNA: The Importance of Initial Complex Stabilization. *J. Phys. Chem. B* **2005**, *109*, 11006–11015. [CrossRef]
5. Romani, A.M.P. Cisplatin in cancer treatment. *Biochem. Pharmacol.* **2022**, *206*, 115323. [CrossRef]
6. Wang, D.; Lippard, S.J. Cellular processing of platinum anticancer drugs. *Nat. Rev. Drug Discov.* **2005**, *4*, 307–320. [CrossRef] [PubMed]
7. Oun, R.; Moussa, Y.E.; Wheate, N.J. The side effects of platinum-based chemotherapy drugs: A review for chemists. *Dalton Trans.* **2018**, *47*, 6645–6653. [CrossRef]
8. Gamberi, T.; Hanif, M. Metal-Based Complexes in Cancer Treatment. *Biomedicines* **2022**, *10*, 2573. [CrossRef]
9. Alessio, E.; Messori, L. NAMI-A and KP1019/1339, Two Iconic Ruthenium Anticancer Drug Candidates Face-to-Face: A Case Story in Medicinal Inorganic Chemistry. *Molecules* **2019**, *24*, 1995. [CrossRef]
10. Liu, H.-K.; Sadler, P.J. Metal Complexes as DNA Intercalators. *Acc. Chem. Res.* **2011**, *44*, 349–359. [CrossRef]
11. Alagesan, M.; Sathyadevi, P.; Krishnamoorthy, P.; Bhuvanesh, N.S.P.; Dharmaraj, N. DMSO containing ruthenium(ii) hydrazone complexes: In vitro evaluation of biomolecular interaction and anticancer activity. *Dalton Trans.* **2014**, *43*, 15829–15840. [CrossRef] [PubMed]
12. Ganeshpandian, M.; Loganathan, R.; Suresh, E.; Riyasdeen, A.; Akbarsha, M.A.; Palaniandavar, M. New ruthenium(ii) arene complexes of anthracenyl-appended diazacycloalkanes: Effect of ligand intercalation and hydrophobicity on DNA and protein binding and cleavage and cytotoxicity. *Dalton Trans.* **2014**, *43*, 1203–1219. [CrossRef]
13. Kilpin, K.J.; Clavel, C.M.; Edafe, F.; Dyson, P.J. Naphthalimide-Tagged Ruthenium–Arene Anticancer Complexes: Combining Coordination with Intercalation. *Organometallics* **2012**, *31*, 7031–7039. [CrossRef]
14. Uma Maheswari, P.; Palaniandavar, M. DNA binding and cleavage properties of certain tetrammine ruthenium(II) complexes of modified 1,10-phenanthrolines—Effect of hydrogen-bonding on DNA-binding affinity. *J. Inorg. Biochem.* **2004**, *98*, 219–230. [CrossRef]

15. Gasser, G.; Ott, I.; Metzler-Nolte, N. Organometallic anticancer compounds. *J. Med. Chem.* **2011**, *54*, 3–25. [CrossRef] [PubMed]
16. Swaminathan, S.; Karvembu, R. Dichloro Ru(II)-p-cymene-1,3,5-triaza-7-phosphaadamantane (RAPTA-C): A Case Study. *ACS Pharmacol. Transl. Sci.* **2023**, *6*, 982–996. [CrossRef] [PubMed]
17. Debreczeni, J.É.; Bullock, A.N.; Atilla, G.E.; Williams, D.S.; Bregman, H.; Knapp, S.; Meggers, E. Ruthenium Half-Sandwich Complexes Bound to Protein Kinase Pim-1. *Angew. Chem. Int. Ed.* **2006**, *45*, 1580–1585. [CrossRef]
18. Bregman, H.; Williams, D.S.; Meggers, E. Pyrido[2,3-a]pyrrolo[3,4-c]carbazole-5,7(6H)-diones: Synthesis, Cyclometalation, and Protein Kinase Inhibition. *Synthesis* **2005**, *2005*, 1521–1527. [CrossRef]
19. Meggers, E.; Atilla-Gokcumen, G.E.; Bregman, H.; Maksimoska, J.; Mulcahy, S.P.; Pagano, N.; Williams, D.S. Exploring Chemical Space with Organometallics: Ruthenium Complexes as Protein Kinase Inhibitors. *Synlett* **2007**, *2007*, 1177–1189. [CrossRef]
20. Wood, T.E.; Thompson, A. Advances in the Chemistry of Dipyrrins and Their Complexes. *Chem. Rev.* **2007**, *107*, 1831–1861. [CrossRef]
21. Mahanta, C.S.; Ravichandiran, V.; Swain, S.P. Recent Developments in the Design of New Water-Soluble Boron Dipyrromethenes and Their Applications: An Updated Review. *ACS Appl. Bio Mater.* **2023**, *6*, 2995–3018. [CrossRef] [PubMed]
22. Shikha Singh, R.; Prasad Paitandi, R.; Kumar Gupta, R.; Shankar Pandey, D. Recent developments in metal dipyrrin complexes: Design, synthesis, and applications. *Coord. Chem. Rev.* **2020**, *414*, 213269. [CrossRef]
23. Gibault, F.; Bailly, F.; Corvaisier, M.; Coevoet, M.; Huet, G.; Melnyk, P.; Cotelle, P. Molecular Features of the YAP Inhibitor Verteporfin: Synthesis of Hexasubstituted Dipyrrins as Potential Inhibitors of YAP/TAZ, the Downstream Effectors of the Hippo Pathway. *ChemMedChem* **2017**, *12*, 954–961. [CrossRef] [PubMed]
24. Kurzwernhart, A.; Kandioller, W.; Bächler, S.; Bartel, C.; Martic, S.; Buczkowska, M.; Mühlgassner, G.; Jakupec, M.A.; Kraatz, H.-B.; Bednarski, P.J.; et al. Structure–Activity Relationships of Targeted RuII(η6-p-Cymene) Anticancer Complexes with Flavonol-Derived Ligands. *J. Med. Chem.* **2012**, *55*, 10512–10522. [CrossRef]
25. Kurzwernhart, A.; Kandioller, W.; Bartel, C.; Bächler, S.; Trondl, R.; Mühlgassner, G.; Jakupec, M.A.; Arion, V.B.; Marko, D.; Keppler, B.K.; et al. Targeting the DNA-topoisomerase complex in a double-strike approach with a topoisomerase inhibiting moiety and covalent DNA binder. *Chem. Commun.* **2012**, *48*, 4839–4841. [CrossRef] [PubMed]
26. Kurzwernhart, A.; Kandioller, W.; Enyedy, É.A.; Novak, M.; Jakupec, M.A.; Keppler, B.K.; Hartinger, C.G. 3-Hydroxyflavones vs. 3-hydroxyquinolinones: Structure–activity relationships and stability studies on RuII(arene) anticancer complexes with biologically active ligands. *Dalton Trans.* **2013**, *42*, 6193–6202. [CrossRef]
27. Kandioller, W.; Balsano, E.; Meier, S.M.; Jungwirth, U.; Göschl, S.; Roller, A.; Jakupec, M.A.; Berger, W.; Keppler, B.K.; Hartinger, C.G. Organometallic anticancer complexes of lapachol: Metal centre-dependent formation of reactive oxygen species and correlation with cytotoxicity. *Chem. Commun.* **2013**, *49*, 3348–3350. [CrossRef]
28. Arion, V.B.; Dobrov, A.; Göschl, S.; Jakupec, M.A.; Keppler, B.K.; Rapta, P. Ruthenium- and osmium-arene-based paullones bearing a TEMPO free-radical unit as potential anticancer drugs. *Chem. Commun.* **2012**, *48*, 8559–8561. [CrossRef]
29. Meier, S.M.; Hanif, M.; Adhireksan, Z.; Pichler, V.; Novak, M.; Jirkovsky, E.; Jakupec, M.A.; Arion, V.B.; Davey, C.A.; Keppler, B.K.; et al. Novel metal(ii) arene 2-pyridinecarbothioamides: A rationale to orally active organometallic anticancer agents. *Chem. Sci.* **2013**, *4*, 1837–1846. [CrossRef]
30. Nazarov, A.A.; Meier, S.M.; Zava, O.; Nosova, Y.N.; Milaeva, E.R.; Hartinger, C.G.; Dyson, P.J. Protein ruthenation and DNA alkylation: Chlorambucil-functionalized RAPTA complexes and their anticancer activity. *Dalton Trans.* **2015**, *44*, 3614–3623. [CrossRef]
31. Gupta, R.K.; Pandey, R.; Sharma, G.; Prasad, R.; Koch, B.; Srikrishna, S.; Li, P.-Z.; Xu, Q.; Pandey, D.S. DNA Binding and Anti-Cancer Activity of Redox-Active Heteroleptic Piano-Stool Ru(II), Rh(III), and Ir(III) Complexes Containing 4-(2-Methoxypyridyl)phenyldipyrromethene. *Inorg. Chem.* **2013**, *52*, 3687–3698. [CrossRef] [PubMed]
32. Paitandi, R.P.; Gupta, R.K.; Singh, R.S.; Sharma, G.; Koch, B.; Pandey, D.S. Interaction of ferrocene appended Ru(II), Rh(III) and Ir(III) dipyrrinato complexes with DNA/protein, molecular docking and antitumor activity. *Eur. J. Med. Chem.* **2014**, *84*, 17–29. [CrossRef] [PubMed]
33. Gupta, R.K.; Sharma, G.; Pandey, R.; Kumar, A.; Koch, B.; Li, P.-Z.; Xu, Q.; Pandey, D.S. DNA/Protein Binding, Molecular Docking, and in Vitro Anticancer Activity of Some Thioether-Dipyrrinato Complexes. *Inorg. Chem.* **2013**, *52*, 13984–13996. [CrossRef]
34. Paitandi, R.P.; Singh, R.S.; Mukhopadhyay, S.; Sharma, G.; Koch, B.; Vishnoi, P.; Pandey, D.S. Synthesis, characterization, DNA binding and cytotoxicity of fluoro-dipyrrin based arene ruthenium(II) complexes. *Inorganica Chim. Acta* **2017**, *454*, 117–127. [CrossRef]
35. Yadav, M.; Singh, A.K.; Maiti, B.; Pandey, D.S. Heteroleptic Arene Ruthenium Complexes Based on meso-Substituted Dipyrrins: Synthesis, Structure, Reactivity, and Electrochemical Studies. *Inorg. Chem.* **2009**, *48*, 7593–7603. [CrossRef]
36. Smart, B.E. Fluorine substituent effects (on bioactivity). *J. Fluor. Chem.* **2001**, *109*, 3–11. [CrossRef]
37. Lund, K.-l.A.R.; Thompson, A. Synthesis of Symmetric meso-H-Dipyrrin Hydrobromides from 2-Formyl-pyrroles. *Synlett* **2014**, *25*, 1142–1144. [CrossRef]

38. Smith, K.M.; Pandey, R.K. Syntheses of isomers of protoporphyrin-IX with permuted propionic side-chains. *J. Heterocycl. Chem.* **1985**, *22*, 1041–1044. [CrossRef]
39. Wilcken, R.; Zimmermann, M.O.; Lange, A.; Joerger, A.C.; Boeckler, F.M. Principles and Applications of Halogen Bonding in Medicinal Chemistry and Chemical Biology. *J. Med. Chem.* **2013**, *56*, 1363–1388. [CrossRef]
40. Auffinger, P.; Hays, F.A.; Westhof, E.; Ho, P.S. Halogen bonds in biological molecules. *Proc. Natl. Acad. Sci. USA* **2004**, *101*, 16789–16794. [CrossRef]
41. Ahmed, B.; Barukial, P.; Bezbaruah, B. Some halogenated anticancer agents, role of halide (-X) and their interaction mechanism with AT/GC base pairs: A computational study. *J. Indian Chem. Soc.* **2023**, *100*, 101027. [CrossRef]
42. MacDonald, S.F. DIPYRRYLMETHANES. *J. Am. Chem. Soc.* **1957**, *79*, 2659. [CrossRef]
43. Treibs, A.; Häberle, N. Über die Synthese und die Elektronenspektren ms-substituierter Porphine. *Justus Liebigs Ann. Chem.* **1968**, *718*, 183–207. [CrossRef]
44. Singh, N.K.; Kumar, Y.; Paitandi, R.P.; Tiwari, R.K.; Kumar, A.; Pandey, D.S. DNA/protein binding and anticancer activity of ruthenium (II) arene complexes based on quinoline dipyrrin. *Inorganica Chim. Acta* **2023**, *545*, 121241. [CrossRef]
45. Abada, Z.; Cojean, S.; Pomel, S.; Ferrié, L.; Akagah, B.; Lormier, A.T.; Loiseau, P.M.; Figadère, B. Synthesis and antiprotozoal activity of original porphyrin precursors and derivatives. *Eur. J. Med. Chem.* **2013**, *67*, 158–165. [CrossRef]
46. Gill, H.S.; Finger, I.; Božidarević, I.; Szydlo, F.; Scott, M.J. Preparation of α,β-unsubstituted meso-arylbidipyrrins via metal-templated, oxidative coupling of dipyrrins. *New J. Chem.* **2005**, *29*, 68–71. [CrossRef]
47. Thoi, V.S.; Stork, J.R.; Magde, D.; Cohen, S.M. Luminescent Dipyrrinato Complexes of Trivalent Group 13 Metal Ions. *Inorg. Chem.* **2006**, *45*, 10688–10697. [CrossRef] [PubMed]
48. Marchetti, F.; Pettinari, C.; Pettinari, R.; Cerquetella, A.; Cingolani, A.; Chan, E.J.; Kozawa, K.; Skelton, B.W.; White, A.H.; Wanke, R.; et al. Areneruthenium(II) 4-Acyl-5-pyrazolonate Derivatives: Coordination Chemistry, Redox Properties, and Reactivity. *Inorg. Chem.* **2007**, *46*, 8245–8257. [CrossRef]
49. Murillo, M.I.; Restrepo-Acevedo, A.C.; Rocha-Roa, C.; Zacchino, S.; Svetaz, L.; Hernández-Ortega, S.; Abonia, R.; Le Lagadec, R.; Cuenú-Cabezas, F. Evaluation of the Antifungal Properties of Azomethine-Pyrazole Derivatives from a Structural Perspective. *ChemistryOpen* **2025**, e202500132. [CrossRef]
50. Chakraborty, D.; Chattaraj, P.K. Conceptual density functional theory based electronic structure principles. *Chem. Sci.* **2021**, *12*, 6264–6279. [CrossRef]
51. Olsztynska, S.; Komorowska, M. (Eds.) *Biomedical Engineering, Trends, Research and Technologies*; IntechOpen: London, UK, 2011. [CrossRef]
52. Liu, Y.-J.; Chao, H.; Tan, L.-F.; Yuan, Y.-X.; Wei, W.; Ji, L.-N. Interaction of polypyridyl ruthenium (II) complex containing asymmetric ligand with DNA. *J. Inorg. Biochem.* **2005**, *99*, 530–537. [CrossRef]
53. Mbugua, S.N.; Sibuyi, N.R.S.; Njenga, L.W.; Odhiambo, R.A.; Wandiga, S.O.; Meyer, M.; Lalancette, R.A.; Onani, M.O. New Palladium(II) and Platinum(II) Complexes Based on Pyrrole Schiff Bases: Synthesis, Characterization, X-ray Structure, and Anticancer Activity. *ACS Omega* **2020**, *5*, 14942–14954. [CrossRef] [PubMed]
54. Vardevanyan, P.O.; Antonyan, A.P.; Parsadanyan, M.A.; Davtyan, H.G.; Karapetyan, A.T. The binding of ethidium bromide with DNA: Interaction with single- and double-stranded structures. *Exp. Mol. Med.* **2003**, *35*, 527–533. [CrossRef]
55. Aslanoglu, M. Electrochemical and Spectroscopic Studies of the Interaction of Proflavine with DNA. *Anal. Sci.* **2006**, *22*, 439–443. [CrossRef]
56. Sirajuddin, M.; Ali, S.; Haider, A.; Shah, N.A.; Shah, A.; Khan, M.R. Synthesis, characterization, biological screenings and interaction with calf thymus DNA as well as electrochemical studies of adducts formed by azomethine [2-((3,5-dimethylphenylimino)methyl)phenol] and organotin(IV) chlorides. *Polyhedron* **2012**, *40*, 19–31. [CrossRef]
57. Zelonka, R.A.; Baird, M.C. Benzene Complexes of Ruthenium(II). *Can. J. Chem.* **1972**, *50*, 3063–3072. [CrossRef]
58. *CrysAlisPro Oxford Diffraction*; Agilent Technologies UK Ltd.: Yarnton, UK, 2017.
59. Dolomanov, O.V.; Bourhis, L.J.; Gildea, R.J.; Howard, J.A.K.; Puschmann, H. OLEX2: A complete structure solution, refinement and analysis program. *J. Appl. Crystallogr.* **2009**, *42*, 339–341. [CrossRef]
60. Impact, C. Diamond–crystal and molecular structure visualization. *Kreuzherrenstr* **2014**, *102*, 53227.
61. Macrae, C.F.; Bruno, I.J.; Chisholm, J.A.; Edgington, P.R.; McCabe, P.; Pidcock, E.; Rodriguez-Monge, L.; Taylor, R.; van de Streek, J.; Wood, P.A. Mercury CSD 2.0—New features for the visualization and investigation of crystal structures. *J. Appl. Crystallogr.* **2008**, *41*, 466–470. [CrossRef]
62. Xcalibur, C. *System, CrysAlis CCD and CrysAlis RED*, Versions 1.171. 36.32; Oxford Diffraction Ltd.: Abingdon, UK, 2013.
63. Clark, R.C.; Reid, J.S. The analytical calculation of absorption in multifaceted crystals. *Acta Crystallogr. Sect. A* **1995**, *51*, 887–897. [CrossRef]
64. Sheldrick, G. SHELXT—Integrated space-group and crystal-structure determination. *Acta Crystallogr. Sect. A* **2015**, *71*, 3–8. [CrossRef] [PubMed]
65. Sheldrick, G. Crystal structure refinement with SHELXL. *Acta Crystallogr. Sect. C* **2015**, *71*, 3–8. [CrossRef] [PubMed]

66. Farrugia, L. WinGX and ORTEP for Windows: An update. *J. Appl. Crystallogr.* **2012**, *45*, 849–854. [CrossRef]
67. Lee, C.; Yang, W.; Parr, R.G. Development of the Colle-Salvetti correlation-energy formula into a functional of the electron density. *Phys. Rev. B* **1988**, *37*, 785–789. [CrossRef]
68. Roy, L.E.; Hay, P.J.; Martin, R.L. Revised Basis Sets for the LANL Effective Core Potentials. *J. Chem. Theory Comput.* **2008**, *4*, 1029–1031. [CrossRef] [PubMed]
69. Marenich, A.V.; Cramer, C.J.; Truhlar, D.G. Universal Solvation Model Based on Solute Electron Density and on a Continuum Model of the Solvent Defined by the Bulk Dielectric Constant and Atomic Surface Tensions. *J. Phys. Chem. B* **2009**, *113*, 6378–6396. [CrossRef]
70. Marenich, A.V.; Jerome, S.V.; Cramer, C.J.; Truhlar, D.G. Charge Model 5: An Extension of Hirshfeld Population Analysis for the Accurate Description of Molecular Interactions in Gaseous and Condensed Phases. *J. Chem. Theory Comput.* **2012**, *8*, 527–541. [CrossRef]
71. Frisch, M.J.; Trucks, G.W.; Schlegel, H.B.; Scuseria, G.E.; Robb, M.A.; Cheeseman, J.R.; Scalmani, G.; Barone, V.; Petersson, G.A.; Nakatsuji, H.; et al. *Gaussian 16 Rev. C.01*; Gaussian, Inc.: Wallingford, CT, USA, 2016.
72. Takahara, P.M.; Rosenzweig, A.C.; Frederick, C.A.; Lippard, S.J. Crystal structure of double-stranded DNA containing the major adduct of the anticancer drug cisplatin. *Nature* **1995**, *377*, 649–652. [CrossRef]
73. Hakkennes, M.L.A.; Buda, F.; Bonnet, S. MetalDock: An Open Access Docking Tool for Easy and Reproducible Docking of Metal Complexes. *J. Chem. Inf. Model.* **2023**, *63*, 7816–7825. [CrossRef]
74. Morris, G.M.; Huey, R.; Lindstrom, W.; Sanner, M.F.; Belew, R.K.; Goodsell, D.S.; Olson, A.J. AutoDock4 and AutoDockTools4: Automated docking with selective receptor flexibility. *J. Comput. Chem.* **2009**, *30*, 2785–2791. [CrossRef]
75. Morris, G.M.; Goodsell, D.S.; Halliday, R.S.; Huey, R.; Hart, W.E.; Belew, R.K.; Olson, A.J. Automated docking using a Lamarckian genetic algorithm and an empirical binding free energy function. *J. Comput. Chem.* **1998**, *19*, 1639–1662. [CrossRef]
76. Pettersen, E.F.; Goddard, T.D.; Huang, C.C.; Couch, G.S.; Greenblatt, D.M.; Meng, E.C.; Ferrin, T.E. UCSF Chimera—A visualization system for exploratory research and analysis. *J. Comput. Chem.* **2004**, *25*, 1605–1612. [CrossRef] [PubMed]
77. *Maestro*; Schrödinger LLC: New York, NY, USA, 2023.
78. Sirajuddin, M.; Ali, S.; Badshah, A. Drug–DNA interactions and their study by UV–Visible, fluorescence spectroscopies and cyclic voltametry. *J. Photochem. Photobiol. B Biol.* **2013**, *124*, 1–19. [CrossRef] [PubMed]
79. Stevanović, N.; Mazzeo, P.P.; Bacchi, A.; Matić, I.Z.; Đorđić Crnogorac, M.; Stanojković, T.; Vujčić, M.; Novaković, I.; Radanović, D.; Šumar-Ristović, M.; et al. Synthesis, characterization, antimicrobial and cytotoxic activity and DNA-binding properties of d-metal complexes with hydrazones of Girard's T and P reagents. *JBIC J. Biol. Inorg. Chem.* **2021**, *26*, 863–880. [CrossRef]
80. Cuartas, V.; Aragón-Muriel, A.; Liscano, Y.; Polo-Cerón, D.; Crespo-Ortiz, M.d.P.; Quiroga, J.; Abonia, R.; Insuasty, B. Anticancer activity of pyrimidodiazepines based on 2-chloro-4-anilinoquinazoline: Synthesis, DNA binding and molecular docking. *RSC Adv.* **2021**, *11*, 23310–23329. [CrossRef]

inorganics

MDPI

Article

Synthesis, Characterisation, DFT Study and Biological Evaluation of Complexes Derived from Transition Metal and Mixed Ligands

Enas H. Mohammed [1], Eman R. Mohammed [1], Eman M. Yahya [1] and Mohammed Alsultan [2,*]

[1] Department of Science, College of Basic Education, University of Mosul, Mosul 41002, Iraq; enashazim@uomosul.edu.iq (E.H.M.); eman.r.ghazal@uomosul.edu.iq (E.R.M.); eman.m.alkhashashab@uomosul.edu.iq (E.M.Y.)

[2] Department of New and Renewable Energy, College of Science, University of Mosul, Mosul 41002, Iraq

* Correspondence: mohamadfkaleel@uomosul.edu.iq

Abstract

This research prepared and characterised novel mixed coordination complexes derived from escitalopram with eugenol and curcumin to form (L_1) and (L_2), respectively. The complexes were prepared via Williamson ether synthesis and analysed by FTIR, UV–Vis, ^{1}H-NMR spectroscopy, elemental analysis, molar conductivity, and magnetic susceptibility. The results confirmed their octahedral geometries. Magnetic investigation reported high-spin configurations for Mn(II), Co(II), and Ni(II) complexes, whereas Cu(II) exhibited a distorted octahedral arrangement with characteristic d–d transitions. In addition, the calculation of Density functional theory (DFT) provided more insight into the detailed structural and electronic properties of the new ligand and its complexes. Antimicrobial compounds were evaluated against *Escherichia coli*, *Staphylococcus aureus*, and *Candida albicans* through the agar well diffusion method. The reported results revealed that Cobalt complexes showed antimicrobial activity followed by Copper (Cu), Nickel (Ni) and Manganese(Mn) complexes, respectively, due to an increase in Co-lipophilicity, which leads to improved diffusion through microbial cell membranes. The research findings confirmed that escitalopram-based mixed ligands coordinate with transition metals and could have significant biological applications.

Keywords: escitalopram; eugenol; curcumin; Williamson ether synthesis; coordination complexes; biological activity; DFT study

1. Introduction

Recently, coordination complexes that involve biologically active ligands have been highlighted in much research due to their potential therapeutic applications. Eugenol, Escitalopram, and Curcumin have enormous interstitial medical applications, such as antidepressant, antioxidant, reducing heart diseases, anti-inflammatory, antimicrobial properties, reducing pain and anticancer factor. Furthermore, they have interesting antibacterial effects [1–5].

Eugenol (Eug) is a phenolic compound that has an allylchain-substituted guaiacol and belongs to the phenylpropanoid chemical compound family. It is a colourless to pale yellow oil. It is simply extracted from plant leaves. The polyphenol (eugenol) can damage the membranes of negative and positive Gram bacteria due to its antibacterial properties [6]. Eugenol is a bidentate ligand recently applied in coordination chemistry with some transition metal elements (Figure 1a). Then the resulting complexes exhibited valuable bacterial activities [7]. At the same time, escitalopram (Esc) is classified under the medical class Selective Serotonin Reuptake Inhibitor (SSRI). Its Core chemical structure has phenylbutylamine, a nitrile group and a dihydrobenzofuran ring. It can be chemically linked to form multifunctional ligands with chelating ability [8] (Figure 1b).

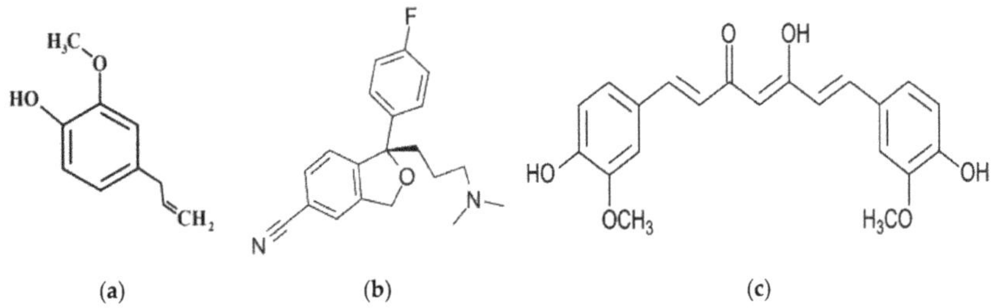

Figure 1. Chemical structure of (**a**) Eugenol, (**b**) Escitalopram and (**c**) Curcumin.

Finally, Curcumin (Cur) (diferuloylmethane) is a natural hydrophobic polyphenol derived from turmeric powder of Curcuma longa. It has been extensively investigated for pharmacological and biological effects. Curcumin can form strong complexes with most of the central transition metal ions through the enolic group [9] (Figure 1c).

Transition metals like Mn(II), Cu(II), Zn(II), Fe(II) and Co(II) could form a stable coordination bond with Eug, Cur and Esc through the N atom in amine or amide, O atom related to O in β-diketone moiety and/or phenolic oxygen. This new modulation of electronic properties can potentially enhance therapeutic efficacy [10]. In addition, some theoretical studies noted that these types of ligands have effective roles against advanced viruses such as COVID-19 [11], especially when these ligands can be mixed together.

To the best of our knowledge, there is no study that examines yet the effect of mixed Eug_Esc or Cur_Esc mixed ligand complexes against some types of bacteria, such as Gram-positive, which is *Staphylococcus aureus* type and Gram-negative, like *E. coli*. The mixed Eug_Esc and Cur_Esc mixed ligands were prepared via Williamson ether synthesis. The ligand and complexes were characterised by FTIR and ^{1}HNMR spectroscopy and Uv-vis absorbance. The molar conductivity of complexes was tested in DMSO solvent. Carbon(C), hydrogen(H), nitrogen(N) and central ion quantities were also investigated. The resulting complexes showed octahedral geometry. All complexes reported very high activity towards these types of bacteria.

2. Materials and Methods

2.1. Material and Measurement Technique

All chemical compound; manganese Chloride($MnCl_2{\cdot}5H_2O$), Cobalt Chloride ($CoCl_2{\cdot}6H_2O$), Nickel Chloride($NiCl_2{\cdot}6H_2O$) Copper chloride($CuCl_2{\cdot}2H_2O$), Absolute Ethanol(Etol), Dimethyl sulfoxide (DMSO), Chloroform (CCl_4), Eugenol (4-Allyl-2-methoxyphenol), Escitalopram ((S)-1-[3-(Dimethylamino)propyl]-1-(4-fluorophenyl)-1,3-dihydro isobenzofuran-5-carbonitrile) and Curcumin 1E,6E)-1,7-bis (4-hydroxy-3-methoxyphenyl)-1,6-heptadiene-3,5-dione were provided either from Merck or Fluka company (Darmstadt, Germany) without further purification.

A Fourier-transform infrared (FTIR) spectra Module 8300 Shimadzu spectrophotometer (Kyoto, Japan) was utilised to examine the prepared mixed ligands and complexes using caesium iodide discs. The transmitted absorbance of samples was measured using an Ultraviolet light-visible (Uv-vis) spectrophotometer module 1600 Shimadzu (Kyoto, Japan). in the range of (200–800 nm). To determine the melting point (MP) of the prepared complexes, a digital display MP temperature Stuart apparatus module SMP30 (Staffordshire, UK)was used. For electrical molar conductivity measurements of complexes, they were carried out at room temperature (25 °C) using a 4510–Jenway in (10^{-3} mol·L^{-1}) DMF solution. Bruker Bio Spin GmbH, 400 MHz, (Rheinstetten, Germany) was used to determine ^{1}H NMR spectra of complexes using tetramethylsilane as internal reference solvent, while the chemical shifts were quoted in δ (ppm). Atomic absorption module SensAA (Melbourne, Australia) was utilised to determine central metal ions (Mn(II), Ni(II), Co(II) and Cu(II). While the effective magnetic moments measured by MSB MK1 Sherwood (Cambridge, UK).

2.2. Preparation of the Mixed Escitalopram and Clove Oil (Eugenol) (Esc-Eug) Ligand

The cyclic (Esc-Eug) ligand was prepared through Williamson Ether Synthesis [12–14] by reacting 0.01 M (equivalent to 3.24 g) escitalopram with Eugenol (0.01 M (equivalent to 1.64 g). The chemical reaction was carried out in a 100 mL round-bottom flask. Chloroform was used as a solvent in a basic medium (commonly using bases like NaOH or K_2CO_3). The chemical mixture was then refluxed for 8 h under continuous stirring to ensure the reaction completion and ether formation. At the end of refluxing, the resulting precipitate was then filtered, washed, and dried thoroughly. And recrystallised with ethanol to purify the product. Scheme 1 shows the chemical mechanism of cyclic ether formation.

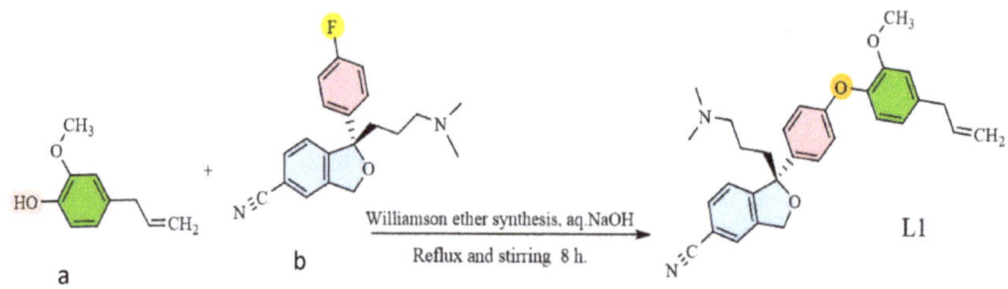

Scheme 1. Illustrated preparation of Eugenol-propyl-escitalopram ether (L_1) from reacting (**a**) Eugenol, with (**b**) Escitalopram via Williamson Ether Synthesis.

2.3. Preparation of the Mixed Escitalopram and Curcumin (Esc_Cur) Ligand

The cyclic (Esc_Cur) ligand was prepared through Williamson Ether Synthesis, similar to Section 2.2, by reacting 0.01 M (equivalent to 3.24 g) escitalopram with Curcumin (0.01 M (equivalent to 3.68 g). The final precipitate was then filtered, washed, and dried thoroughly and recrystallised with ethanol to purify the product. The Scheme 2 shows the chemistry of cyclic ether formation.

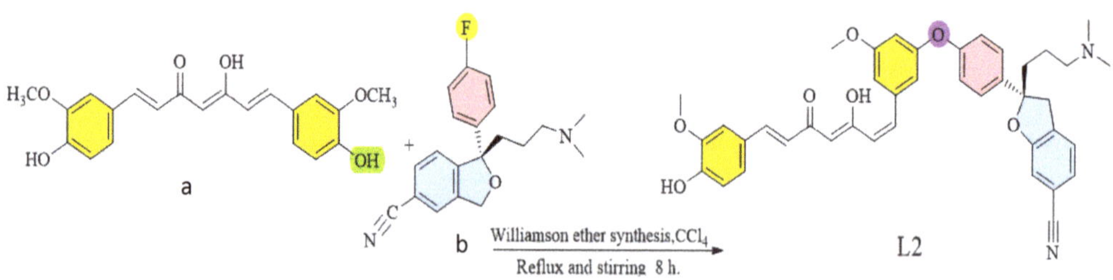

Scheme 2. Iluustrates preparation of Esc_Curc Either conjugate (L_2) from reacting (**a**) Eugenol, with (**b**) Escitalopram via Williamson Ether Synthesis.

2.4. General Preparation of the Mixed Ligand-Transition Metal Complexes

Solutions of the metal chlorides (0.01 M) like $MnCl_2 \cdot 5H_2O$, $CoCl_2 \cdot 6H_2O$, $NiCl_2 \cdot 6H_2O$ and $CuCl_2 \cdot 2H_2O$ were mixed with an equivalent molar amount (0.01 M) of the prepared ligand in Section 2.2 or Section 2.3 using chloroform as solvent. The reaction mixture was then refluxed for 4 h to ensure the formation of transition metal complexes. At the end of refluxing, the formed precipitates were then collected by drying and recrystallised using ethanol to obtain purified crystalline complexes.

3. Results and Discussion

3.1. Fourier Transfer Infrared Spectroscopy (FTIR)

As shown in Figure 2. The ligand (L_1) exhibits peaks at (3339) cm^{-1} that belong to the ν(N-H) bond stretching. Other peaks at (3061, 3028) are attributed to the ν(C-H) aromatic stretch. While Aliphatic ν(C-H) stretching peaks appeared at (2913, 2853, 2806) cm^{-1}, respectively.

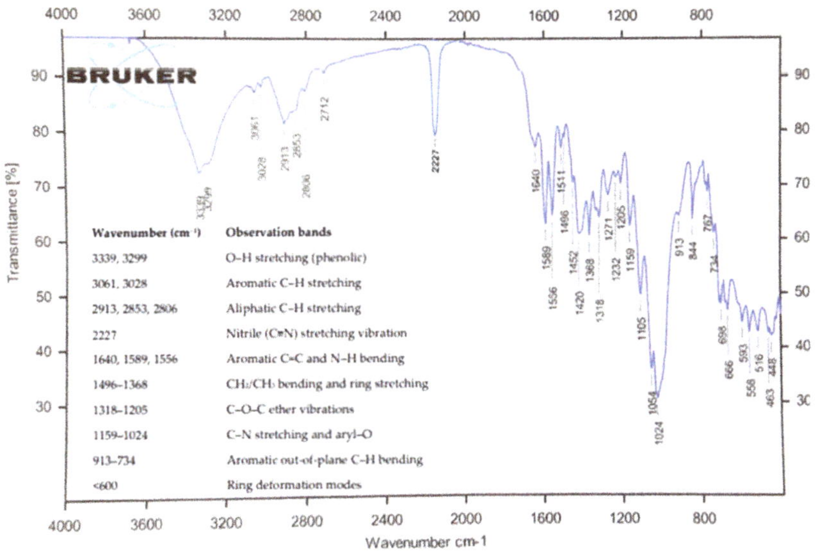

Figure 2. FTIR spectroscopy of L_1. The inset table shows the interpretation of the highlighted absorbed bands.

The Peaks at (1640, 1589, 1556) (cm^{-1}) belong to aromatic ν(C=C) and ν(N-H) bending. ν(CH_2/CH_3) bending and ring stretching displayed peaks at 1496–1368 cm^{-1}, respectively. ν(C-O-C) either vibration cited in 1318–1205 cm^{-1}. C–N stretching and aryl–O displayed bands at 1159–1024 cm^{-1}. Aromatic ν(C-H) out-of-plane ν(C-H) bending displayed peaks at 913–734 cm^{-1}. A band was observed at 2227 cm^{-1}, attributed to the ν(C≡N) stretching vibration. This band shifted lower when L_1 coordinated with the central ion due to π-back-donation into the nitrile. Which π* weakens C≡N to the red shift, as seen in Table 1 [15]. Finally, the peaks below 600 cm^{-1} are attributed to the deformation of ring modes [9,16–18].

FTIR of complexes [M(L_1)Cl]Cl, where M can represent Mn(II), Ni(II), Cu(II) and Co(II) as central coordination ions, showed successful metal ion coordination. [M(L_1)Cl]Cl formula suggested after FTIR and HNMR, and conductivity study as explained in Sections 3.1–3.3. For more details, in complexes [X(L_1)Cl]Cl, slightly new bands appear in the range of (524–601) cm^{-1}, (412–464) cm^{-1} could be attributed to M-O and M-N, respectively. The strong band of ν(C=N) or ν(C=O) stretching, which is shifted to a lower value if compared to L1, suggests coordination through nitrogen or oxygen as donor atoms. The ν(N-H) bond has been shifted down when the N atom is coordinated in complexes. Thus, the new ν(N-H) appeared at a range of (3331–3338) cm^{-1}. Finally, in all [M(L_1)Cl] complexes appeared bands (2905–2918) cm^{-1} and (2995–3007) cm^{-1}, corresponding to ν(C-H) aromatic and aliphatic [18,19].

Table 1. Infrared Spectroscopic Data (ν/cm^{-1}) for N–H, ν(C≡N, C=O, C–N, C–O–C, C=C, M–N, and M–O) Modes of L_1, L_2 and Their Metal Complexes.

Sample	ν(N–H)	ν(C≡N)	ν(C=O)	ν(C–N)	ν(C–O–C)	ν(C=C) (Aryl)	ν(M–N)	ν(M–O)
L_1	3339	2227	1589	1368	1105	1640	—	—
L_2	2963	—	1642	1363	1100	1642	—	—
[Co(L_1)Cl]Cl	2963	—	1555	1421	1106	1614	522	553
[Ni(L_1)Cl]Cl	2901	—	1574	1376	1108	1552	463	556
[Mn(L_1)Cl]Cl	2902	—	1505	1417	1105	1663	447	557
[Cu(L_1)Cl]Cl	2995	—	1516	1409	1103	1644	494	558
[Co(L_2)Cl]	3334	2196	1515	1410	1105	1643	491	554
[Ni(L_2)Cl]	3333	2205	1516	1419	1104	1625	524	565
[Mn(L_2)Cl]	3335	2212	1559	1406	1106	1642	526	571
[Cu(L_2)Cl]	3332	2185	1509	1419	1105	1636	491	572

In a similar pattern, L_2 displayed bands as plotted in Figure 3. A broad absorption was observed at 3334 cm^{-1} and 3284 cm^{-1} corresponding to ν(N-H) and ν(O-H) stretching vibrations, respectively. The bands at 3010 cm^{-1} and 2918 cm^{-1} are attributed to aromatic and aliphatic ν(C-H) stretches. The peaks at (951, 846, and 701) cm^{-1} are attributed to aromatic ν(C-H) vibration. The presented bands at 1515, 1466, and 1406 cm^{-1} confirm the ν(C=C) stretches of integrity of the aromatic backbone. Strong bands appeared at (1368, 1316, 1296, 1226, 1202, 1151, 1103, 1078, and 1005) cm^{-1}, confirming ν(C–O–C) attributed to C–O–C stretching vibrations, which confirm formation of ether linkage through Williamson ether synthesis. The bands 951, 846, and 701 cm^{-1} correspond to aromatic ν(C-H) out-of-plane bend vibrations [20].

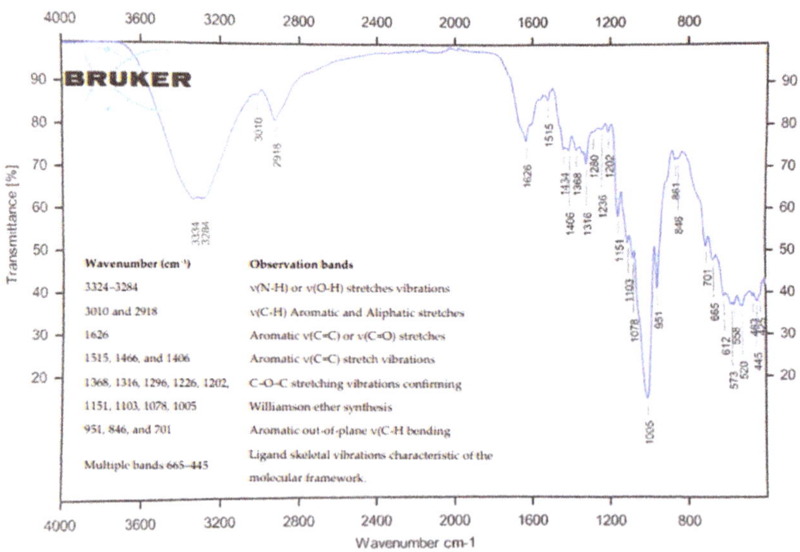

Figure 3. FTIR spectroscopy of L_2. The inset table shows the interpretation of the highlighted absorbed bands.

FTIR of complexes [X(L_2)Cl], new bands appear in the range of (526–601) cm^{-1}, (404–478) cm^{-1} correspond to M-O and M-N, respectively. ν(C=N) or ν(C=O) stretching band in complexes shifted lower wavenumber. This suggested the coordination of X central ion(II) with nitrogen or oxygen as donor atoms. It was noted also that the ν(N-H) bond has been shifted down when the N atom is coordinated in complexes. Thus, the new ν(N-H) appeared at a range of (3331–3338) cm^{-1}. Finally, in all [X(L_1)Cl]Cl complexes, bands appeared at (2905–2918) cm^{-1} and (2995–3007) $cm^{-1,}$ corresponding to ν(C-H) aromatic and aliphatic absorbance regions.

When L_1 and L_2 coordinate with Co(II), Ni(II), Cu(II), and Mn(II), they are commonly approved by new low-frequency bands of the IR spectrum. Weak to medium absorptions are often reported in the range (490–530) cm^{-1}. which is corresponding to ν(M–N) stretching. These values are consistent with ligand nitrogen atoms coordination (azomethine or nitrile donors) to the metal central ions. Additional bands observed near (555–580) cm^{-1} are attributed to ν(M–O) stretching vibrations. It could arise from the bonding of phenolic or ether oxygen atoms to the central ions. These low-energy modes are absent in the free L_1 and L_2. They appeared only after coordination, confirming the formation of true chelate structures. The values are in high agreement with other reports for octahedral M(II) Schiff base and nitrile complexes. When ν(M–N) typically sited at (450–530) cm^{-1} and ν(M–O) at (520–600) cm^{-1} [21,22].

3.2. ^{1}H-NMR Spectra of L_1 and L_2

Figure 4 shows ^{1}H-NMR of L_1; as seen, the shifted peak at δ 8.05–6.78 ppm (m, ~12H) is attributed to aromatic protons that are related to naphthoxy and phenoxy moieties; (naphthyl ring can appear at δ 7.8–8.1, and phenoxy/anisole resonances can show peaks near δ 7.5 and 6.8. While allyl fragment can be observed at δ 5.9–5.1 region: δ 5.86 (m, 1H, CH=), δ 5.25 (d, J = 17 Hz, 1H, =CH_2 trans), and δ 5.12 (d, J = 10 Hz, 1H, =CH_2 cis). The diarylmethyl methine displayed peaks at δ 5.45 (d, J ≈ 6 Hz, 1H), consistent with coupling to the adjacent methylene. In addition, benzylic methylene can be sited near the allyl group to obtain peaks at δ 3.32 (s, 2H). Ether linkages observed at δ 4.25 (t, J ≈ 6.8 Hz, 2H, Ph–O–CH_2–) and δ 4.12 (t, J ≈ 6.8 Hz, 2H, Nap–O–CH_2–). A sharp singlet peak reported at δ 3.78 (s, 3H) belongs to O–CH_3. The dimethylamino arm observed as δ 2.85 (t, J ≈ 6.8 Hz, 2H, –CH_2–NMe_2) and δ 2.75 (t, J ≈ 6.8 Hz, 2H, –CH_2– between O and N), with the terminal NMe_2 methyls at δ 2.30 (s, 6H). No phenolic OH is detected (no broad signal at ~δ 4.5–5.0), consistent with etherification of the parent eugenol unit. In ^{1}HNMR Summary was δ 8.05–6.78 (m, ~12H, Ar–H); 5.86 (m, 1H, CH= of allyl); 5.25 (d, J = 17 Hz, 1H, =CH_2 trans); 5.12 (d, J = 10 Hz, 1H, =CH_2 cis); 5.45 (d, J ≈ 6 Hz, 1H, diarylmethyl CH); 4.25 (t, J ≈ 6.8 Hz, 2H, Ph–O–CH_2–); 4.12 (t, J ≈ 6.8 Hz, 2H, Nap–O–CH_2–); 3.78 (s, 3H, O–CH_3); 3.32 (s, 2H, benzylic CH_2 to allyl); 2.85 (t, J ≈ 6.8 Hz, 2H, –CH_2–NMe_2); 2.75 (t, J ≈ 6.8 Hz, 2H, –CH_2– between O and N); 2.30 (s, 6H, NMe_2. The absence of a broad singlet at ~4.5–5.0 ppm supports successful etherification and deprotonation of eugenol –OH [23,24].

While Figure 5 shows the ^{1}H NMR spectrum of L_2. The Phenolic OH peaks showed Distinct peaks, singlets s at δ 10.20 and 9.80 ppm (1H each), attributed to free two aromatic. Aromatic region at (δ 8.05–6.70 ppm with ~18–20 protons covers the naphthoxy, phenoxy, and anisole aromatic rings. However, the most deshielded protons near δ 8.0 ppm arise from peri-protons of the naphthyl fragment; other aromatic signals appear in the range of δ 7.6–6.7 ppm. Diarylmethyl CH: A showed a small doublet at δ 5.42 (J ≈ 6 Hz, 1H), belonging to the central benzylic methine linking the two aryl fragments. Ether linkages: Two nonequivalent –OCH_2– groups appear as triplets at δ 4.20 and 4.05 (J ≈ 6.8 Hz, 2H each), reflecting as two sharp singlets at δ 3.80 and 3.76 (3H each) are assigned to anisole O–CH_3 groups. The methylene was sites next to nitrogen resonates at δ 2.90 (t, J ≈ 6.8 Hz, 2H), while the terminal NMe_2 methyls appear as a singlet at δ 2.28 (6H). The δ 10.20 (s, 1H, OH); 9.80 (s, 1H, OH); 8.05–6.70 (m, ~18–20H, Ar–H); 5.42 (d, J ≈ 6 Hz, 1H, diarylmethyl CH); 4.20 (t, J ≈ 6.8 Hz, 2H, O–CH_2–); 4.05 (t, J ≈ 6.8 Hz, 2H, O–CH_2–); 3.80 (s, 3H, O–CH_3); 3.76 (s, 3H, O–CH_3); 2.90 (t, J ≈ 6.8 Hz, 2H, –CH_2–N); 2.28 (s, 6H, NMe_2). The clear presence of phenolic OH signals and sharp methoxy/dimethylamino resonances differentiates L_2 from L_1, where etherification eliminates OH peaks and modifies the side-chain environment [24,25].

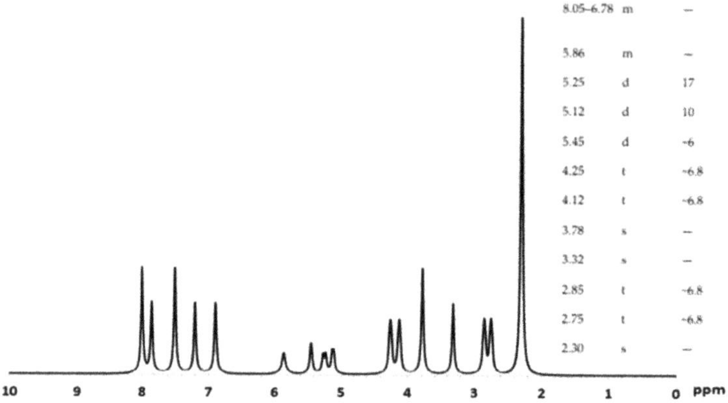

δ (ppm)	Multip.	J (Hz)	Integ.	Notes
8.05–6.78	m	–	~12H	Aromatic protons (naphthoxy and phenoxy); peri protons δ 7.8–8.1, anisole/phenoxy δ 7.5, 6.8
5.86	m	–	1H	Internal vinylic proton (–CH= of allyl)
5.25	d	17	1H	=CH₂ trans of allyl
5.12	d	10	1H	=CH₂ cis of allyl
5.45	d	~6	1H	Benzylic CH (diarylmethyl)
4.25	t	~6.8	2H	Ph–O–CH₂– (ether linker to allyl)
4.12	t	~6.8	2H	Nap–O–CH₂– (ether linker to amino arm)
3.78	s	–	3H	OCH₃ (para-methoxy on phenoxy)
3.32	s	–	2H	CH₂ adjacent to allyl
2.85	t	~6.8	2H	–CH₂–NMe₂
2.75	t	~6.8	2H	–CH₂– between O and N
2.30	s	–	6H	Terminal NMe₂ (dimethylamino)

Figure 4. ^{1}H NMR of Eugenol-propyl-escitalopram ether (L_1) spectroscopy. The inset table shows the interpretation of the highlighted absorbed band.

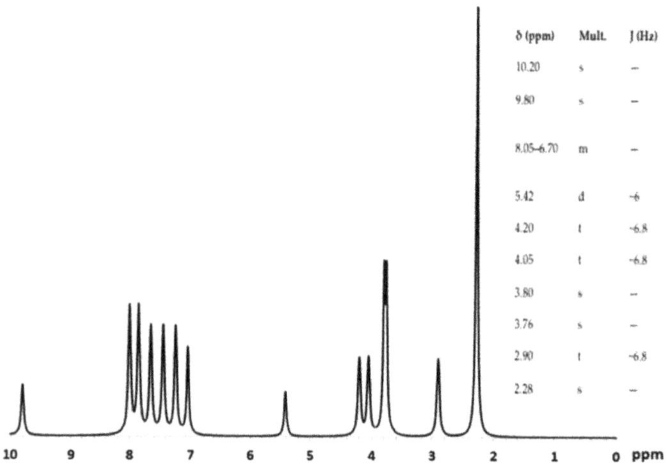

δ (ppm)	Mult.	J (Hz)	Integ.	Notes
10.20	s	–	1H	Phenolic OH (para-methoxyphenyl)
9.80	s	–	1H	Phenolic OH (para-methoxyphenyl)
8.05–6.70	m	–	~18–20H	Aromatic protons (naphthoxy, phenoxy, anisole); peri protons near δ 8.0
5.42	d	~6	1H	Benzylic methine (chiral diarylmethyl centre)
4.20	t	~6.8	2H	O–CH₂– (ether linkage 1)
4.05	t	~6.8	2H	O–CH₂– (ether linkage 2)
3.80	s	–	3H	OCH₃ (anisole methoxy)
3.76	s	–	3H	OCH₃ (anisole methoxy)
2.90	t	~6.8	2H	–CH₂–N (dimethylamino arm)
2.28	s	–	6H	NMe₂ (terminal dimethylamino)

Figure 5. ^{1}H NMR of Escitalopram_curcumin (Esc_Curr) and curcumin Either conjugate (L_2) spectroscopy. The inset table shows the interpretation of the highlighted absorbed bands.

3.3. Molar Conductivity, (d-d) Spectrum and Expected Geometry of the Prepared [M(L_1)Cl]Cl and [M(L_2)Cl] Complexes

Table 2 displayed some highlighted physical properties such as molar conductivity of their complexes measurement that carried out at (1×10^{-3}) M in Dimethyl sulfoxide (DMSO) solvent, colour, melting point (M.P), effective magnetic moment in Bohr magnetons (B.M) and found (calculated CHN and/or M) quantitative analysis values of [M(L_1)Cl]Cl and of [M(L_2)Cl] complexes. As seen, all L_1-coordination complexes exhibit conductivity in the range of (22–32) Ω^{-1} $cm^2{\cdot}mol^{-1}$, revealing 1:1 electrolytic behaviour. This indicates the prepared complexes are partially dissociating in DMSO solvent to result in mono-cationic complex species and free anion. At the same time, L_2 complexes showed non-conductive activity.

Table 2. Highlighted data of Physical and analytical characterisation of L_1 and L_2, as well as their complex properties.

Compound	Colour	Yield	M.P (°C)	Conductivity (Ω^{-1} $cm^2 \cdot mol^{-1}$)	μeff (B.M)	Found (Calc.) %			
						C	H	N	M
ESC	white	-	150	-	-	**73.98**	**6.47**	**8.63**	-
CUR	yellow	-	183	-	-	**68.4**	**5.42**	-	-
EUG	colourless	-	liquid	-	-	**73.17**	**7.31**	-	-
L_1	yellow	71%	218	-	-	**74.4** (73.9)	**6.21** (6.6)	**6.67** (6.35)	-
L_2	Off-white	73%	223	-	-	**74.6** (77.9)	**6.65** (6.9)	**6.16** (5.8)	-
[Co(L_1)Cl]Cl	green	82%	240 °C	25	5.10	**67.84** (66.9)	**5.84** (5.78)	**5.45** (5.5)	**11.48** (11.5)
[Ni(L_1)Cl]Cl	green	81%	280 °C	29	3.31	**67.87** (66.98)	**5.85** (5.80)	**5.46** (5.49)	**11.44** 11.00)
[Mn(L_1)Cl]Cl	brown	88%	268 °C	32	3.98	**(68.36)** (67.90)	**5.89** (5.77)	**5.50** (5.48)	**10.80** (10.35)
[Cu(L_1)Cl]Cl	Brown	75%	257 °C	28	2.80	**67.24** (66.89)	**5.79** (5.93)	**5.41** (5.22)	**12.27** (12.98)
[Co(L_2)Cl]	Blue	68%	184 °C	8	5.10	**65.60** (64.93)	**5.33** (5.29)	**3.73** (2.89)	**7.85** (7.88)
[Ni(L_2)Cl]	green	75%	209 °C	7	3.38	**65.62** (66.09)	**5.33** (5.41)	**3.73** (2.91)	**7.82** (7.66)
[Mn(L_2)Cl]	Brown	84%	186 °C	11	3.98	**65.95** (65.44)	**5.36** (5.78)	**3.75** (2.89)	**7.37** (6.89)
[Cu(L_2)Cl]	green	79%	192 °C	10	2.80	**65.20** (64.87)	**5.30** (5.26)	**3.71** (3.65)	**8.41** (8.36)

Measurement that carried out at (1×10^{-3}) M in Dimethyl sulfoxide (DMSO) solvent, colour, melting point (M.P), effective magnetic moment in Bohr magnetons (B.M) and found (calculated CHN and/or M) quantitative analysis values of [M(L_1)Cl]Cl and of [M(L_2)Cl] complexes.

As seen, all L_1-coordination complexes exhibit conductivity in the range of (22–32) Ω^{-1} $cm^2 \cdot mol^{-1}$, revealing 1:1 electrolytic behaviour. This indicates the prepared complexes are partially dissociating in DMSO solvent to result in mono-cationic complex species and free anion, while L_2 complexes showed non-conductive activity.

The magnetic moment of both Ni(II) complexes [Ni(L_1)Cl] and [Ni(L_2)Cl] reported (3.31–3.38) (B.M) indicated an octahedral geometry (high spin) with electron configuration of $3d^8$. This electron configuration can be split into $t_2g^6eg^2$ with 2 unpaired electrons. Thus, the expected (d-d) spectra of Ni(II) d^8 high spin can show three allowed spins from the (^3A_2g) as the ground state term. The allowed-spin are ($^3A_2g \rightarrow {}^3T_2g$) cm^{-1}, ($^3A_2g \rightarrow {}^3T_1g_{(F)}$) and $^3A_2g \rightarrow {}^3T_1g_{(P)}$. They often cited at region (~9000–10,500), (~14,300–15,400) and (~20,000–22,500) cm^{-1}, respectively. Consistently, in practical experiment d-d spectra of Ni(II) d^8 high-spin showed three main regions at (9225–9615), (14,477–15,025) and (21,045–22,300) cm^{-1}

In a similar pattern, [Co(L_1)Cl]Cl and [Co(L_2)Cl] complexes exhibit a magnetic moment of (4.61–5.10) (B.M). They also indicated an octahedral geometry (high spin) with an electron configuration of $3d^7$. Their electron configuration can be split to $t_2g^5eg^2$ with 3 unpaired electrons. (d-d) spectra of Co(II) d^7 high spin has three allowed-spin $^4T_1g_{(F)} \rightarrow {}^4T_2g_{(F)}$, $^4T_1g_{(F)} \rightarrow {}^4A_2g_{(F)}$ and $^4T_1g_{(F)} \rightarrow {}^4T_1g_{(P)}$ at regions (~8000–9000), (~16,000–18,000) and (~19,000–22,000) cm^{-1}, respectively. Practically, regions of allowed the spin of Co(II) d^7 recorded at (8220–9924), 16,220–17,564) and (19,893–21,636,363) cm^{-1}.

[Mn(L_1)Cl] and [Mn(L_2)Cl] complexes showed a magnetic moment (5.91–5.94) B.M. They expected to have d^5 splitting to $t_2g^3eg^2$ with 5 unpaired electrons. While the (d-d) spectrum of prepared Mn(II) complexes assigned weak (d–d) absorbed bands at (20,800–24,200) and (15,380–17,210) cm^{-1}. These bands attributed to $^6A_1g \rightarrow {}^4T_1g_{(G)}$ and $^6A_1g \rightarrow {}^4T_2g_{(G)}$. These results are consistent with previous spectral data reported for octahedral (high-spin) Mn(II) complexes. The weak d-d absorption corresponded to the forbidden-spin nature of these transition bands, due to vibronic coupling and spin–orbit interaction. That confirms the octahedral high-spin configuration of Mn(II).

Finally, (d-d) transition spectrum of [Cu(L_1)Cl]Cl and [Cu(L_2)Cl] d^9 ($t_2g^6eg^3$) showed weak and broad bands in the range (12,600–16,870) cm^{-1}. In spite of its d-d transition allowed-spin $^2Eg \rightarrow {}^2T_1g$, the band was weak and broad. This is attributed to LaPorte's centrosymmetric field of [Cu(L)Cl]. As well as the electrons-vibronic coupling.

From all previous investigations, the suggested scheme formula structure of complexes of [ML_1)Cl]Cl and [ML_2)Cl] are shown in Figures 6a and 6b, respectively.

(**a**)

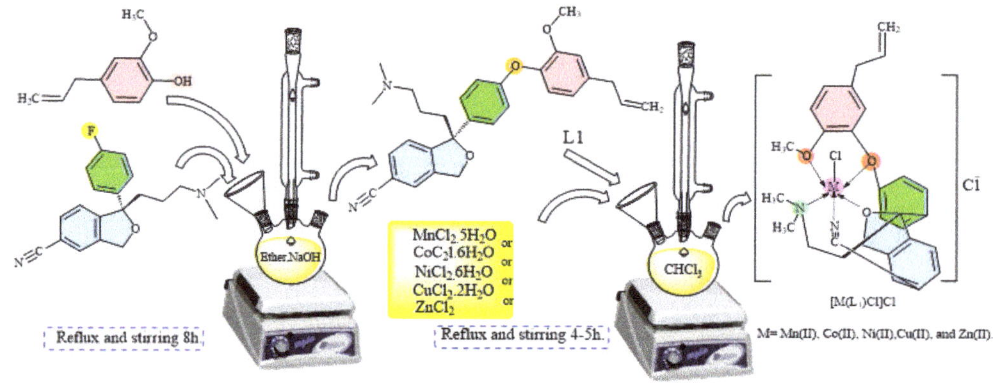

(**b**)

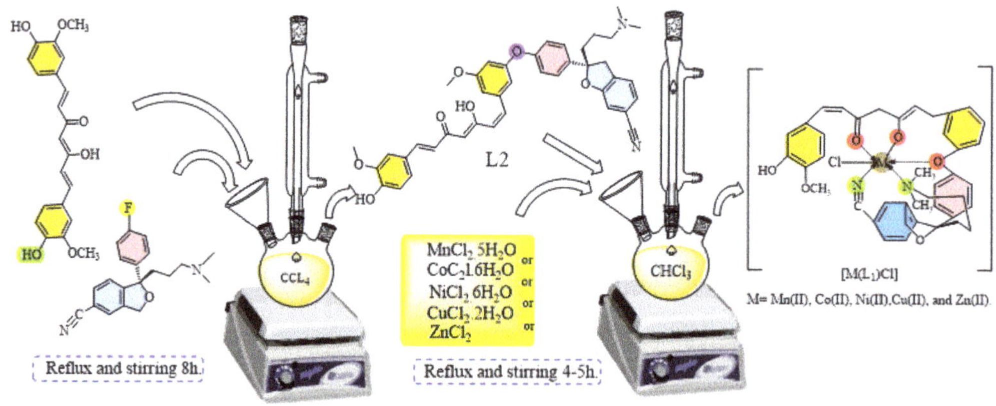

Figure 6. Scheme structure formula of (**a**) [ML_1]Cl and (**b**) [ML_2]Cl.

The successful synthesis of [M(L_1)Cl] and b [M(L_2)Cl] complexes was confirmed through FTIR and ^{1}H NMR spectroscopy. FTIR spectra of metal–ligand bands were confirmed in the range of 420–550 cm^{-1}. ν(N-H) and ν(C=O) stretches were shifted down due to the coordination bond forming of (M--N) and (M--O), respectively. At the same time, ^{1}H NMR showed ether formation as well as aromatic and aliphatic protons in the 6.8–7.1 ppm range. This is a complex method procedure. It is opening pathways for developing metal-based therapeutic candidates with dual bioactivity.

3.4. Antimicrobial Activities of [M(L_1)Cl] and [M(L_2)Cl] Complexes

The antimicrobial activities of prepared complexes [M(L_1)Cl], [M(L_2)Cl] and free ligand were investigated vs. *Escherichia coli* (St = 24), *Staphylococcus aureus*, and *Candida albicans* (St = 21) using agar well diffusion method at a range concentrations of (100, 50, 25, and 12.5) mg/mL. The results can be seen in Table 3.

Table 3. Antimicrobial activities of prepared metal complexes against *Candida albicans*, *Staphylococcus aureus*, and *Escherichia coli* in different concentrations expressed as inhalation zone area (St.mm).

Complexes	Conc. mg/mL	*Escherichia coli* St = 24	*Staphylococcus aureus* St = 25	*Candida albicans* St = 21
[Cu(L_1)Cl]	100	16.5	23	13
	50	12.5	18.5	9
	25	R	14.5	R
	12.5	R	R	R
[Cu(L_2)Cl]	100	17.5	24	14
	50	13.5	19.5	10
	25	R	15.5	R
	12.5	R	R	R
[Ni(L_1)Cl]	100	11.5	15	15
	50	R	12.5	R
	25	R	R	R
	12.5	R	R	R
[Ni(L_2)Cl]	100	12.5	16	16
	50	R	13.5	R
	25	R	R	R
	12.5	R	R	R
[Mn(L_1)Cl]	100	R	R	R
	50	R	R	R
	25	R	R	R
	12.5	R	R	R
[Mn(L_2)Cl]	100	R	R	R
	50	R	R	R
	25	R	R	R
	12.5	R	R	R
[Co(L_1)Cl]	100	17.5	32.5	26.1
	50	13	28.2	21
	25	10.5	21	13
	12.5	R	17	R
[Co(L_2)Cl]	100	19.5	31.5	25
	50	14	27.2	20
	25	11	20	12
	12.5	R	16	R

Table 3. *Cont.*

Complexes	Conc. mg/mL	*Escherichia coli* St = 24	*Staphylococcus aureus* St = 25	*Candida albicans* St = 21
L_1	100	R	R	12.5
	50	R	R	10.5
	25	R	R	R
	12.5	R	R	R
L_2	100	R	R	11.5
	50	R	R	11
	25	R	R	R
	12.5	R	R	R

Copper complexes [Cu(L_1or$_2$)Cl] at 100 mg/mL exhibited a moderate activity against *E. coli*, with inhibition zones of 17.5 mm, higher activity against *S. aureus* in 23 mm against and lower activity against *C. albicans*. However, the activity decreased with a decrease in complex concentration. [Ni(L_1or$_2$)Cl] and [Mg(L_1or$_2$)Cl] exhibit negligible to non-activity behaviour against the tested microorganisms.

Copper complexes [Cu(L_1or$_2$)Cl] at 100 mg/mL exhibited a moderate activity against *E. coli*, with inhibition zones of 17.5 mm, higher activity against *S. aureus* in 23 mm against and lower activity against *C. albicans*. However, the activity decreased with a decrease in complex concentration. [Ni(L_1or$_2$)Cl] and [Mg(L_1or$_2$)Cl] exhibit negligible to non-activity behaviour against the tested microorganisms.

In contrast, Cobalt complexes [Co(L_1)Cl] demonstrated broad-spectrum activity at 100 mg/mL, reaching up to (32.5–31.5) mm vs. *S. aureus*, (26.1–25) mm vs. *C. albicans* and (17.5–18.5) mm vs. *E. coli*, showing superior antimicrobial behaviour among other complexes. In a similar pattern of Cu-complexes, the activity became less at lower Co-complex conc. Finally, the ligands L_1 and L_2 exhibit weak to acceptable inhibition at higher concentrations.

Overall, Co(II) complexes demonstrated the highest antimicrobial activity, potentially due to an increase in lipophilicity and an improvement of diffusion via microbial cell membranes upon Co-coordination, as shown in Figure 7.

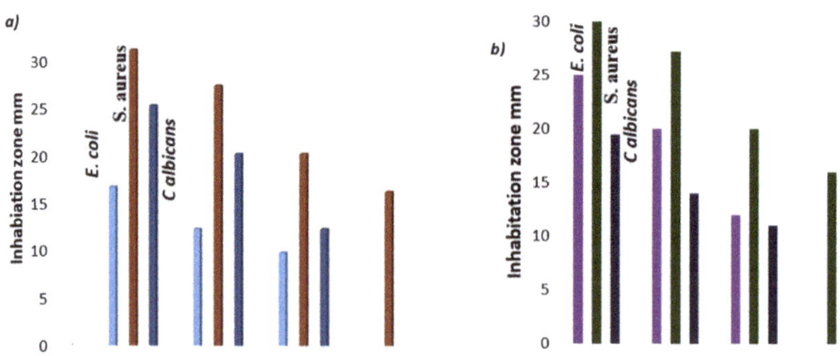

Figure 7. Antimicrobial Activity of (**a**) [Cu(L_1)Cl]Cl and (**b**) [Cu(L_1)Cl] at (100, 50, 25, and 12.5) mg/mL from left to right.

3.5. Density Functional Theory (DFT)

At times, the lack of a crystal structure necessitated computer investigations to gain a better understanding of the molecular structures of the complexes. The GW9 Gaussian 16 program was utilised to conduct geometric optimisations. The structures of the investigated compounds were optimised at the B3LYP/6-311G(d,p) level, as applied in the Gaussian 09W software (version number A.02) [26]. The GaussView 6 package was used to draw the primary structures and visually display the results. The computational optimisation was performed in the gas phase at the ground state. DFT calculations were employed to determine the 3D optimised structures of the compounds, as shown in Figure 9, as well as to analyse the HOMO-LUMO diagram, as shown in Figures 10 and 11. These describe the ultimate electron transfer interactions within the chemical compound, including electronegativity (χ), chemical potential (μ), global hardness (η), global softness (S), and the global electrophilicity index (ω) [27,28], which are listed in Table 4.

Table 4. The calculated energy values of ligands and their Mn(II) and Co(II) complexes using the DFT/B3LYP method.

Compound	EHOMO (eV)	ELUMO (eV)	ΔE (eV)	μ (Debye)	χ (eV)	η (eV)	σ (eV)	ω (eV)	ΔNmax
L_1	−5.429	−1.285	4.144	−3.357	−3.357	2.072	0.241	2.719	2.551
Mn(II)-complex	−5.533	−1.360	4.173	−3.446	− 3.446	2.086	0.240	2.847	1.652
Co(II)-complex	−6.130	−2.350	3.780	−4.780	−4.240	1.880	0.520	4.763	2.246
L_2	−5.641	−1.473	4.168	−3.557	−3.557	2.084	0.240	3.036	1.620
Mn(II)-complex	−5.627	−1.460	4.167	−3.544	−3.544	2.082	0.240	3.013	1.652
Co(II)-complex	−5.831	−2.547	3.284	−4.189	−4.189	1.642	0.305	5.343	1.701

$I = -E_{HOMO}$; $A = -E_{LOMO}$; $(\Delta E) = E_{LUMO} - E_{HOMO}$; $\mu = -(I + A)/2$; $\chi = (I + A/2)$; $\eta = (I - A)/2$; $Pi = -\chi$; $\sigma = 1/\eta$; $S = 1/2\eta$; $\omega = Pi2/2\eta$; $\Delta N\ max = Pi/\eta$.

DFT Calculations

The synthesised complexes were optimised using the density functional theory (DFT) method with the 6–31G (d, p) basis set at the B3LYP level in a DMSO medium. The HOMO and LUMO, along with their energy gap, of some synthesised complexes are illustrated in Figure 8 and listed in Table 3. The B3LYP/6-311G(d,p) module type was used to theoretically create the geometric structures of compounds as seen in Figure 9. Chemical descriptors such as hardness and softness, derived from these energies, provide insight into reactivity. Figures 9 and 10 display the molecular orbital structure of ligands and their Mn(II) and Co(II) complexes, illustrating the significance of the HOMO and LUMO in electron donation and acceptance. The energy gap between these orbitals influences the molecule's stability and reactivity, with smaller gaps indicating higher polarizability and larger gaps indicating greater stability but lower reactivity. Parameters like electrophilicity (ω) and hardness (η) further evaluate stability and reactivity [29]. The high electrophilicity value of Co(II) complexes suggests strong biological activity. Molecular stability and reactivity can be inferred from the calculations of both absolute softness (σ) and absolute hardness (η) parameters. Additional parameters such as global electrophilicity (ω), global softness (S), electrophilicity index (χ), and chemical potential (μ) have been calculated using HOMO and LUMO energies [30]. The high electrophilicity value of Co(II) complexes suggests strong biological activity. Molecular stability and reactivity can be inferred from the calculations of both absolute softness (σ) and absolute hardness (η) parameters. Additional parameters such as global electrophilicity (ω), global softness (S), electrophilicity index (χ), and chemical potential (μ) have been calculated using HOMO and LUMO energies [31], as previously shown in Table 3. At the same time, the electronic charge (ΔN max) was calculated for molecules (Table 4). The high ω value of two Co(II) complexes indicates a strong potential for biological activity, which is further confirmed by the experimental data.

The minimum electrophilicity theory can be used to estimate the reactivity of the ligand and metal complexes. According to the theory of minimum electrophilicity (ω), compounds with the lowest electrophilicity exhibit the highest stability [31,32], as previously shown in Table 4.

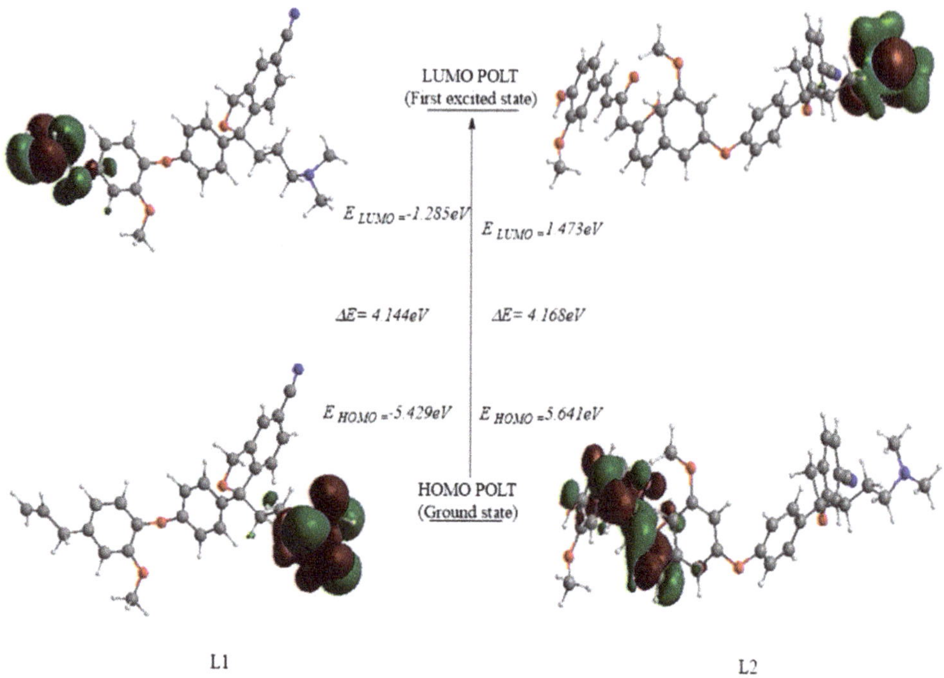

Figure 8. HOMO and LUMO orbitals of the synthesised ligands.

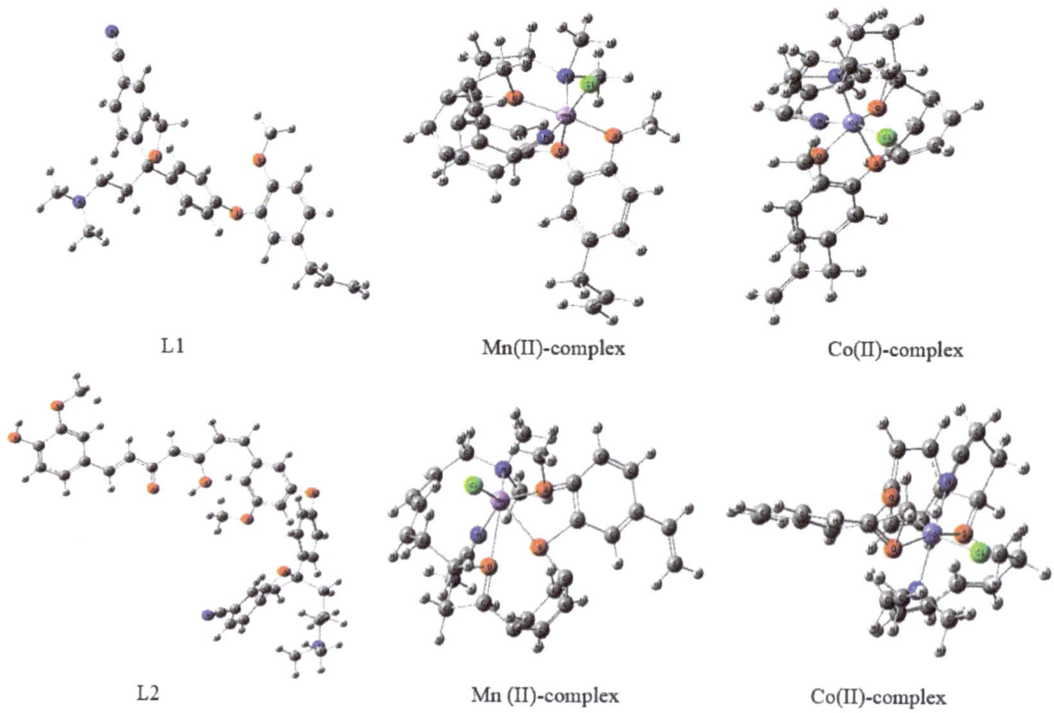

Figure 9. Optimized geometry of some synthesized L_1 and L_2-Coordination compounds.

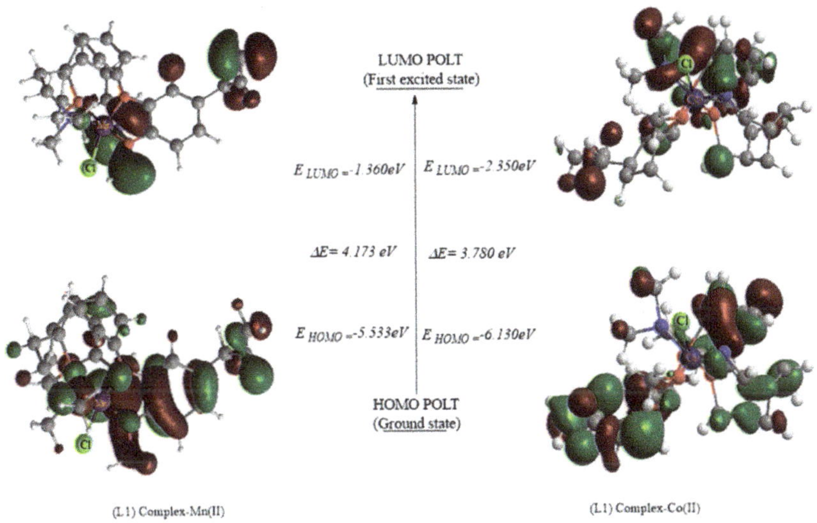

Figure 10. HOMO and LUMO orbitals of some synthesised complexes Mn(II) and Co(II) through L_1.

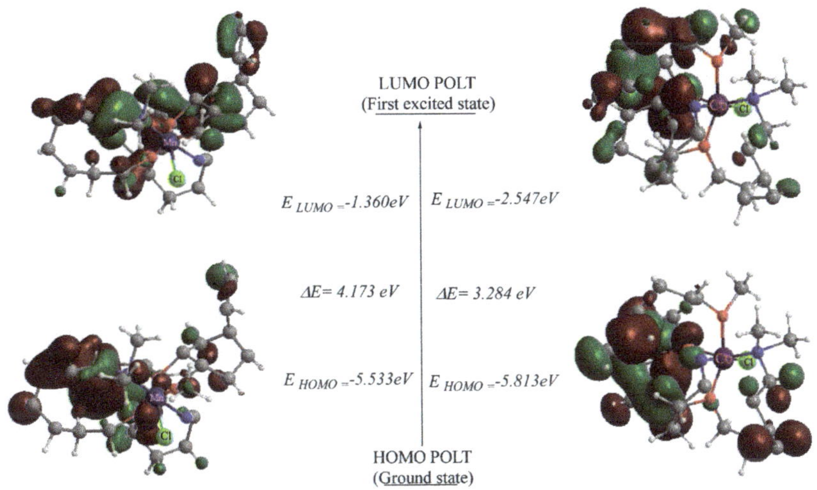

Figure 11. HOMO and LUMO orbitals of some synthesised complexes Mn(II) and Co(II) through L_2.

4. Conclusions

Biological mixed ligands based on escitalopram–eugenol and escitalopram–curcumin were prepared, coordinated with transition metal ions and characterised successfully. The coordination Mn(II), Cu(II), Zn(II), Fe(II), and Co(II) complexes showed predominantly octahedral geometries. The stable coordination occurs through oxygen and nitrogen as donor sites. The magnetic measurement exhibits high-spin states for Mn(II), Co(II), and Ni(II) complexes. This is consistent with their electronic configurations, while Cu(II) complexes showed distorted octahedral geometry with broad d–d bands. Biological investigation demonstrated that cobalt(II) complexes have promising antimicrobial activity, which enhances their potential as therapeutic agents. DFT reported results provide additional structural information with minimised energy confirmation. It is also revealed to be a strong foundation for future investigations into the pharmacological applications. Further investigations are recommended, including in vivo practical experiments and comprehensive biological scanning, to confirm their therapeutic potential or their clinical applicability.

Author Contributions: Conceptualisation, E.H.M.; methodology, and software, E.R.M.; validation and formal analysis, E.M.Y.; investigation, resources, data curation, and writing—original draft preparation, M.A.; writing—reviewing and editing, E.H.M.; supervision, E.M.Y.; project administration, E.R.M. All authors have read and agreed to the published version of the manuscript.

Funding: This research received no external funding.

Institutional Review Board Statement: Not applicable.

Informed Consent Statement: Not applicable.

Data Availability Statement: The original contributions presented in this study are included in the article. Further inquiries can be directed to the corresponding author.

Acknowledgments: We would like to thank the University of Mosul for its support.

Conflicts of Interest: The authors declare no conflicts of interest.

References

1. Banerjee, S.; Chakravarty, A.R. Metal Complexes of Curcumin for Cellular Imaging, Targeting, and Photoinduced Anticancer Activity. *Acc. Chem. Res.* **2015**, *48*, 2075–2083. [CrossRef]
2. Sarkar, T.; Bera, A.; Upadhyay, A.; Jain, N.; Are, V.; Eedara, A.; Prakashchandra, R.D.; Panneerselvam, S.; Nanubolu, J.B.; Andugulapati, S.B.; et al. Photostable Mn(II) Complex of Curcumin for Effective Photodynamic Therapy and Precise Three-Dimensional In Vivo Tumor Imaging. *ACS Appl. Mater. Interfaces* **2025**, *17*, 13660–13675. [CrossRef]
3. Coelho, J.R.A.; Pacheco, A.R.F.; Domingues, D.C.; Rodrigues, A.R.O.; Temitope, A.A.; Coutinho, P.J.G.; Fernandes, M.J.G.; Castanheira, E.M.S.; Gonçalves, M.S.T. New Cation Sensors Based on Eugenol-Derived Azo Dyes. *Molecules* **2025**, *30*, 2788. [CrossRef]
4. Al-Noor, T.H.; Mohapatra, R.K.; Azam, M.; Karem, L.K.A.; Mohapatra, P.K.; Ibrahim, A.A.; Parhi, P.K.; Dash, G.C.; El-Ajaily, M.M.; Al-Resayes, S.I.; et al. Mixed-ligand complexes of ampicillin derived Schiff base ligand and Nicotinamide: Synthesis, physico-chemical studies, DFT calculation, antibacterial study and molecular docking analysis. *J. Mol. Struct.* **2021**, *1229*, 129832. [CrossRef]
5. Albotani, A.S.; Mohammed, E.R.; Al-Dulaimi, M.S.; Saied, S.M.; Saleh, M.Y. Synthesis, Characterization, In Silico Screening for Molecular Docking and Computational Exploration of Some Novel Metal Complexes Derived from Ampicillin-Isatine Schiff Bases. *New Mater. Compd. Appl.* **2025**, *9*, 109–122. [CrossRef]
6. Kowalewska, A.; Majewska-Smolarek, K. Eugenol-Based Polymeric Materials—Antibacterial Activity and Applications. *Antibiotics* **2023**, *12*, 1570. [CrossRef]
7. Fujisawa, S.; Murakami, Y. Eugenol and its role in chronic diseases. In *Drug Discovery from Mother Nature*; Gupta, S.C., Prasad, S., Aggarwal, B.B., Eds.; Springer International Publishing: Cham, Switzerland, 2016; pp. 45–66.
8. Kanwal, A.; Afzal, U.; Zubair, M.; Imran, M.; Rasool, N. Synthesis of anti-depressant molecules via metal-catalyzed reactions: A review. *RSC Adv.* **2024**, *14*, 6948–6971. [CrossRef] [PubMed]
9. Refat, M.S. Synthesis and characterization of ligational behavior of curcumin drug towards some transition metal ions: Chelation effect on their thermal stability and biological activity. *Spectrochim. Acta Part A Mol. Biomol. Spectrosc.* **2013**, *105*, 326–337. [CrossRef]
10. Prasad, S.; DuBourdieu, D.; Srivastava, A.; Kumar, P.; Lall, R. Metal-Curcumin Complexes in Therapeutics: An Approach to Enhance Pharmacological Effects of Curcumin. *Int. J. Mol. Sci.* **2021**, *22*, 7094. [CrossRef] [PubMed]
11. Dhar, S.; Bhattacharjee, P. Promising role of curcumin against viral diseases emphasizing COVID-19 management: A review on the mechanistic insights with reference to host-pathogen interaction and immunomodulation. *J. Funct. Foods* **2021**, *82*, 104503. [CrossRef]
12. Parikh, A.; Parikh, H.; Parikh, K. (Eds.) Williamson Ether Synthesis. In *Name Reactions in Organic Synthesis*; Foundation Books: New Delhi, India, 2006; pp. 430–434.
13. Lu, K.-Q.; Li, Y.-H.; Tang, Z.-R.; Xu, Y.-J. Roles of Graphene Oxide in Heterogeneous Photocatalysis. *ACS Mater. Au* **2021**, *1*, 37–54. [CrossRef] [PubMed]

14. Baranová, Z.; Amini, H.; Neupane, M.; Garrett, S.C.; Ehnbom, A.; Bhuvanesh, N.; Reibenspies, J.H.; Gladysz, J.A. Syntheses, Structural Studies, and Copper Iodide Complexes of Macrocycles Derived from Williamson Ether Syntheses Involving 2, 9-Bis (4-hydroxyphenyl)-1, 10-phenanthroline, α, ω-Dibromides, and Resorcinol or 2, 7-Dihydroxynaphthalene. *Aust. J. Chem.* **2016**, *70*, 373–386. [CrossRef]
15. Bresciani, G.; Biancalana, L.; Pampaloni, G.; Zacchini, S.; Ciancaleoni, G.; Marchetti, F. A comprehensive analysis of the metal–nitrile bonding in an organo-diiron system. *Molecules* **2021**, *26*, 7088. [CrossRef] [PubMed]
16. Wong, K.C. Review of *Spectrometric Identification of Organic Compounds*, 8th edition. *J. Chem. Educ.* **2015**, *92*, 1602–1603. [CrossRef]
17. Călinescu, M.; Fiastru, M.; Bala, D.; Mihailciuc, C.; Negreanu-Pîrjol, T.; Jurcă, B. Synthesis, characterization, electrochemical behavior and antioxidant activity of new copper(II) coordination compounds with curcumin derivatives. *J. Saudi Chem. Soc.* **2019**, *23*, 817–827. [CrossRef]
18. Malik, S.; Mondal, U.; Jana, N.C.; Banerjee, P.; Saha, A. Using eugenol scaffold to explore the explosive sensing properties of Cd(ii)-based coordination polymers: Experimental studies and real sample analysis. *Dalton Trans.* **2024**, *53*, 12995–13011. [CrossRef]
19. Kumar, A.; Singh, B. Synthesis and Characterization of Transition Metal Complexes along with their Antibacterial Activities. *Int. J. Curr. Eng. Technol.* **2014**, *4*, 1950–1954.
20. Mohammed, E.H.; AlSultan, M.; Alasalli, S.M.; Sabah, A.A. Syntheses, Characterization and Thermal Stability Study of New Co (II), Cu (II) AND Zn (II) Complexes Deviates from Schiff Base 4-(Dimethylamino) Benzaldehyde. *Kim. Probl.* **2025**, *23*, 395–406. [CrossRef]
21. Abdulkarem, A.A. Synthesis and antibacterial studies of metal complexes of Cu (II), Ni (II) and Co (II) with tetradentate ligand. *J. Biophys. Chem.* **2017**, *8*, 13. [CrossRef]
22. Kumar, S.; Dhar, D.N.; Saxena, P.N. Applications of metal complexes of Schiff bases—A review. *J. Sci. Ind. Res.* **2009**, *68*, 181–187.
23. Wang, X.-q.; He, M.-h.; Wang, Q. Synthesis of Escitalopram Oxalate. *Chin. Pharm. J.* **2009**, *44*, 1429–1431.
24. Tahay, P.; Parsa, Z.; Zamani, P.; Safari, N. A structural and optical study of curcumin and curcumin analogs. *J. Iran. Chem. Soc.* **2022**, *19*, 3177–3188. [CrossRef]
25. Poorghorban, M.; Karoyo, A.H.; Grochulski, P.; Verrall, R.E.; Wilson, L.D.; Badea, I. A ^{1}H NMR Study of Host/Guest Supramolecular Complexes of a Curcumin Analogue with β-Cyclodextrin and a β-Cyclodextrin-Conjugated Gemini Surfactant. *Mol. Pharm.* **2015**, *12*, 2993–3006. [CrossRef]
26. Ibrahim, F.; Hammza, R.; Fadhil, D. Synthesis and characterization of Trimethoprim metal complexes used as corrosion inhibitors for carbon steel in acid media. *Int. J. Corros. Scale Inhib.* **2019**, *8*, 733–742. [CrossRef]
27. Mustafa, N.A.; Ahmed, S.A. Design, synthesis, characterization, DFT, molecular docking, and in vitro screening of metal chelates incorporating Schiff base. *Bull. Chem. Soc. Ethiop.* **2024**, *38*, 1625–1638. [CrossRef]
28. Mohammed, B.A.-K.; Ahmed, S.A.; Al-Healy, F. Design, characterizations, DFT, molecular docking and antibacterial studies of some complexes derived from 4-aminpantipyrine with glycine amino acid and ligand. *Bull. Chem. Soc. Ethiop.* **2024**, *38*, 1609–1624. [CrossRef]
29. Al Quaba, R.A. Synthesis, Characterization, DFT and Antibacterial, Azo Ligand Derived From 2-Amino pyrimidine with Antipyrine Mixed ligand Complexes involving 1,10-phenanthroline. *J. Phys. Conf. Ser.* **2021**, *1999*, 012008.
30. Tegegn, D.F.; Belachew, H.Z.; Salau, A.O. DFT/TDDFT calculations of geometry optimization, electronic structure and spectral properties of clevudine and telbivudine for treatment of chronic hepatitis B. *Sci. Rep.* **2024**, *14*, 8146. [CrossRef]
31. Prabakaran, A.; Maheswari, C.U.; Issaoui, N.; Al-Dossary, O.M.; Rajamani, T.; Gnanasambandan, T.; Saravanan, P.; Vimalan, M.; Manikandan, A. Computational insight into the spectroscopic and molecular docking analysis of estrogen receptor with ligand 2, 3-dimethyl-N [2-(hydroxy) benzylidene] aniline. *J. King Saud Univ. Sci.* **2024**, *36*, 103196. [CrossRef]
32. Zhang, L.; Xie, R.; Zhao, W.; Ren, T.; Wang, Q.; Wang, W. Coumarin-Based Arylpyrazolines: Highly Selective Cu2+ Fluorescent Probes and their AIEE and MFC Properties. *J. Fluoresc.* **2025**, 1–12. [CrossRef] [PubMed]

inorganics

MDPI

Article

Structurally Characterized Cobalt and Nickel Complexes of Flavonoid Chrysin as Potential Radical Scavenging Compounds

Eleftherios Halevas [1,2,*], Barbara Mavroidi [1], Despoina Varna [3], Georgia Zahariou [4], George Litsardakis [5], Maria Pelecanou [2] and Antonios G. Hatzidimitriou [3,*]

1 Institute of Biosciences & Applications, National Centre for Scientific Research "Demokritos", 15310 Athens, Greece; bmavroidi@bio.demokritos.gr
2 Laboratory of Biochemistry, Department of Chemistry, Aristotle University of Thessaloniki, 54124 Thessaloniki, Greece; pelmar@bio.demokritos.gr
3 Laboratory of Inorganic Chemistry, Department of Chemistry, Aristotle University of Thessaloniki, 54124 Thessaloniki, Greece; varndesp@chem.auth.gr
4 Institute of Nanoscience and Nanotechnology, National Centre for Scientific Research "Demokritos", 15310 Athens, Greece; zahariougeorgia@gmail.com
5 Laboratory of Materials for Electrotechnics, Department of Electrical and Computer Engineering, Aristotle University of Thessaloniki, 54124 Thessaloniki, Greece; lits@auth.gr
* Correspondence: lefterishalevas@gmail.com (E.H.); hatzidim@chem.auth.gr (A.G.H.); Tel.: +30-210-6503558 (E.H.); +30-2310-997748 (A.G.H.)

Abstract

Polyphenolic compounds, such as flavonoids, possess important structural and physico-chemical characteristics that in combination with their biological properties render them an important class of natural compounds with medicinal prospects. Chrysin is a well-known flavone with antioxidant activity and a multitude of other beneficial properties. The potential of flavonoids to coordinate with metal ions leads to derivatives with enhanced biological profile. Within this framework, four novel heteroleptic complexes of cobalt and nickel with chrysin and the aromatic bidentate chelating agents 2,2′-bipyridine and 1,10-phenanthroline were synthesized, as well as physico-chemically and structurally characterized. The *in vitro* antioxidant efficiency of the synthesized complexes was examined via the 2,2-diphenyl-1-picrylhydrazyl (DPPH) assay. All complexes showed notable radical scavenging capacity comparable to that of ascorbic acid, providing the incentive for further investigation of their therapeutic potential.

Keywords: cobalt and nickel complexes; flavonoid chrysin; antioxidant potential; radical scavenging activity

1. Introduction

Polyphenolic compounds, such as flavonoids, are the main components of several dietary supplements, fruits, plants, flowers, and medicinal herbs [1,2]. The presence of free hydroxyl moieties in their structure renders them beneficial in the management of oxidative stress, which accompanies many diseases. Further to the antioxidant properties [3], polyphenols possess significant biological activities including anticancer [4], antiallergic [5], anti-inflammatory [6], neuroprotective [7], and hepatoprotective ones [8], and, as a result, their impact in nutrition is constantly being upgraded. An important polyphenol with remarkable biological properties is the flavonoid chrysin (5,7-dihydroxyflavone), which has the flavone general structure with two hydroxyl groups at the 5 and 7 carbons of the A ring (Figure 1) [9]. Chrysin is copiously found in many plant extracts, as well as in propolis and

 https://doi.org/10.3390/inorganics13070230

honey extracts [10,11]. It is also present in several medicinal plants and fruits, including *Passiflora caerulea* (blue passion flower), *Momordica charantia* (bitter melon) or *Pyrus pashia* (the wild Himalayan pear) [12,13].

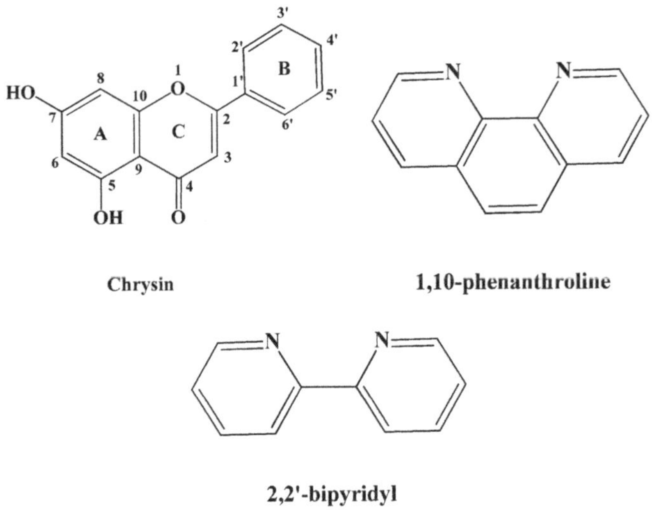

Figure 1. Schematic representation of the employed ligands chrysin (HL), 1,10-phenanthroline (phen) and 2,2′-bipyridyl (bpy).

Chrysin exhibits a well-investigated pharmacological profile. Further to its antioxidant and anti-inflammatory properties, chrysin has a well-established anti-cancer activity against a broad spectrum of cancers, including prostate, breast, lung, liver, colon and pancreatic cancer, and also human uveal melanoma, glioblastoma, and epidermoid carcinoma [14,15]. Furthermore, it has been welldocumented that the utilization of chrysin in applications of combination therapy enhances the efficiency of several chemotherapeutic drugs such as docetaxel [16], cisplatin, and camptothecin [17]. This can be explained by the intense interactions of chrysin metabolites with human serum albumins (HAS), affecting the albumin-binding properties and thus the biological activity, pharmacokinetics, and half-life of these drugs [18]. Moreover, chrysin has strong hepatoprotective effects protecting the liver cells from toxic substances [19], enhanced skin protective activity attenuating psoriasis-like skin lesions [20], significant neuroprotective properties improving memorization processes [21], and important antiviral activity against influenza H1N1 [22], coxsackie virus B3 (CVB3) [23], human rhinoviruses (HRV) [24], and others.

The bioactivity of chrysin is strongly related to its efficiency to interact with plasma proteins [18] and cell membranes [25], scavenge radicals [26], and chelate transition metal ions that produce radicals [27]. However, its fast metabolism, limited aqueous solubility, low systemic availability, and absorption rates have led researchers to develop metallo-chrysin complexes as substitute biologically active compounds with upgraded physico-chemical and possibly, biological properties [13–15].

In our on-going research endeavors to generate innovative, bioactive metal complexes based on naturally derived chelators, the properties of chrysin were integrated with those of cobalt (Co) and nickel (Ni) metal ions. Ni is considered to be an important trace element for biological systems. Ni ions serve as enzyme cofactors of important enzymes including nickel superoxide dismutase, urease, carbon monoxide dehydrogenase, methyl-coenzyme M reductase, acetyl-coenzyme A synthase, acireductone dioxygenase, [NiFe]-hydrogenase, glyoxylase I and lactate racemase [28–30]. Furthermore, Ni(II) complexes display beneficial activities, such as antimicrobial, antioxidant, anticancer, anticonvulsant, anti-inflammatory, and antileishmanial [31,32], while several Ni(II) complexes serve as DNA intercalators [33], cleaving agents [34], and potential binders to serum albumins [35]. On the other hand, Co is also a crucial trace element which shows interest, mainly due to its existence in the vitamin B12 active center. It can be found in both cobalamin-dependent enzymes and non-corrin cobalt-containing enzymes [36]. Several Co(II) and Co(III)-based metallo-complexes display remarkable antitumor, anti-angiogenic, antimicrobial, antiviral and antioxidant properties [37–40], and act as hydrolytic agents for DNA cleavage [41] and binding [42].

In light of the preceding facts, in this work we demonstrate the synthetic procedure, physico-chemical characterization, structural elucidation, and the DPPH radical scavenging activity of four novel Co(II), Co(III), and Ni(II) heteroleptic complexes with chrysin (HL) and the auxiliary *N,N*-donor ligands 1,10-phenanthroline (phen) and 2,2′-bipyridine (bpy) (Figure 1). Until today, there are no structurally elucidated, through single-crystal X-ray diffractometry, chrysin-based complexes of Co and Ni reported in the literature.

2. Results

2.1. Synthesis

A comprehensive examination of the reaction parameters including pH, temperature, metal-to-ligand-to-chelator molar ratios, solvent system, and crystallization procedures, led to the generation of heteroleptic Co(III)-chrysin-bpy, Co(II)-chrysin-phen, Ni(II)-chrysin-bpy, and Ni(II)-chrysin-phen complexes in alcoholic media and in sufficient yields. A detailed presentation of the examined reaction and (re)crystallization parameters and observations is presented in Supplementary Information Table S1. The stoichiometric equations that led to complexes **1–4** are presented in Scheme 1 below:

$$4\,[Co(NO_3)_2{\cdot}6H_2O] + 4\,HL + 8\,bpy + 6\,EtOH + O_2 \xrightarrow[Co^{+2} \rightarrow Co^{+3}]{N(C_2H_5)_3} 4\,\{[Co(L)(bpy)_2]{\cdot}(NO_3)_2{\cdot}1.5EtOH\} + 26\,H_2O$$

Complex 1

$$2\,Co(NO_3)_2{\cdot}6H_2O + 2\,HL + 2\,phen + 2\,N(C_2H_5)_3 + 6\,MeOH \longrightarrow [Co_2(L)_2(phen)_2(MeOH)_2]{\cdot}(NO_3)_2{\cdot}4MeOH + 2\,\{[(C_2H_5)_3NH](NO_3)\} + 12\,H_2O$$

Complex 2

$$2\,Ni(NO_3)_2{\cdot}6H_2O + 2\,HL + 2\,bpy + 2\,N(C_2H_5)_3 + 4\,EtOH \longrightarrow [Ni_2(L)_2(bpy)_2(EtOH)_2]{\cdot}(NO_3)_2{\cdot}2EtOH + 2\,\{[(C_2H_5)_3NH](NO_3)\} + 12\,H_2O$$

Complex 3

$$2\,Ni(NO_3)_2{\cdot}6H_2O + 2\,HL + 2\,phen + 2\,N(C_2H_5)_3 + 5\,EtOH \longrightarrow [Ni_2(L)_2(phen)_2(EtOH)_2]{\cdot}(NO_3)_2{\cdot}3EtOH + 2\,\{[(C_2H_5)_3NH](NO_3)\} + 12\,H_2O$$

Complex 4

Scheme 1. Stoichiometric equations of complexes **1–4**.

Ethanol was selected as the most adequate reaction solvent for the generation of stable crystalline materials in the cases of complexes **1**, **3**, and **4**. However, in the case of complex **2**, methanol was proved the most appropriate reaction solvent for the formation of a crystalline product of sufficient yield and with the ability to maintain its crystallinity right after its isolation from the mother liquor and for fairly long periods of time. Diethylether was employed as the precipitation inducing solvent to all the reaction solutions. Dark brown and dark red crystals emerged in the reactions of Co(III) and Co(II), respectively, whereas green crystals emerged in the case of Ni(II) reactions. The verification of the generated crystalline products was performed via elemental analysis, X-ray crystallography, and FT-IR spectroscopy. All materials retain their crystallinity when exposed to air and are insoluble in dichloromethane, acetone and acetonitrile, soluble in dimethylacetamide (DMA), dimethylsulfoxide (DMSO), dimethylformamide (DMF), MeOH and EtOH, and moderately soluble in H_2O under ambient conditions.

2.2. X-Ray Crystallographic Analysis

Figures 2A, 3A, 4A and 5A illustrate the solid-state lattices of **1**–**4**. The crystal collection data are given in Table 1, while the interatomic bond angles and lengths are provided in Table 2 [43–47]. In all cases, the monoionic chrysin is coordinated to the metal centers (Co(III), Co(II), and Ni(II)) in a bidentate mode via the deprotonated phenolic and carbonyl oxygens, leading to a six-membered ring rigid formation. In the case of the divalent metal ion complexes, the deprotonated phenolic oxygen is also acting as bridge between two metal centers forming a $M^{II}{}_2O_2$ four-membered ring.

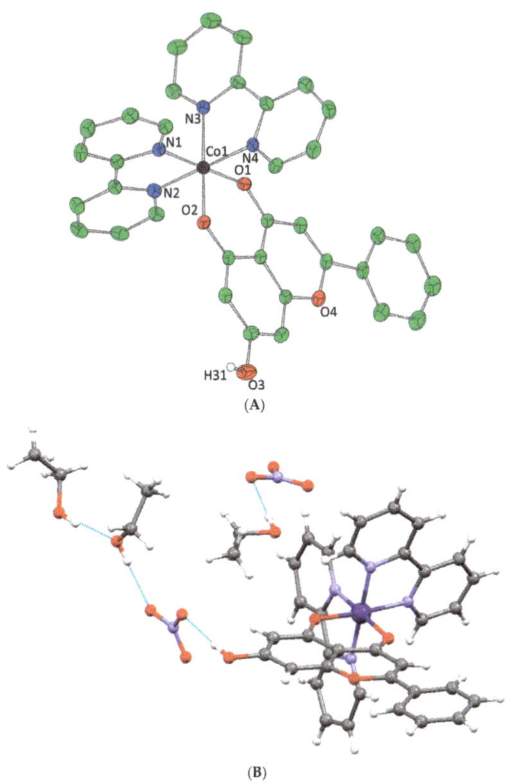

Figure 2. (**A**) Cameron scheme of $[Co(L)(bpy)_2]^{2+}$ complex assembly in compound **1**. Aromatic hydrogen atoms omitted for clarity. (**B**) Hydrogen bonding in complex **1** (light blue dotted lines).

Complex **1** crystallizes in the monoclinic space group $P2_1/n$ and its unit cell contains four molecules. The crystal asymmetric unit consists of two nitrate (NO_3^-) counter ions, one dicationic $[Co(L)(bpy)_2]^{2+}$ complex assembly, and one and a half lattice EtOH molecules. The molecular structure of **1** is composed of a Co(III) ion that is coordinated with one deprotonated chrysin and two bpy chelators, giving rise to a $Co^{III}N_4O_2$ chromophore with a distorted octahedral arrangement (Figure 2A). The lattice of **1** includes local intermolecular hydrogen bonds that are developed between the hydroxyl groups of the lattice EtOH molecules, the oxygens of the NO_3^- groups, and the non-coordinated phenolic group that derives from the deprotonated chrysin giving stability to the formed structure (Figure 2B, Supplementary Information Table S2).

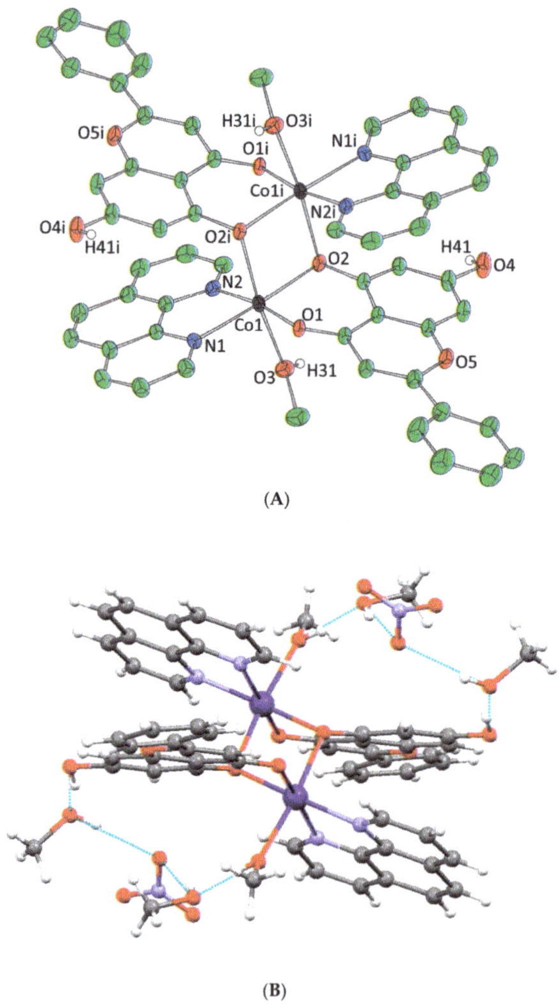

Figure 3. (**A**) Cameron scheme of $[Co_2(L)_2(phen)_2(MeOH)_2]^{2+}$ complex assembly in compound **2**. Aromatic and methyl hydrogen atoms are omitted for clarity. (**B**) Hydrogen bonding in complex **2** (azure dotted lines).

Complex **2** crystallizes in the triclinic space group $P\bar{1}$ and its unit cell consists of four lattice MeOH molecules, one $[Co_2(L)_2(phen)_2(MeOH)_2]^{2+}$ cation, and two NO_3^- counter ions. Each Co(II) cation is bound to the oxygen of a MeOH molecule, two oxygens from one monoanionic *O,O*-donor chrysin, two nitrogen atoms of one phen molecule as well as the bridging deprotonated phenolic oxygen of a second chrysin from the symmetrically developed neighboring $[Co(L)(phen)(MeOH)]^{+}$ formation. Thus, the final dimer dinuclear complex dication is formed. Each metal center presents a $Co^{II}N_2O_4$ chromophore and a coordination arrangement with a distorted octahedral geometry (Figure 3A). The lattice of **2** includes four intermolecular hydrogen bonding interactions. The unbound phenolic oxygens of chrysin ligands bind to the oxygen of a lattice MeOH molecule. The oxygen atom of the ligated MeOH molecule binds the oxygen of the second lattice MeOH molecule. Finally, both oxygen atoms of the two lattice MeOH molecules bind the same oxygen of the NO_3^- anion thus forming a local network of hydrogen bonding interactions giving further stability to the lattice formed (Figure 3B, Supplementary Information Table S2).

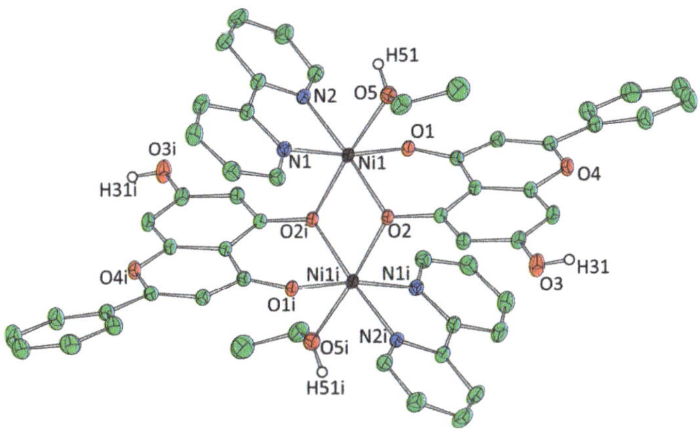

(A)

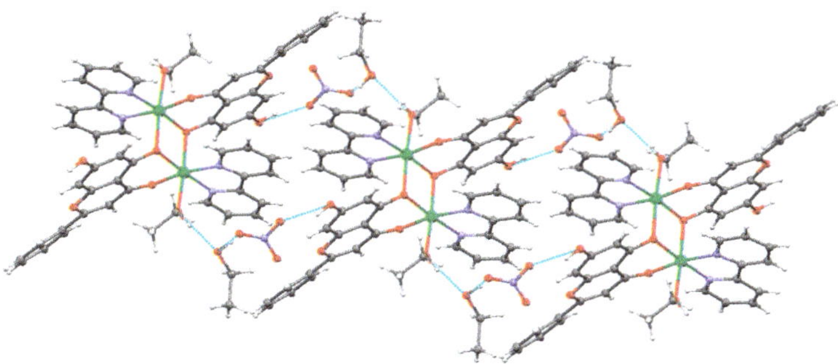

(B)

Figure 4. **(A)** Cameron scheme of $[Ni_2(L)_2(bpy)_2(EtOH)_2]^{2+}$ complex assembly in compound **3**. Aromatic and ethyl hydrogen atoms are omitted for clarity. **(B)** Hydrogen bonding in complex **3** (azure dotted lines).

Complexes **3** and **4** crystallize in the triclinic space group *P*-1. The unit cell of **3** contains two lattice EtOH molecules, two NO_3^- counter ions, and one $[Ni_2(L)_2(bpy)_2(EtOH)_2]^{2+}$ cation. In the case of **4**, the unit cell contains three EtOH molecules, one $[Ni_2(L)_2(phen)_2(EtOH)_2]^{2+}$ cation, and two NO_3^- counter ions. Both binuclear complexes result from the dimerism of the mononuclear complex $[Ni(L)(N,N\text{-donor})(EtOH)]^{+}$, present in the asymmetric unit, as to the center of symmetry located on the center of Ni-Ni mid distance. The deprotonated phenolic oxygen atoms are acting not only as chelators for each metal ion but also as bridges between them. Moreover, in both complexes, each Ni(II) ion is coordinated to one EtOH molecule, one monoionic chrysin, one bpy or phen molecule and the deprotonated phenolic oxygen of a second flavonoid molecule from the symmetrically developed assembly, giving rise to a $Ni^{II}N_2O_4$ chromophore around each Ni(II) core and a coordination arrangement with a distorted octahedral geometry (Figures 4A and 5A). Hydrogen bonding interactions in **3** and **4** emerge between the ligated ethanolic oxygen atom to one oxygen atom of a lattice NO_3^- anion while a second oxygen from the same NO_3^- anion interacts with the oxygen of a lattice EtOH. The latter also binds the free phenolic oxygen of a neighboring complex assembly thus forming a polymeric chain of hydrogen bonding interactions giving extra stability to the system and forming an extended final 1D crystal lattice (Figures 4B and 5B, Supplementary Information Table S2).

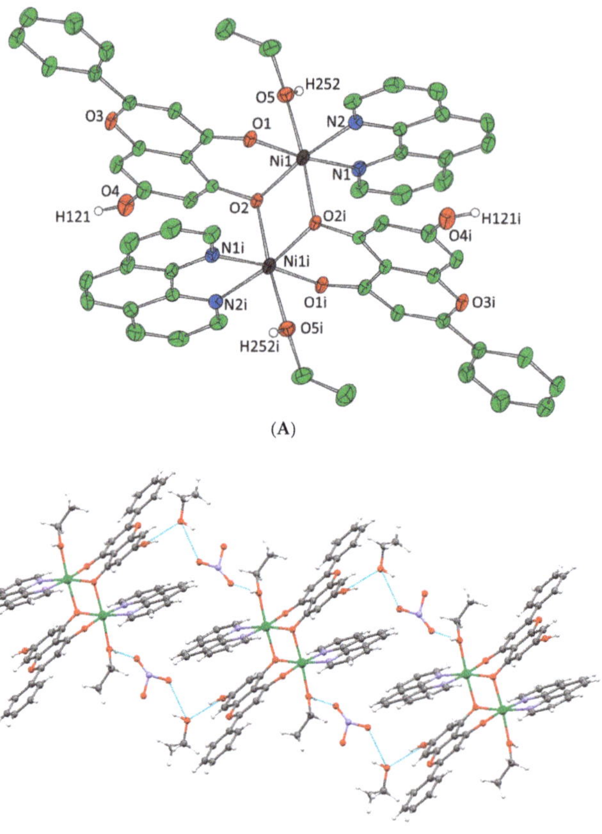

Figure 5. (**A**) Cameron plot of $[Ni_2(L)_2(phen)_2(EtOH)_2]^{2+}$ complex assembly in compound **4**. Aromatic and ethyl hydrogen atoms are omitted for clarity. (**B**) Hydrogen bonding in complex **4** (azure dotted lines).

The Co(III)-N (bpy), Co(III)-O, Co(II)-N (phen), Co(II)-O, Ni(II)-N (bpy), Ni(II)-N (phen) and Ni(II)-O bond lengths are in the range of 1.9099(19)–1.950(2) Å, 1.8445(17)–1.8806(16) Å, 2.101(2)–2.115(2) Å, 2.034(2)–2.147(2) Å, 2.043(2)–2.061(2) Å, 2.0067(19)–2.145(2) Å, 2.039(3)–2.076(3) Å, and 1.998(2)–2.137(2) Å, respectively. The measured lengths are similar to others found in literature reported Co(III), Co(II), and Ni(II) complexes [33,38,48].

Table 1. Summary of crystal, intensity collection and refinement data for $[Co(L)(bpy)_2]\cdot(NO_3)_2\cdot1.5(EtOH)$ (**1**), $[Co_2(L)_2(phen)_2](NO_3)_2\cdot2(MeOH)$ (**2**), $[Ni_2(L)_2(bpy)_2(EtOH)_2](NO_3)_2\cdot2(EtOH)$ (**3**), and $[Ni_2(L)_2(phen)_2(EtOH)_2](NO_3)_2\cdot3(EtOH)$ (**4**), where L is the deprotonated ligand chrysin.

	1	**2**	**3**	**4**
Crystal Data				
Chemical formula sum	$C_{38}H_{34}N_6O_{11.5}Co$	$C_{60}H_{58}N_6O_{20}Co_2$	$C_{58}H_{58}N_6O_{18}Ni_2$	$C_{64}H_{64}N_6O_{19}Ni_2$
Chemical formula moiety	$[Co(L)(bpy)_2]\cdot(NO_3)_2\cdot1.5(EtOH)$	$[Co_2(L)_2(phen)_2](NO_3)_2\cdot2(MeOH)$	$[Ni_2(L)_2(bpy)_2(EtOH)_2](NO_3)_2\cdot2(EtOH)$	$[Ni_2(L)_2(phen)_2(EtOH)_2](NO_3)_2\cdot3(EtOH)$
M_r	817.65	1301.01	1244.55	1338.66
Crystal system, Space group	Monoclinic, $P2_1/n$	Triclinic, *P*-1	Triclinic, *P*-1	Triclinic, *P*-1
Temperature (K)	295	295	295	295
a, *b*, *c* (Å)	13.144 (2), 10.9453 (17), 29.641 (5)	10.4626 (6), 11.2633 (8), 13.5977 (10)	10.6335 (12), 11.3275 (10), 13.8952 (17)	9.3331 (8), 12.6102 (12), 15.3374 (12)
a (°) β (°) γ (°)	98.340 (5)	87.743 (4) 71.447 (3) 76.325 (3)	72.933 (6) 71.685 (6) 66.191 (5)	72.730 (4) 88.713 (4) 68.463 (4)
V (Å³)	4219.3 (12)	1474.94 (18)	1426.7 (3)	1595.7 (2)
Z	4	1	1	1
Radiation type	Mo Kα	Mo Kα	Mo Kα	Mo Kα
μ (mm⁻¹)	0.47	0.65	0.74	0.67
Crystal size (mm)	0.23 × 0.16 × 0.15	0.25 × 0.23 × 0.15	0.15 × 0.07 × 0.06	0.19 × 0.18 × 0.12
Data collection				
Diffractometer	Bruker Kappa Apex2	Bruker Kappa Apex2	Bruker Kappa Apex2	Bruker Kappa Apex2
Absorption correction	Numerical Analytical Absorption [49]	Numerical Analytical Absorption [49]	Numerical Analytical Absorption [49]	Numerical Analytical Absorption [49]
Tmin, Tmax	0.93, 0.93	0.86, 0.91	0.95, 0.96	0.89, 0.92
No. of measured, independent and observed [I > 2.0σ(I)] reflections	41,491, 7992, 5795	19,214, 5549, 4560	17,025, 5406, 3980	27,053, 6093, 4366
Rint	0.028	0.021	0.023	0.030
(sin θ/λ)max (Å⁻¹)	0.613	0.613	0.615	0.613
Refinement				
$R[F^2 > 2\sigma(F^2)]$, $wR(F^2)$, S	0.048, 0.087, 1.00	0.046, 0.091, 1.00	0.039, 0.077, 1.00	0.044, 0.084, 1.00
No. of reflections	5795	4560	3980	4366
No. of parameters	505	397	373	413
No. of restraints	6	0	12	10
H-atom treatment	H-atom parameters constrained	H-atom parameters constrained	H-atom parameters constrained	H atoms treated by a mixture of independent and constrained refinement
$\Delta\rho_{max}$, $\Delta\rho_{min}$ (e Å⁻³)	0.33, −0.48	0.35, −0.90	0.60, −0.49	0.48, −0.30

Table 2. Bond lengths [Å] and angles [deg] for [Co(L)(bpy)$_2$]·(NO$_3$)$_2$·1.5(EtOH) (**1**), [Co$_2$(L)$_2$(phen)$_2$](NO$_3$)$_2$·2(MeOH) (**2**), [Ni$_2$(L)$_2$(bpy)$_2$(EtOH)$_2$](NO$_3$)$_2$·2(EtOH) (**3**), and [Ni$_2$(L)$_2$(phen)$_2$(EtOH)$_2$](NO$_3$)$_2$·3(EtOH) (**4**), where L is the deprotonated ligand chrysin.

Bond Lengths (Å)							
1		**2**		**3**		**4**	
Co(1)—O(1)	1.8806 (16)	Co(1)—O(2)^i	2.137 (2)	Ni(1)—O(2)^i	2.112 (2)	Ni(1)—O(2)^i	2.137 (2)
Co(1)—O(2)	1.8445 (17)	Co(1)—N(1)	2.101 (2)	Ni(1)—O(1)	2.0067 (19)	Ni(1)—O(1)	2.0123 (19)
Co(1)—N(1)	1.9099 (19)	Co(1)—N(2)	2.115 (2)	Ni(1)—O(2)	2.0098 (19)	Ni(1)—O(2)	1.998 (2)
Co(1)—N(2)	1.923 (2)	Co(1)—O(1)	2.0492 (19)	Ni(1)—O(5)	2.145 (2)	Ni(1)—O(5)	2.125 (2)
Co(1)—N(3)	1.950 (2)	Co(1)—O(2)	2.034 (2)	Ni(1)—N(1)	2.061 (2)	Ni(1)—N(1)	2.076 (3)
Co(1)—N(4)	1.925 (2)	Co(1)—O(3)	2.147 (2)	Ni(1)—N(2)	2.043 (2)	Ni(1)—N(2)	2.039 (3)
Angles (°)							
1		**2**		**3**		**4**	
O(1)—Co(1)—O(2)	96.70 (7)	O(2)^i—Co(1)—N(1)	96.05 (9)	O(2)^i—Ni(1)—O(1)	91.97 (8)	O(2)^i—Ni(1)—O(1)	90.25 (8)
O(1)—Co(1)—N(1)	175.32 (8)	O(2)^i—Co(1)—N(2)	89.76 (8)	O(2)^i—Ni(1)—O(2)	82.41 (8)	O(2)^i—Ni(1)—O(2)	82.11 (8)
O(2)—Co(1)—N(1)	86.26 (8)	N(1)—Co(1)—N(2)	79.18 (9)	O(1)—Ni(1)—O(2)	88.60 (8)	O(1)—Ni(1)—O(2)	88.72 (8)
O(1)—Co(1)—N(2)	93.47 (9)	O(2)^i—Co(1)—O(1)	95.81 (8)	O(2)^i—Ni(1)—O(5)	175.16 (8)	O(2)^i—Ni(1)—O(5)	175.52 (9)
O(2)—Co(1)—N(2)	87.33 (8)	N(1)—Co(1)—O(1)	94.19 (8)	O(1)—Ni(1)—O(5)	87.11 (8)	O(1)—Ni(1)—O(5)	87.80 (9)
N(1)—Co(1)—N(2)	83.03 (9)	N(2)—Co(1)—O(1)	171.78 (9)	O(2)—Ni(1)—O(5)	92.81 (8)	O(2)—Ni(1)—O(5)	93.81 (9)
O(1)—Co(1)—N(3)	86.90 (8)	O(2)^i—Co(1)—O(2)	80.82 (8)	O(2)^i—Ni(1)—N(1)	93.46 (9)	O(2)^i—Ni(1)—N(1)	94.88 (9)
O(2)—Co(1)—N(3)	174.66 (8)	N(1)—Co(1)—O(2)	176.29 (9)	O(1)—Ni(1)—N(1)	170.80 (9)	O(1)—Ni(1)—N(1)	173.61 (10)
N(1)—Co(1)—N(3)	90.41 (9)	N(2)—Co(1)—O(2)	98.74 (9)	O(2)—Ni(1)—N(1)	99.47 (9)	O(2)—Ni(1)—N(1)	95.74 (9)
N(2)—Co(1)—N(3)	96.42 (9)	O(1)—Co(1)—O(2)	88.13 (8)	O(5)—Ni(1)—N(1)	88.08 (9)	O(5)—Ni(1)—N(1)	87.35 (10)
O(1)—Co(1)—N(4)	86.96 (8)	O(2)^i—Co(1)—O(3)	170.87 (8)	O(2)^i—Ni(1)—N(2)	93.21 (9)	O(2)^i—Ni(1)—N(2)	91.29 (9)
O(2)—Co(1)—N(4)	92.14 (9)	N(1)—Co(1)—O(3)	91.92 (9)	O(1)—Ni(1)—N(2)	92.56 (9)	O(1)—Ni(1)—N(2)	95.48 (10)
N(1)—Co(1)—N(4)	96.58 (9)	N(2)—Co(1)—O(3)	87.37 (9)	O(2)—Ni(1)—N(2)	175.50 (10)	O(2)—Ni(1)—N(2)	172.21 (9)
N(2)—Co(1)—N(4)	179.35 (9)	O(1)—Co(1)—O(3)	88.01 (8)	O(5)—Ni(1)—N(2)	91.59 (10)	O(5)—Ni(1)—N(2)	92.91 (10)
N(3)—Co(1)—N(4)	84.09 (9)	O(2)—Co(1)—O(3)	91.05 (8)	N(1)—Ni(1)—N(2)	79.73 (10)	N(1)—Ni(1)—N(2)	80.62 (11)
				Ni(1)^i—O(2)—Ni(1)	97.59 (8)	Ni(1)^i—O(2)—Ni(1)	97.89 (8)

Symmetry code: (i) −x+1, −y+1, −z+1.

2.3. FT-IR Studies

The FT-IR spectroscopy data were in agreement with the structure of complexes **1–4** and display significant modifications of vibrational patterns of the ligands upon metal ion complexation (Supplementary Information Figure S2). In the free chrysin spectrum, an intense band was observed at 1652 cm^{-1}, demonstrating the C=O bond stretching vibration of the chromone ring. The strong signals at 1448 cm^{-1}, 1575 cm^{-1} and 1608 cm^{-1} are indicative of the ν(C=C) stretching in the γ-pyrone and benzene rings. The δ(O–H) and ν(C−O) linked vibrational modes are spotted at 1312 cm^{-1} and 1354 cm^{-1}, respectively. The v(C–O–C) stretching signals are spotted at 1244 cm^{-1}, whereas the phenolic stretches ν(C–H) and v(O–H) are detected between 2627 cm^{-1} and 3150 cm^{-1} [50,51].

In the spectra of the complex assemblies, the intense signals at 1632 cm^{-1} for **1**, 1629 cm^{-1} for **2**, 1625 cm^{-1} for **3** and 1629 cm^{-1} for **4**, are indicative of the v(C=O) stretches and exhibit a shift to lower energy in comparison to plain flavonoid, in agreement with coordination to the metal cores [50,51]. The bands at 1312 cm^{-1} and 1338 cm^{-1} for **1**, 1279 cm^{-1} and 1325 cm^{-1} for **2**, 1266 cm^{-1} and 1302 cm^{-1} for **3** and 1280 cm^{-1} and 1324 cm^{-1} for **4**, indicative of the coupled ν(C–O) and δ(O–H) vibrations, show notable shifts to lower energy in comparison to plain chrysin. The bands located at 1402 cm^{-1} for **1**, 1386 cm^{-1} for **2**, 1354 cm^{-1} for **3** and 1404 cm^{-1} for **4**, are due to the degenerate $\nu 3(E')$ stretches of the planar ($D3h$) NO^{3-} anion [52]. In all complexes, the weak but distinguishable bands in the 3100–2800 cm^{-1} region, are due to the v(O–H) and aromatic $v(C–H)$ stretches. The bands at 1243 cm^{-1} for **1**, 1279 cm^{-1} for **2**, 1266 cm^{-1} for **3** and 1245 cm^{-1} for **4**, are attributed to the v(C–O–C) vibrational modes. The signals located at 574 cm^{-1} for **1**, 562 cm^{-1} for **2**, 559 cm^{-1} for **3** and 652 cm^{-1} for **4**, can be attributed to the v(M–O) stretches [53].

In the spectra of **1** and **3**, the v(C–N) stretches indicating the bpy moiety, can be spotted at 1520 cm^{-1} and 1538 cm^{-1}, respectively, exhibiting shifts to lower frequencies due to the coordination to the metal ions [54]. A similar observation applies for the stretches of the phen molecule in the spectra of **2** and **4** that can be found at around 1538 cm^{-1}. The indicative out-of-plane C–H deformation motion on the heterocyclic rings can be found as strong bands between 832–728 cm^{-1} for **1**, 860–726 cm^{-1} for **2**, 864–735 cm^{-1} for **3** and 862–725 cm^{-1} for **4**, respectively [54].

2.4. UV-Vis Studies

The UV–Vis spectra of pure chrysin, **1** and **2** (Figure 6A), as well as **3** and **4** (Figure 6B), were recorded in EtOH (10^{-5} M). The chrysin spectrum demonstrates a high intensity absorption signal at 270 nm (ε~28,490 $M^{-1}\cdot cm^{-1}$), characteristic of the $\pi \rightarrow \pi^*$ electronic transitions over Band II (A and C rings), and a broader lower signal around 315 nm, indicative of $\pi \rightarrow \pi^*$ transitions over Band I (B-ring) [50,51]. The UV-Vis spectra of **1–4** display an intense hyperchromic shift of the Band II maximum absorption compared to pure flavonoid. A strong shift of this band to longer wavelengths (315 nm) was observed in the spectra of **1** and **3**. These shifts are characteristic of the rigid flavonoid coordination to the metal core in combination with ligand-to-metal charge transfer (LMCT) interactions [50–52,55]. More accurately, Band II shows up at 295 nm (ε~70,030 $M^{-1}\cdot cm^{-1}$) for **1** and at 287 nm (ε~80,100 $M^{-1}\cdot cm^{-1}$) for **3**, showing shifts of 25 nm and 17 nm, respectively, compared to pure flavonoid, and at 270 nm for **2** (ε~67,141 $M^{-1}\cdot cm^{-1}$) and **4** (ε~88,138 $M^{-1}\cdot cm^{-1}$). Band I can be observed as a very weak shoulder in the spectrum of complexes **2**, **3** and **4** at around 328 nm, confirming the absence of B ring from the complexation process. The broad bands at around 395 nm (ε~10,979 $M^{-1}\cdot cm^{-1}$) for **1**, 380 nm (ε~6010 $M^{-1}\cdot cm^{-1}$) for **2**, 389 nm for **3** (ε~9000 $M^{-1}\cdot cm^{-1}$) and 388 nm for **4** (ε~5952 $M^{-1}\cdot cm^{-1}$) can be assigned to the LMCT process [50–52,55]. A weak shoulder observed at around 289 nm (ε~42,640 $M^{-1} \cdot cm^{-1}$) for complexes **2** and **4** is due to the $\pi \rightarrow \pi^*$ charge transfer of the chelated phen molecule [56]. Nevertheless, in the spectra of complexes **1** and **3**, a similar signal cannot be easily distinguished.

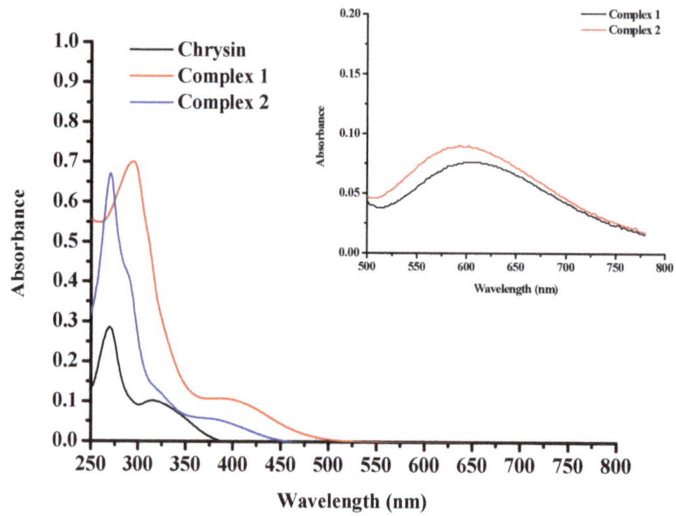

(A)

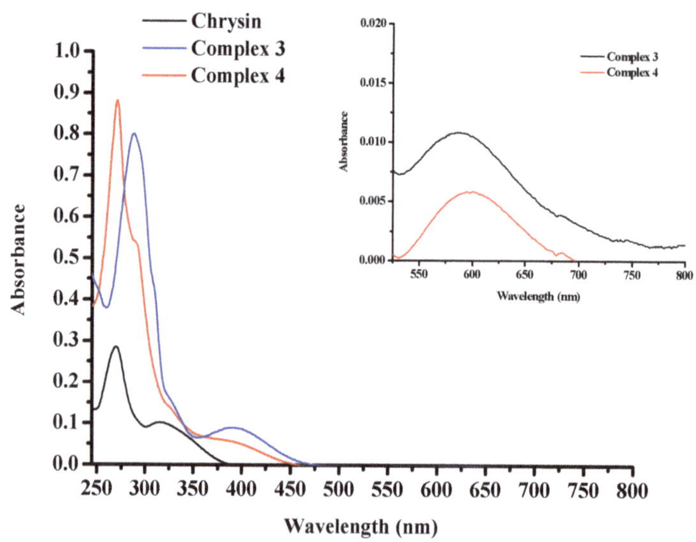

(B)

Figure 6. UV-Vis spectra of (**A**) **1** (red line), **2** (blue line), and pure chrysin (black line); and (**B**) **3** (blue line), **4** (red line), and plain chrysin (black line), in EtOH at 10^{-5} M. The inset graphs in both spectra present the d–d transitions detected in EtOH at 10^{-3} M.

No d–d transitions are detected in the spectra of complexes **1–4** due to the low concentration of the solutions (10^{-5} M). Additional measurements at 10^{-3} M showed broad bands at 606 nm for **1**, 593 nm for **2**, 584 nm for **3** and 597 for **4**, respectively, characteristic of d–d transitions (Figure 6A, B inset graphs).

2.5. Fluorescence Studies

Emission spectra of the free flavonoid and complexes **1–4** were measured in EtOH solutions (10^{-5} M), at ambient temperature (Figure 7A, B). In particular, chrysin in its free form exhibits two emission bands with emission maxima at 414 and 436 nm, upon excitation at 270 nm. Complexes **1** and **3** show emission maxima at 437 and 436 nm, respectively, when excited at 295 nm. An additional emission band of lower intensity was observed at 414 nm, in the spectrum of **3**. Emission spectra of **2** and **4** show reduced emission maxima in comparison to plain chrysin whereas complexes **1** and **3** present enhanced emission. Complex **2** displays an emission maximum at 436 nm and a weaker signal at 413 nm whereas for complex **4**, the corresponding signals are observed at 436 and 415 nm, upon excitation at 270 nm. The intensity variations of the emission signals of **1–4** compared to the plain flavonoid may reflect the rigidity of coordination of chrysin to the metal cores [50–52]. In general, the emission at higher energies is attributed to the contribution of chrysin in the fluorescence of the complexes. However, at lower energies all complexes exhibit emission maxima that are assigned to the d–d transitions in the metal center. Due to the presence of potential aggregates in solution that may lead to decrease of fluorescence, the emission spectra at 600–800 nm were recorded in solid state and are presented in Supplementary Information Figure S3. The corresponding excitation spectra of complexes **1–4** are presented in Supplementary Information Figure S4. Table 3 summarizes the emission band maxima λ_{max}(em) (nm) in EtOH solutions (10^{-5} M) and solid state.

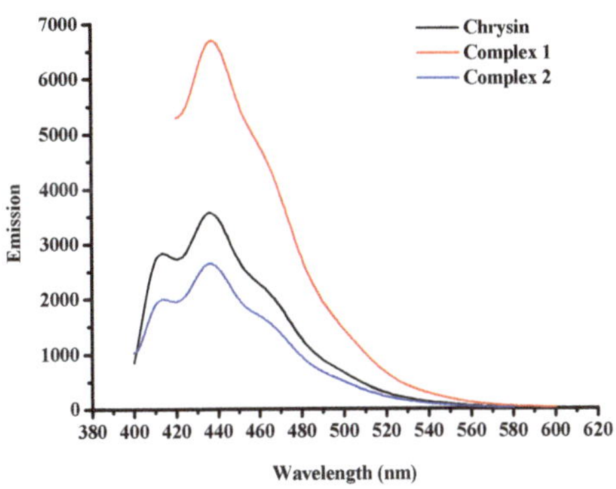

(**A**)

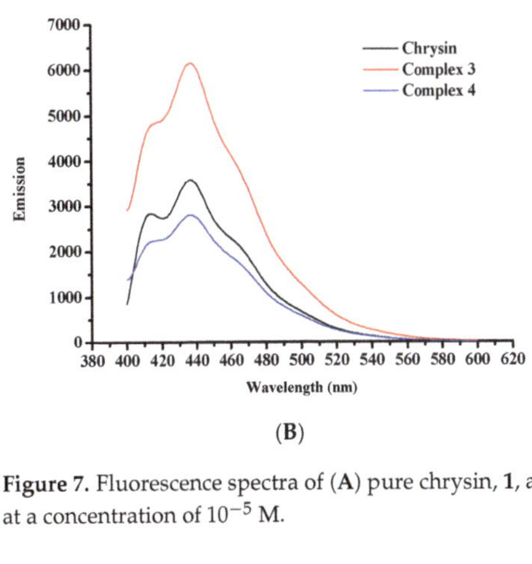

(**B**)

Figure 7. Fluorescence spectra of (**A**) pure chrysin, **1**, and **2**, and (**B**) pure chrysin, **3**, and **4**, in EtOH at a concentration of 10^{-5} M.

Table 3. Emission band maxima λ_{max}(em) (nm) in EtOH solutions (10^{-5} M) and solid state.

Complex	λ_{max}(em)/(nm)
1	437, 649, 752
2	436, 649, 752
3	436, 708, 748
4	415, 436, 708, 748

2.6. Thermogravimetric Studies

The thermal degradation of complexes **1–4** was assessed via thermogravimetric analysis (TGA), in air (Figure 8A,B).

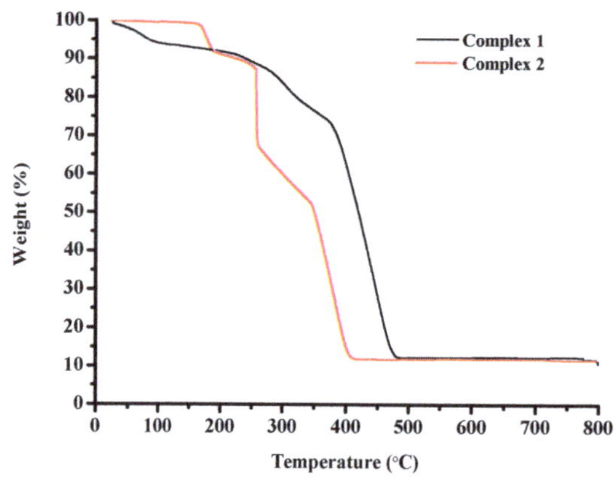

(A)

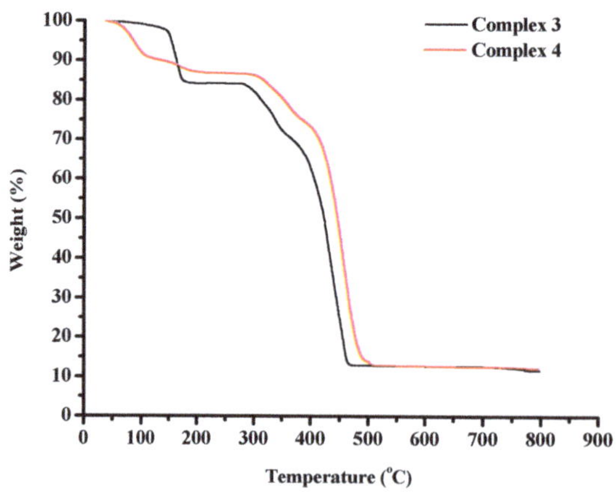

(B)

Figure 8. TGA diagrams of (**A**) complexes **1** and **2**, and (**B**) complexes **3** and **4**.

The thermal decomposition of complex **1** begins at 25 °C and proceeds rapidly until 100 °C potentially due to the discharge of volatile compounds. Between 100 °C and 215 °C the rate decreases. After 215 °C and until 485 °C the decomposition curve descends very rapidly and a significant weight loss can be observed that could be attributed to the destruction of the organic constituents of the starting molecule. A horizontal portion is located between 485 °C and 800 °C.

Complex **2** is thermally stable until 165 °C. From this point a rapid weight loss is detected that may be attributed to the degradation of the organic matter of the starting molecule. The curve descends continuously accompanied by abrupt changes in direction until 422 °C. A perfectly horizontal portion is spotted between 422 °C and 800 °C.

A negligible weight loss is noticed for complex **3** between 25 °C and 145 °C indicative of the release of volatile substances. From this point on, a rapid weight loss is noticed until 185 °C. After that, a horizontal portion of the decomposition curve is recorded between 185 °C and 277 °C. The decomposition then proceeds very rapidly up to 475 °C pointing to the destruction of the organic constituents of the starting molecule. Subsequently, a plateau appears that remains stable until 800 °C.

The thermal decomposition of complex **4** begins at 25 °C and proceeds rapidly until 205 °C, potentially due to the release of volatile substances. After that, a horizontal portion of the decomposition curve is located between 205 °C and 290 °C. Then the complex decomposes very rapidly until 510 °C pointing to the destruction of the organic substances of the starting molecule. Subsequently, a horizontal portion appears and remains until 800 °C.

The performed analysis is consistent with documented TGA data on Co(II), Co(III) and Ni(II)-containing compounds [57].

2.7. EPR Analysis

The continuous wave (cw) X-band EPR spectra [58] of frozen dimethyl sulfoxide (DMSO) solutions of nickel complexes **3** and **4** in are depicted in Figure 9A. The spectra of both complexes do not present any EPR signal, indicating that both compounds are diamagnetic. Figure 9B shows the X-band EPR spectra of cobalt complexes **1** and **2**. The spectrum of complex **1** presents a very wide EPR signal which indicates the paramagnetic nature of this compound. However, the determination of the spin value is impossible, due to the broadness of the EPR signal. In the case of complex **2**, the absence of an EPR signal indicates that it is diamagnetic.

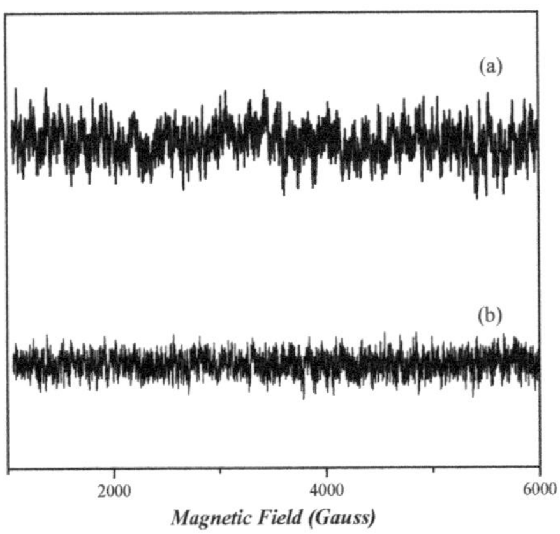

(A)

Figure 9. *Cont.*

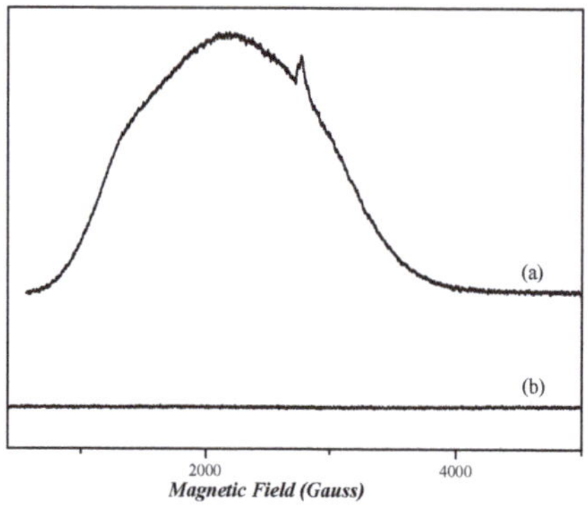

(**B**)

Figure 9. CW X-band EPR spectra of complexes **1**, **2**, **3**, and **4**, all diluted in DMSO at ~1 mM. (**A**) Complex **4** (trace a) and complex **3** (trace b). (**B**) Complex **1** (trace a) and complex **2** (trace b). Measurement conditions: T = 6 K, modulation amplitude 12.5 Gpp, microwave frequency 9.41 GHz, microwave power 30 mW.

2.8. Antioxidant Activity

The antioxidant activity of complexes **1**–**4** was determined *in vitro* through their potential to scavenge free DPPH radicals [59] and is depicted in Figure 10. The well-known antioxidant ascorbic acid (AA) was employed as reference substance. Plain chrysin showed weak antioxidant activity at all tested concentrations. However, all the complexes under evaluation presented dose-dependent antioxidant potential, higher than the pure flavonoid. More specifically, Ni complexes **3** and **4** displayed higher antioxidant activity than cobalt complexes **1** and **2** ($p < 0.001$), plain chrysin ($p < 0.01$) and the control samples $Ni(NO_3)_2{\cdot}6H_2O$, phen and bpy ($p < 0.01$). Co complexes **1** and **2** scavenged DPPH radicals to a greater degree than plain chrysin ($p > 0.05$), $Co(NO_3)_2{\cdot}6H_2O$ ($p > 0.05$) and the corresponding co-ligands ($p < 0.001$). It is worth noting that no significant antioxidant activity was detected in a similar set of experiments utilizing the plain precursors $Ni(NO_3)_2{\cdot}6H_2O$, $Co(NO_3)_2{\cdot}6H_2O$ and co-ligands phen, bpy (Supplementary Information Figure S5).

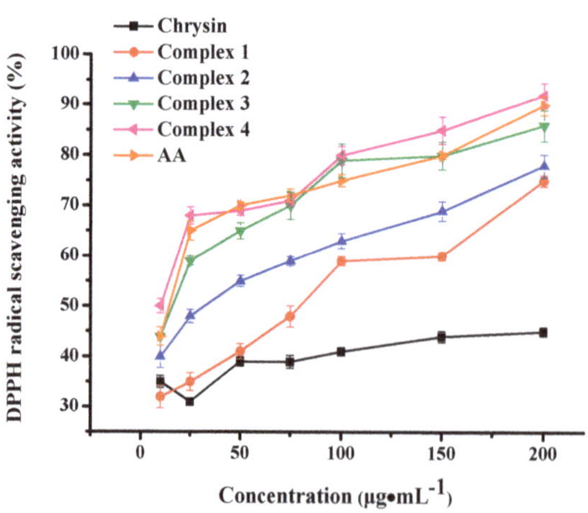

Figure 10. DPPH radical scavenging activity of complexes **1**–**4** and chrysin. AA was employed as standard antioxidant. Data are shown as means ± SEM after three independent measurements.

3. Discussion

The interaction of flavonoids with metal ions has gathered significant attention in the last two decades, in the search of flavonoid-metal complexes with enhanced features and potential drugability. Metal complexation has been shown to strengthen the biological activity of flavonoids and contribute to their stability, solubility, and bioavailability. Although many studies on the synthetic process and physico-chemical characterization of metal-flavonoid complexes are reported in the literature [60,61], yet their structure is scarcely elucidated via X-ray crystallography, mainly because of their negligible solubility and crystallinity.

Our research team has an extended experience in the development of novel crystalline, physico-chemically, structurally, and biologically characterized metallo-flavonoid complexes with a variety of flavonoids and metal ions including lanthanides. Until today, only a few fully characterized via X-ray crystallography metal complexes of chrysin are present in the literature [50–52,62,63], and most of them have been developed by our research team. In this work, a systematic investigation of the experimental conditions led to the development of four crystalline chrysin-based heteroleptic Co(II), Co(III) and Ni(II) complexes with a coordination motif elucidated via spectroscopic, analytical, and X-ray crystallographic techniques.

After careful examination of all reaction parameters, the four complexes were synthesized in alcoholic media employing triethylamine as deprotonating agent. EtOH was chosen as the most appropriate reaction solvent in the cases of complexes **1**, **3**, and **4**, whereas, in the case of complex **2**, MeOH proved to be the most adequate solvent for the formation of stable crystalline material. As observed by our research group in our previous works [50–52], the employment of *N,N*-donor chelators, such as bpy and phen, along with the main flavonoid ligand, leads to the formation of stable heteroleptic complexes with exceptional crystallinity and in some cases, remarkable structural differentiation. Interestingly, in this work, the combination of EtOH, triethylamine, and bpy led to the formation of a Co(III) monomeric mononuclear complex **1**, whereas in the case of complex **2**, the combination of MeOH, triethylamine, and phen led to the development of a dimer dinuclear Co(II) complex. The observed differences in the oxidation state of cobalt and structures of these complexes could be attributed to the combined effect of the solvent and *N,N*-donor chelator. In the case of complex **1**, it is possible that the lower coordination ability of EtOH for Co(II) [64], compared to MeOH, leaves the metal center more exposed to atmospheric oxygen, resulting in the oxidation of Co(II) $\rightarrow$ Co(III). Moreover, the strong ligand field of the incorporated bpy may stabilize low-spin Co(III) whilst its non-bridging nature and smaller steric demand, compared to that of phen, promote the formation of a monomeric structure. In the case of complex **2**, the combination of MeOH, with the stronger coordination capacity for Co(II), and the co-ligand phen with its rigid, planar structure promoting π-stacking and bridging interactions, facilitated the formation of the final dimeric Co(II) assembly [65,66]. In the case of the Ni(II) complexes, the greater resistance to oxidation of Ni(II) compared to Co(II), may explain the formation of the dimeric Ni(II) complexes **3** and **4**, irrespective of solvent, or type of *N,N*-chelator.

As previously reported [67], plain chrysin is a weak DPPH scavenger. Herein, the Ni(II) complexation of chrysin resulted in great enhancement of its radical scavenging activity, that became comparable to that of AA. No significant difference in activity was detected between the bpy and phen-bearing complexes. A comparison to the literature reported Ni(II) complexes revealed that complexes **3** and **4** are better radical scavengers compared to Ni(II)-indomethacin [31], Ni(II)-diflunisal [32], Ni(II)-mefenamato [33], and Ni(II)-naproxen [68] mixed-ligand compounds.

Cobalt complexes **1** and **2** also increased the DPPH scavenging efficiency of chrysin but in a lesser degree compared to Ni(II) complexes **3** and **4**. However, the phen-ligated complex **2** appeared to be a better scavenger than the bpy-ligated complex **1**, a result that is in agreement with documented results on phen or bpy-bearing Co complexes [69]. Herein, the observed scavenging ability is of the same magnitude as recently reported cobalt complexes of non-steroidal anti-inflammatory drugs in the presence or absence of nitrogen-donor heterocyclic ligands [37,38].

The results in this work are in accord with our previous studies on the synthesis and evaluation of novel metal-flavonoid complexes. All complexes exhibited enhanced biological activities such as cytotoxic and ROS generating potential in a series of diverse cancer cell lines [51,70], induction of sister chromatid exchanges, reduction in proliferation rate index and mitotic index values in healthy human lymphocytes and cytotoxic activity against human non-small cell lung carcinoma and lung adenocarcinoma cell lines [50], increased radical scavenging activity [52,71], antioxidant and antibacterial activity [72], and DNA binding potential [73], hence demonstrating the strong effect that metal complexation has on the enhancement of the biological potential of flavonoids.

4. Materials and Methods

A detailed report of the employed equipment, methods and materials is provided in Supplementary Information.

4.1. Synthesis

4.1.1. Synthesis of $[Co(L)(bpy)_2]\cdot(NO_3)_2\cdot1.5EtOH$ (1)

Ethanolic solutions of $Co(NO_3)_2\cdot6H_2O$ (0.15 g, 0.5 mmol in 10 mL EtOH) and HL (0.13 g, 0.5 mmol in 10 mL EtOH) were stirred under steady reflux conditions at 60 °C for 90 min. To the occurring bright red clear solution, an ethanolic solution of bpy (0.08 g, 0.5 mmol in 5 mL EtOH) was added under constant stirring. A dark red transparent solution was obtained that was cooled until reaching ambient conditions. To that, triethylamine (55 μL, 0.4 mmol) was added. After 60 min a crimson colored clear solution resulted. To the resulting solution, 1 mL of diethyl ether was gradually added applying the solvent diffusion/layering technique and, after sealing with a rubber septum, the reaction container was refrigerated. This procedure was repeated daily.Twenty days later, dark brown plate-like crystals emerged that were filtrated and vacuum-dried. Yield: 0.06 g (15%). Anal. calcd. for **1**, $[Co(L)(bpy)_2]\cdot(NO_3)_2\cdot1.5EtOH$ ($C_{38}H_{34}N_6O_{11.5}Co$, Mr: 817.65): C, 55.82; H, 4.19; N, 10.28%. Found: C, 55.73; H, 4.14; N, 10.27%. HR-ESI-MS (positive mode), calcd. for $[Co(L)_2(bpy)]^+$ m/z= 721.1021, found m/z = 721.1011.

4.1.2. Synthesis of $[Co_2(L)_2(phen)_2(MeOH)_2]\cdot(NO_3)_2\cdot4MeOH$ (2)

Methanolic solutions of $Co(NO_3)_2\cdot6H_2O$ (0.15 g, 0.5 mmol in 10 mL MeOH) and HL (0.13 g, 0.5 mmol in 10 mL MeOH) were stirred under steady reflux conditions at 60 °C for 90 min. To the bright red clear solution that resulted, a methanolic solution of phen hydrate (0.1 g, 0.5 mmol in 5 mL MeOH) was added under constant stirring. A dark red transparent solution was obtained that was cooled until reaching ambient conditions.To that, triethylamine (70 μL, 0.5 mmol) was added. After 60 min, a crimson colored clear solution resulted. To the resulting solution, diethyl ether was gradually added applying the solvent diffusion/layering technique and, after sealing with a rubber septum, the reaction container was refrigerated. The following day, the reaction vessel was unsealed and left to evaporate slowly at 4 °C. Eight days later, dark red plate like crystalline material emerged that was filtrated and vacuum-dried. Yield: 0.13 g (39%). Anal. calcd. for **2**, $[Co_2(L)_2(phen)_2(MeOH)_2]\cdot(NO_3)_2\cdot4MeOH$ ($C_{60}H_{58}N_6O_{20}Co_2$, Mr: 1301.01): C, 55.39; H,

4.49; N, 6.46%. Found: C, 55.34; H, 4.44; N, 6.45%. HR-ESI-MS (positive mode), calcd. for $[Co(L)(phen)]^+$ m/z = 492.0520, found m/z = 492.0540.

4.1.3. Synthesis of $[Ni_2(L)_2(bpy)_2(EtOH)_2]\cdot(NO_3)_2\cdot 2EtOH$ (3)

Ethanolic solutions of $Ni(NO_3)_2\cdot 6H_2O$ (0.15 g, 0.5 mmol in 10 mL EtOH) and HL (0.13 g, 0.5 mmol in 10 mL EtOH) were stirred under steady reflux conditions at 60 °C for 90 min. To the bright green clear solution that resulted, an ethanolic solution of bpy (0.08 g, 0.5 mmol in 5 mL EtOH) was added under constant stirring. A green transparent solution was obtained that was cooled until reaching ambient conditions. To that, triethylamine (70 μL, 0.5 mmol) was added. After 60 min, an emerald green homogeneous solution was obtained. To the resulting solution, diethyl ether was gradually added applying the solvent diffusion/layering technique and, after sealing with a rubber septum, the reaction container was refrigerated. Eight days later, green needlelike crystalline precipitate emerged which was filtrated and vacuum-dried.Yield: 0.11 g (35%). Anal. calcd. for **3**, $[Ni_2(L)_2(bpy)_2(EtOH)_2]\cdot(NO_3)_2\cdot 2EtOH$ ($C_{58}H_{58}N_6O_{18}Ni_2$, Mr: 1244.55): C, 55.98; H, 4.70; N, 6.75%. Found: C, 55.98; H, 4.67; N, 6.71%. HR-ESI-MS (positive mode), calcd. for $[Ni(L)(bpy)]^+$ m/z = 467.0542, found m/z = 467.0550.

4.1.4. Synthesis of $[Ni_2(L)_2(phen)_2(EtOH)_2]\cdot(NO_3)_2\cdot 3EtOH$ (4)

Ethanolic solutions of $Ni(NO_3)_2\cdot 6H_2O$ (0.15 g, 0.5 mmol in 10 mL EtOH) and HL (0.13 g, 0.5 mmol in 10 mL EtOH) were stirred under steady reflux conditions at 60 °C for 90 min. To the bright green clear solution that resulted, an ethanolic solution of phen hydrate (0.1 g, 0.5 mmol in 5 mL EtOH) was added under constant stirring. A green homogeneous solution was obtained that was allowed to cool to ambient temperature.To that, triethylamine (70 μL, 0.5 mmol) was added. After 60 min, a deep green transparent solution resulted. To the resulting solution, diethyl ether was gradually added applying the solvent diffusion/layering technique and, after sealing with a rubber septum, the reaction container was refrigerated. Ten days later, green cubic crystalline material emerged that was filtrated and vacuum-dried. Yield: 0.08 g (24%). Anal. calcd. for **4**, $[Ni_2(L)_2(phen)_2(EtOH)_2]\cdot(NO_3)_2\cdot 3EtOH$ ($C_{64}H_{64}N_6O_{19}Ni_2$, Mr: 1338.66): C, 57.42; H, 4.82; N, 6.28%. Found: C, 57.38; H, 4.77; N, 6.27%. HR-ESI-MS (positive mode), calcd. for $[Ni(L)(phen)]^+$ m/z = 491.0542, found m/z = 491.0563.

4.2. X-Ray Crystal Structure Determination

The X-ray monocrystal methodology for the elucidation of the crystal structures of complexes **1**–**4** is provided in Supplementary Information.

4.3. Antioxidant Activity

The DPPH˙ protocol was utilized for the *in vitro* estimation of the antioxidant potential of complexes **1**–**4**, the precursors $Co(NO_3)_2\cdot 6H_2O$ and $Ni(NO_3)_2\cdot 6H_2O$, and the participating ligands chrysin, bpy, and phen, as previously described [71]. Additional information can be found in Supplementary Information.

5. Conclusions

The four novel Co(III)-chrysin-bpy, Co(II)-chrysin-phen, Ni(II)-chrysin-bpy, and Ni(II)-chrysin-phen complexes presented herein demonstrate the effective complexation of flavonoid chrysin with the bioactive Co(II), Co(III) and Ni(II) ions into stable formations. The four complexes were physico-chemically and structurally characterized in solution phase and solid form, indicating a distinct coordination motif around the metal cores with a flavonoid ligand and the ancillary phen and bpy chelators. The *in vitro* biological evaluation revealed notable enhancement of radical scavenging activity after metal complexation, a fact that paves the way for advanced investigation of their pharmacological potential.

Supplementary Materials: The following supporting information can be downloaded at https://www.mdpi.com/article/10.3390/inorganics13070230/s1, Figure S1 HR-ESI-MS spectra of **1**(**A**), **2**(**B**), **3**(**C**), and **4**(**D**); Table S1 Presentation of the examined reaction and (re)crystallization parameters and observations; Table S2 Hydrogen bonds in **1–4**; Figure S2 FT-IR spectra of: chrysin (black line), complex **1** (red line), complex **2** (blue line), complex **3** (purple line), complex **4** (green line), bpy (orange line), and phen (brown line); Figure S3 Emission spectra of complexes **1** (**A**), **2** (**B**), **3** (**C**), and **4** (**D**) upon excitation at 540–600 nm; Figure S4 Excitation spectra of complexes **1–4** corresponding to the observed fluorescence spectra; Figure S5 Effect of $Ni(NO_3)_2{\cdot}6H_2O$, $Co(NO_3)_2{\cdot}6H_2O$ and co-ligands phen, bpy on DPPH radical scavenging activity. Data were expressed as means ± SEM obtained in three independent experiments.

Author Contributions: Conceptualization, E.H.; methodology, E.H., B.M., D.V., G.Z. and A.G.H.; software, G.Z. and A.G.H.; validation, E.H., B.M., D.V., G.Z. and A.G.H.; formal analysis, E.H., B.M., D.V., G.Z. and A.G.H.; investigation, E.H., B.M., D.V., G.Z. and A.G.H.; resources, E.H.; data curation, E.H., B.M., D.V., G.Z. and A.G.H.; writing—original draft preparation, E.H., B.M., D.V., G.Z. and A.G.H.; writing—review and editing, E.H., B.M., D.V., G.Z., G.L., M.P. and A.G.H.; visualization, E.H., B.M., D.V., G.Z. and A.G.H.; supervision, E.H. and A.G.H.; project administration, E.H.; funding acquisition, E.H. All authors have read and agreed to the published version of the manuscript.

Funding: This research received no external funding.

Institutional Review Board Statement: Not applicable.

Informed Consent Statement: Not applicable.

Data Availability Statement: The raw data supporting the conclusions of this article will be made available by the authors on request. CCDC 2448473 (1), 2448474 (2), 2448475 (3) and 2448476 (4) contain the supplementary crystallographic data for this paper. These data can be obtained free of charge via https://www.ccdc.cam.ac.uk/structures/ (accessed on 3 May 2025) (or from the Cambridge Crystallographic Data Centre, 12 Union Road, Cambridge CB21EZ, UK; fax: (+44) 1223-336-033; or deposit@ccde.cam.ac.uk).

Acknowledgments: E. Halevas acknowledges the Foundation for Education and European Culture (IPEP) founded by Nicos and Lydia Tricha. G. Zahariou gratefully acknowledges helpful discussions with Y. Sanakis on the simulated analysis of the EPR spectra.

Conflicts of Interest: The authors declare no conflicts of interest.

Abbreviations

The following abbreviations are used in this manuscript:

ROS	Reactive Oxygen Species
HL	Chrysin
bpy	2,2′-bipyridine
phen	1,10-phenanthroline
MeOH	Methanol
$Co(NO_3)_2{\cdot}6H_2O$	Cobalt nitrate hexahydrate
$Ni(NO_3)_2{\cdot}6H_2O$	Nickel nitrate hexahydrate
LMCT	Ligand-to-Metal Charge Transfer
AA	Ascorbic Acid
DPPH	2,2-diphenyl-1-picrylhydrazyl
FT-IR	Fourier Transform-Infrared spectroscopy
GC	Gas Chromatography
HR-ESI-MS	High Resolution-ElectroSpray Ionization Mass Spectra
UV-Vis	Ultraviolet-Visible

TGA	Thermogravimetric Analysis
cw	Continuous Wave
EPR	Electron Paramagnetic Resonance
SEM	Standard Error of Mean
ANOVA	Analysis of variance
EtOH	Ethanol
DMA	Dimethylacetamide
DMF	Dimethylformamide
DMSO	Dimethyl sulfoxide

References

1. Chen, S.; Liu, J.; Dong, G.; Zhang, X.; Liu, Y.; Sun, W.; Liu, A. Flavonoids and caffeoylquinic acids in *Chrysanthemum morifolium* Ramat flowers: A potentially rich source of bioactive compounds. *Food Chem.* **2021**, *344*, 128733. [CrossRef]
2. Zhang, T.; Qiu, F.; Chen, L.; Liu, R.; Chang, M.; Wang, X. Identification and in vitro anti-inflammatory activity of different forms of phenolic compounds in *Camellia oleifera* oil. *Food Chem.* **2021**, *344*, 128660. [CrossRef] [PubMed]
3. Wang, N.; Wang, L.; Yang, J.; Wang, Z.; Cheng, L. Quercetin promotes osteogenic differentiation and antioxidant responses of mouse bone mesenchymal stem cells through activation of the AMPK/SIRT1 signaling pathway. *Phytother. Res.* **2021**, *35*, 2639–2650. [CrossRef]
4. Shendge, A.K.; Chaudhuri, D.; Mandal, N. The natural flavones, acacetin and apigenin, induce Cdk-Cyclin mediated G2/M phase arrest and trigger ROS-mediated apoptosis in glioblastoma cells. *Mol. Biol. Rep.* **2021**, *48*, 539–549. [CrossRef] [PubMed]
5. Liu, T.; Xu, J.; Wu, Y.; Li, X.; Ding, D.; Ma, D.; Yao, M.; Wei, W.; Zhang, W.; Wang, S.; et al. Beneficial effects of baicalein on a model of allergic rhinitis. *Acta Pharm.* **2020**, *70*, 35–47. [CrossRef] [PubMed]
6. Wu, Z.; Fan, Q.; Miao, Y.; Tian, E.; Ishfaq, M.; Li, J. Baicalin inhibits inflammation caused by coinfection of *Mycoplasma gallisepticum* and *Escherichia coli* involving IL-17 signaling pathway. *Poult. Sci.* **2020**, *99*, 5472–5480. [CrossRef]
7. Ding, Y.M.; Lin, J.T.; Fang, L.M.; Lou, Z.Q.; Liang, G.N.; Zhang, X.Y.; Li, A.Q.; Zhang, X. The neuroprotective effect of apigenin against OGD/R injury in rat hippocampal neurons. *Pak. J. Pharm. Sci.* **2020**, *33*, 1527–1533.
8. Yue, S.; Xue, N.; Li, H.; Huang, B.; Chen, Z.; Wang, X. Hepatoprotective effect of apigenin against liver injury *via* the non-canonical NF-κB pathway in vivo and in vitro. *Inflammation* **2020**, *43*, 1634–1648. [CrossRef]
9. Stompor-Gorący, M.; Bajek-Bil, A.; Machaczka, M. Chrysin: Perspectives on contemporary status and future possibilities as pro-health agent. *Nutrients* **2021**, *13*, 2038. [CrossRef]
10. Chan, C.W.; Deadman, B.J.; Manley-Harris, M.; Wilkins, A.L.; Alber, D.G.; Harry, E. Analysis of the flavonoid component of bioactive New Zealand mānuka (*Leptospermum scoparium*) honey and the isolation, characterisation and synthesis of an unusual pyrrole. *Food Chem.* **2013**, *141*, 1772–1781. [CrossRef]
11. Wożniak, M.; Mrówczyńska, L.; Kwaśniewska-Sip, P.; Waśkiewicz, A.; Nowak, P.; Ratajczak, I. Effect of the solvent on propolis phenolic profile and its antifungal, antioxidant, and in vitro cytoprotective activity in human erythrocytes under oxidative stress. *Molecules* **2020**, *25*, 4266. [CrossRef]
12. Lopez, A.P.; Galuch, M.B.; Petenuci, M.E.; Oliveira, J.H.; Canesin, E.A.; Schneider, V.V.A.; Visentainer, J.V. Quantification of phenolic compounds in ripe and unripe bitter melons (*Momordica charantia*) and evaluation of the distribution of phenolic compounds in different parts of the fruit by UPLC–MS/MS. *Chem. Pap.* **2020**, *74*, 2613–2625. [CrossRef]
13. Sharma, P.; Kumari, A.; Gulati, A.; Krishnamurthy, S.; Hemalatha, S. Chrysin isolated from *Pyruspashia* fruit ameliorates convulsions in experimental animals. *Nutr. Neurosci.* **2019**, *22*, 569–577. [CrossRef] [PubMed]
14. Xie, Y.; Peng, X. Effect of chrysin on apoptosis of oral squamous cell carcinoma KB cells and its mechanism. *J. Cent. South Univ. Med. Sci.* **2019**, *44*, 522–527.
15. Chen, H.Y.; Jiang, Y.W.; Kuo, C.L.; Day, T.; Chou, Y.C.; Chang, Y.S.; Chung, J.G. Chrysin inhibit human melanoma A375.S2 cell migration and invasion via affecting MAPK signaling and NF-κB signaling pathway in vitro. *Environ. Toxicol.* **2019**, *34*, 434–442. [CrossRef]
16. Ghamkhari, A.; Pouyafar, A.; Salehi, R.; Rahbarghazi, R. Chrysin and docetaxel loaded biodegradable micelle for combination chemotherapy of cancer stem cell. *Pharm. Res.* **2019**, *36*, 165. [CrossRef]
17. Wang, J.N.; Li, X.; Chen, M.F.; Wang, F.Y.; Xiong, X.K.; Chen, X.J.; Yang, M.L.; Huang, J.M. Sensitization of chrysin on the apoptosis induced by cisplatin or camptothecin in hepatoma cell lines (Hep G2). *Chin. Pharm. J.* **2016**, *51*, 2088–2093.
18. Mohos, V.; Fliszár-Nyúl, E.; Schilli, G.; Hetènyi, C.; Lemli, B.; Kunsági-Mátè, S.; Bognár, B.; Poór, M. Interaction of chrysin and its main conjugated metabolites chrysin-7-sulfate and chrysin-7-glucuronide with serum albumin. *Int. J. Mol. Sci.* **2018**, *19*, 4073. [CrossRef]

19. Song, Y.; Wu, W.; Sheng, L.; Jiang, B.; Li, X.; Cai, K. Chrysin ameliorates hepatic steatosis induced by a diet deficient in methionine and choline by inducing the secretion of hepatocyte nuclear factor 4α-dependent very low-density lipoprotein. *J. Biochem. Mol. Toxicol.* **2020**, *34*, e22497. [CrossRef]
20. Li, H.J.; Wu, N.L.; Pu, C.M.; Hsiao, C.Y.; Chang, D.C.; Hung, C.F. Chrysin alleviates imiquimod-induced psoriasis-like skin inflammation and reduces the release of CCL20 and antimicrobial peptides. *Sci. Rep.* **2020**, *10*, 2932. [CrossRef]
21. Bortolotto, V.C.; Araujo, S.M.; Pinheiro, F.C.; Poetini, M.R.; de Paula, M.T.; Meichtry, L.B.; de Almeida, F.P.; Musachio, E.A.S.; Guerra, G.P.; Prigol, M. Modulation of glutamate levels and Na^+, K^+-ATPase activity contributes to the chrysin memory recovery in hypothyroidism mice. *Physiol. Behav.* **2020**, *222*, 112892. [CrossRef] [PubMed]
22. Sadati, S.M.; Gheibi, N.; Ranjbar, S.; Hashemzadehm, M.S. Docking study of flavonoid derivatives as potent inhibitors of influenza H1N1 virus neuraminidase. *Biomed. Rep.* **2019**, *10*, 33–38. [CrossRef] [PubMed]
23. Song, J.H.; Kwon, B.E.; Jang, H.; Kang, H.; Cho, S.; Park, K.; Ko, H.J.; Kim, H. Antiviral activity of chrysin derivatives against Coxsackie virus B3 invitro and in vivo. *Biomol. Ther.* **2015**, *23*, 465–470. [CrossRef]
24. Kwon, M.J.; Shin, H.M.; Perumalsamy, H.; Wang, X.; Ahn, Y.J. Antiviral effects and possible mechanisms of action of constituents from Brazilian propolis and related compounds. *J. Apicult. Res.* **2020**, *59*, 413–425. [CrossRef]
25. Tsuji, P.A.; Walle, T. Cytotoxic effects of the dietary flavones chrysin and apigenin in a normal trout liver cell line. *Chem Biol. Interact.* **2008**, *171*, 37–44. [CrossRef]
26. Harris, G.K.; Qian, Y.; Leonard, S.S.; Sbarra, D.C.; Shi, X. Luteolin and chrysin differentially inhibit cyclooxygenase-2 expression and scavenge reactive oxygen species but similarly inhibit prostaglandin-E_2 formation in raw 264.7 cells. *J. Nutr.* **2006**, *136*, 1517–1521. [CrossRef] [PubMed]
27. Rana, A.C.; Gulliya, B. Chemistry and pharmacology of flavonoids—A review. *Indian J. Pharm. Educ. Res.* **2019**, *53*, 8–20. [CrossRef]
28. Boer, J.L.; Mulrooney, S.B.; Hausinger, R.P. Nickel-dependent metalloenzymes. *Arch. Biochem. Biophys.* **2014**, *544*, 142–152. [CrossRef]
29. Ragsdale, S.W. Nickel-based enzyme systems. *J. Biol. Chem.* **2009**, *284*, 18571–18575. [CrossRef]
30. Zamble, D. Introduction to the biological chemistry of nickel. In *The Biological Chemistry of Nickel*; Zamble, D., Rowińska-Żyrek, M., Kozlowski, H., Eds.; Royal Society of Chemistry: Cambridge, UK, 2017; pp. 1–11.
31. Perontsis, S.; Tialiou, A.; Hatzidimitriou, A.G.; Papadopoulos, A.N.; Psomas, G. Nickel(II)-indomethacin mixed-ligand complexes: Synthesis, characterization, antioxidant activity and interaction with DNA and albumins. *Polyhedron* **2017**, *138*, 258–269. [CrossRef]
32. Tserkezidou, C.; Hatzidimitriou, A.G.; Psomas, G. Nickel(II) complexes of flufenamic acid: Characterization, structure and interaction with DNA and albumins. *Polyhedron* **2016**, *117*, 184–192. [CrossRef]
33. Totta, X.; Papadopoulou, A.A.; Hatzidimitriou, A.G.; Papadopoulos, A.; Psomas, G. Synthesis, structure and biological activity of nickel(II) complexes with mefenamato and nitrogen-donor ligands. *J. Inorg. Biochem.* **2015**, *145*, 79–93. [CrossRef]
34. Gupta, S.K.; Anjana, C.; Sen, N.; Butcher, R.J.; Jasinski, J.P.; Golen, J.A. An unusual hydroxy-substituted mononuclear nickel(II) complex with a tetradentate Schiff base: Synthesis, spectroscopy, electrochemistry, crystallography, DNA binding, and theoretical investigation. *Polyhedron* **2015**, *89*, 219–231. [CrossRef]
35. Kyropoulou, M.; Raptopoulou, C.P.; Psycharis, V.; Psomas, G. Ni(II) complexes withnon-steroidal anti-inflammatory drugdiclofenac: Structure and interaction with DNA and albumins. *Polyhedron* **2013**, *61*, 126–136. [CrossRef]
36. Okamoto, S.; Eltis, L.D. The biological occurrence and trafficking of cobalt. *Metallomics* **2011**, *3*, 963–970. [CrossRef]
37. Tsiliou, S.; Kefala, L.A.; Hatzidimitriou, A.G.; Kessissoglou, D.P.; Perdih, F.; Papadopoulos, A.N.; Turel, I.; Psomas, G. Cobalt(II) complexes with non-steroidal anti-inflammatory drugs and α-diimines. *J. Inorg. Biochem.* **2016**, *160*, 125–139. [CrossRef]
38. Dimiza, F.; Papadopoulos, A.N.; Tangoulis, V.; Psycharis, V.; Raptopoulou, C.P.; Kessissoglou, D.P.; Psomas, G. Biological evaluation of cobalt(II) complexes with non-steroidal anti-inflammatory drug naproxen. *J. Inorg. Biochem.* **2012**, *107*, 54–64. [CrossRef] [PubMed]
39. Thamilarasan, V.; Sengottuvelan, N.; Sudha, A.; Srinivasan, P.; Chakkaravarthi, G. Cobalt(III) complexes as potential anticancer agents: Physicochemical, structural, cytotoxic activity and DNA/protein interactions. *J. Photochem. Photobiol. B Biol.* **2016**, *162*, 558–569. [CrossRef]
40. Ambika, S.; Manojkumar, Y.; Arunachalam, S.; Gowdhami, B.; Kumar, K.; Sundaram, M.; Solomon, R.V.; Venuvanalingam, P.; Akbarsha, M.A.; Sundararaman, M. Biomolecular interaction, anti-cancer and anti-angiogenic properties of cobalt(III) schiffbase complexes. *Sci. Rep.* **2019**, *9*, 2721. [CrossRef]
41. Tsiliou, S.; Kefala, L.A.; Perdih, F.; Turel, I.; Kessissoglou, D.P.; Psomas, G. Cobalt(II) complexes with non-steroidal anti-inflammatory drug tolfenamic acid: Structure and biological evaluation. *Eur. J. Med. Chem.* **2012**, *48*, 132–142. [CrossRef]
42. Pages, B.J.; Ang, D.L.; Wright, E.P.; Aldrich-Wright, J.R. Metal complex interactions with DNA. *Dalton Trans.* **2015**, *44*, 3505–3526. [CrossRef] [PubMed]
43. Bruker Analytical X-Ray Systems, Inc. *Apex2, Version 2 User Manual*; M86-E01078; Bruker Analytical X-Ray Systems, Inc.: Madison, WI, USA, 2006.

44. Siemens Industrial Automation, Inc. *SADABS: Area-Detector Absorption Correction*; Siemens Industrial Automation, Inc.: Madison, WI, USA, 1996.
45. Betteridge, P.W.; Carruthers, J.R.; Cooper, R.I.; Prout, K.; Watkin, D.J. CRYSTALS Version 12: Software for guided crystal structure analysis. *J. Appl. Cryst.* **2003**, *36*, 1487. [CrossRef]
46. Palatinus, L.; Chapuis, G. SUPERFLIP—A computer program for the solution of crystal structures by charge flipping in arbitrary dimensions. *J. Appl. Cryst.* **2007**, *40*, 786–790. [CrossRef]
47. Watkin, D.J.; Prout, C.K.; Pearce, L.J. *CAMERON*; Chemical Crystallography Laboratory: Oxford, UK, 1996.
48. Craig, P.R.; Brothers, P.J.; Clark, G.R.; Wilson, W.R.; Denny, W.A.; Ware, D.C. Anionic carbonato and oxalate cobalt(iii) nitrogen mustard complexes. *Dalton Trans.* **2004**, *2004*, 611–618. [CrossRef]
49. De Meulenaer, J.; Tompa, H. The absorption correction in crystal structure analysis. *Acta Cryst.* **1965**, *19*, 1014. [CrossRef]
50. Halevas, E.; Mitrakas, A.; Mavroidi, B.; Athanasiou, D.; Gkika, P.; Antoniou, K.; Samaras, G.; Lialiaris, E.; Hatzidimitriou, A.; Pantazaki, A.; et al. Structurally characterized copper-chrysin complexes display genotoxic and cytotoxic activity in human cells. *Inorg. Chim. Acta* **2021**, *515*, 120062. [CrossRef]
51. Halevas, E.; Mavroidi, B.; Antonoglou, O.; Hatzidimitriou, A.; Sagnou, M.; Pantazaki, A.A.; Litsardakis, G.; Pelecanou, M. Structurally characterized gallium–chrysin complexes with anticancer potential. *Dalton Trans.* **2020**, *49*, 2734–2746. [CrossRef]
52. Halevas, E.; Mavroidi, B.; Zahariou, G.; Pelecanou, M.; Hatzidimitriou, A.G. Structurally characterized copper complexes of flavonoid naringenin with enhanced radical scavenging activity. *Inorg. Chim. Acta* **2023**, *546*, 121325. [CrossRef]
53. Nakamoto, K. *Infrared and Raman Spectra of Inorganic and Coordination Compounds: Part B: Applications in Coordination, Organometallic, and Bioinorganic Chemistry*, 6th ed.; Wiley Inc.: New York, NY, USA, 2009; pp. 96–98.
54. Boudalis, A.K.; Nastopoulos, V.; Perlepes, S.P.; Raptopoulou, C.P.; Terzis, A. Reactions of 2,2′-bipyridine (bpy) and 1,10-phenanthroline (phen) with yttrium(III) nitrate: Preparation, X-ray crystal structures and spectroscopic characterization of the bis-bpy and bis-phen complexes. *Transit. Metal Chem.* **2001**, *26*, 276–281. [CrossRef]
55. Tan, M.; Zhu, J.; Pan, Y.; Chen, Z.; Liang, H.; Liu, H.; Wang, H. Synthesis, cytotoxic activity, and DNA binding properties of copper(II) complexes with hesperetin, naringenin, and apigenin. *Bioinorg. Chem. Appl.* **2009**, *2009*, 347872. [CrossRef]
56. Tosonian, S.; Ruiz, C.J.; Rios, A.; Frias, E.; Eichler, J.F. Synthesis, characterization, and stability of iron(III) complex ions possessing phenanthroline-based ligands. *Open J. Inorg. Chem.* **2013**, *3*, 7–13. [CrossRef] [PubMed]
57. Duval, C. *Inorganic Thermogravimetric Analysis*, 2nd ed.; Elsevier Publishing Co.: Amsterdam, The Netherlands, 1963; pp. 416–420.
58. Stoll, S.; Schweiger, A. EasySpin, a comprehensive software package for spectral simulation and analysis in EPR. *J. Magn. Reson.* **2006**, *178*, 42–55. [CrossRef] [PubMed]
59. Kedare, S.B.; Singh, R.P. Genesis and development of DPPH method of antioxidant assay. *J. Food Sci. Technol.* **2011**, *48*, 412–422. [CrossRef] [PubMed]
60. Khater, M.; Ravishankar, D.; Greco, F.; Osborn, H.M. Metal complexes of flavonoids: Their synthesis, characterization and enhanced antioxidant and anticancer activities. *Future Med. Chem.* **2019**, *11*, 2845–2867. [CrossRef]
61. Kasprzak, M.M.; Erxleben, A.; Ochocki, J. Properties and applications of flavonoid metal complexes. *RSC Adv.* **2015**, *5*, 45853–45877. [CrossRef]
62. Matsia, S.; Lazopoulos, G.; Hatzidimitriou, A.; Reimann, M.K.; Pöttgen, R.; Salifoglou, A. Chemical reactivity profile of rare earth metal ions with flavonoids. From structural speciation to magneto-optical properties. *Polyhedron* **2023**, *230*, 116231. [CrossRef]
63. Matsia, S.; Tsave, O.; Hatzidimitriou, A.; Salifoglou, A. Chromium flavonoid complexation in an antioxidant capacity role. *Int. J. Mol. Sci.* **2022**, *23*, 7171. [CrossRef]
64. Payehghadr, M.; Hashemi, S.E. Solvent effect on complexation reactions. *J. Incl. Phenom. Macrocycl. Chem.* **2017**, *89*, 253–271. [CrossRef]
65. Liu, J.; Lorraine, S.C.; Dolinar, B.S.; Hoover, J.M. Aerobic oxidation reactivity of well-defined cobalt(II) and cobalt(III) aminophenol complexes. *Inorg. Chem.* **2022**, *61*, 6008–6016. [CrossRef]
66. Bencini, A.; Lippolis, V. 1,10-Phenanthroline: A versatile building block for the construction of ligands for various purposes. *Coord. Chem. Rev.* **2010**, *254*, 2096–2180. [CrossRef]
67. Uivarosi, V.; Munteanu, A.C. Flavonoid complexes as promising anticancer metallodrugs. In *Flavonoids—From Biosynthesis to Human Health*; Justino, G.C., Ed.; IntechOpen: London, UK, 2017; pp. 305–334.
68. Totta, X.; Hatzidimitriou, A.G.; Papadopoulos, A.N.; Psomas, G. Nickel(ii)–naproxen mixed-ligand complexes: Synthesis, structure, antioxidant activity and interaction with albumins and calf-thymus DNA. *New J. Chem.* **2017**, *41*, 4478–4492. [CrossRef]
69. Festus, C.; Okafor, S.N.; Ekennia, A.C. Heteroleptic metal complexes of a pyrimidinyl based schiff base ligand incorporating 2,2′-bipyridine moiety: Synthesis, characterization, and biological studies. *Front. Chem.* **2019**, *7*, 862. [CrossRef] [PubMed]
70. Halevas, E.; Mavroidi, B.; Kaplanis, M.; Hatzidimitriou, A.G.; Moschona, A.; Litsardakis, G.; Pelecanou, M. Hydrophilic bis-MPA hyperbranched dendritic scaffolds as nanocarriers of a fully characterized flavonoid morin-Zn(II) complex for anticancer applications. *J. Inorg. Biochem.* **2022**, *232*, 111832. [CrossRef]

71. Halevas, E.; Mavroidi, B.; Pelecanou, M.; Hatzidimitriou, A.G. Structurally characterized zinc complexes of flavonoids chrysin and quercetin with antioxidant potential. *Inorg. Chim. Acta* **2021**, *523*, 120407. [CrossRef]
72. Halevas, E.; Matsia, S.; Hatzidimitriou, A.; Geromichalou, E.; Papadopoulos, T.A.; Katsipis, G.; Pantazaki, A.; Litsardakis, G.; Salifoglou, A. A unique ternary Ce(III)-quercetin-phenanthroline assembly with antioxidant and anti-inflammatory properties. *J. Inorg. Biochem.* **2022**, *235*, 111947. [CrossRef]
73. Halevas, E.; Pekou, A.; Papi, R.; Mavroidi, B.; Hatzidimitriou, A.G.; Zahariou, G.; Litsardakis, G.; Sagnou, M.; Pelecanou, M.; Pantazaki, A.A. Synthesis, physicochemical characterization and biological properties of two novel Cu(II) complexes based on natural products curcumin and quercetin. *J. Inorg. Biochem.* **2020**, *208*, 111083. [CrossRef]

inorganics

MDPI

Article

Structural, Antioxidant, and Protein/DNA-Binding Properties of Sulfate-Coordinated Ni(II) Complex with Pyridoxal-Semicarbazone (PLSC) Ligand

Violeta Jevtovic [1], Luka Golubović [2], Odeh A. O. Alshammari [1], Munirah Sulaiman Alhar [1], Tahani Y. A. Alanazi [1], Aleksandra Radulović [3], Đura Nakarada [2], Jasmina Dimitrić Marković [2], Aleksandra Rakić [2] and Dušan Dimić [2,*]

1 Department of Chemistry, College of Science, University Ha'il, Ha'il 81451, Saudi Arabia
2 Faculty of Physical Chemistry, University of Belgrade, Studentski trg 12-16, 11000 Belgrade, Serbia
3 Department of Prosthodontics, School of Medicine, 78000 Banja Luka, Bosnia and Herzegovina
* Correspondence: ddimic@ffh.bg.ac.rs

Abstract: The pyridoxal-semicarbazone (PLSC) ligand and its transition metal complexes have shown significant biological activity. In this contribution, a novel nickel(II)-PLSC complex, $[Ni(PLSC)(SO_4)(H_2O)_2]$, was obtained, and its structure was determined by X-ray crystallographic analysis, FTIR, and UV-VIS spectroscopy. The sulfate ion is directly coordinated to the central metal ion. The intermolecular stabilization interactions were examined using Hirshfeld surface analysis. The crystal structure was optimized by a B3LYP functional using two pseudopotentials for nickel(II) (LanL2DZ and def2-TZVP) together with a 6-311++G(d,p) basis set for non-metallic atoms. The experimental and theoretical bond lengths and angles were compared, and the appropriate level of theory was determined. The stabilization interactions within the coordination sphere were investigated by the Quantum Theory of Atoms in Molecules (QTAIM). The antioxidant activity towards hydroxyl and ascorbyl radicals was measured by EPR spectroscopy. The interactions between Human Serum Albumin (HSA) and the complex were examined by spectrofluorimetric titration and a molecular docking study. The mechanism of binding to DNA was analyzed by complex fluorescence quenching, potassium iodide quenching, and ethidium bromide displacement studies in conjunction with molecular docking simulations.

Keywords: Ni(II) complexes; pyridoxal-semicarbazone; DFT; molecular docking; EPR; HSA; CT-DNA

1. Introduction

Pyridoxal-semicarbazone (PLSC) is a tridentate ligand formed in the reaction between pyridoxal and semicarbazone [1,2]. The donor atoms of PLSC include phenolic oxygen, hydrazone nitrogen, and carbonyl oxygen. Because pyridoxal moiety is a part of the vitamin B6 structure, this ligand resembles many active biomolecules [3]. Due to the presence of several protonation sites, PLSC can exist in three forms: neutral (H_2L), monoanionic (HL^-), and dianionic (L^{2-}) (Figure 1). When in neutral form, pyridine and hydrazone nitrogen atoms are protonated, while the hydroxyl group attached to the aromatic ring is deprotonated. The monoanionic form is obtained by removing hydrogen atoms from protonated hydrazone nitrogen. When both protons from the mentioned nitrogen atoms are removed, the ligand is dianionic. The presence of several donor atoms in the structure leads to the formation of many transition metal complexes. The overall geometry of complexes depends on the chosen metal and pH of the solution (when pH is above 7, the mono- and dianionic forms are predominantly present).

The first complex compounds incorporating the PLSC ligand were those with copper and platinum, $[Pt(PLSC\text{-}H)Cl_3]$, and $[CuBr_2(PLSC)]$ [4]. Vojinović-Ješić and coworkers described several chromium(III)-PLSC complexes with *mer*-octahedral structure [5].

To the present date, only several complexes between nickel(II) and PLSC ligands have been described in the literature. Complex compounds containing PLSC and nickel(II) described in the literature include $Ni(PLSC)Cl_2{\cdot}3.5H_2O$, $[Ni(PLSC)(H_2O)_3](NO_3)_2$, $Ni(PLSC\text{-}2H)(NCS)_2{\cdot}4H_2O$, and $[Ni(PLSC\text{-}2H)NH_3]{\cdot}1.5H_2O$ [6]. The first three compounds were paramagnetic with an octahedral structure, while the fourth was diamagnetic with a square planar structure. Two complexes were formed between nickel(II) nitrate and PLSC in the presence of NaN_3: binuclear octahedral complex $[Ni_2(PLSC)_2(\mu_{1,1}\text{-}N_3)_2]{\cdot}2H_2O$ and octahedral $[Ni(PLSC\text{-}H)_2]{\cdot}2H_2O$ [7]. An unusual complexation mode of PLSC was examined in the paper by Leovac and coworkers in which an elementary cell consisted of two monomeric complex cations $[Ni(PLSC)(H_2O)_3]^{2+}$, one centrosymmetric dimeric cation $[Ni_2(PLSC)_2(H_2O)_4]^{4+}$, eight Cl^- anions, and four molecules of crystal water [8]. In the paper by Jevtovic and coworkers, the crystal structure and antibacterial activity of $[Ni(PLSC\text{-}H)_2]{\cdot}H_2O$ are presented [9]. Aduri and coworkers published an article describing the synthesis, characterization, DNA binding, and antibacterial studies of $[Ni(PLSC\text{-}H)Cl(H_2O)_2]$ and $[Cu(PLSC\text{-}H)_2]{\cdot}2H_2O$ [10].

Nickel is an important transition metal, and its complex compounds can be used for S-C bond formation [11], hydrogen production [12,13], and control of the polyethylene microstructure [14]. The effects of nickel(II) compounds on SARS-CoV-2 enzymes have been examined theoretically [15] and experimentally [16]. Bhandarkar and coworkers reported moderate inhibitory effects of nickel(II) complexes towards the COVID-19 and hepatitis viruses [17]. Nickel complexes also show anticancer potential towards human prostate cancer cells [18], primary lung cancer cells [19], mammary gland breast carcinoma, humane prostate carcinoma, colorectal carcinoma [20], and chronic myeloid leukemia [21]. Perontsis and coworkers presented results on the significant radical-scavenging activity of nickel(II) complexes with the anti-inflammatory drug diflunisal towards diphenyl-picrylhydrazyl ($DPPH^{\bullet}$), 2,2′-azinobis(3-ethylbenzothiazoline-6-sulfonic acid) and hydroxyl radicals [22]. The radical-scavenging activity of nickel(II)-rutin complex towards $DPPH^{\bullet}$ was higher than the respective activity of free rutin in reference [23]. The antibacterial activity of the nickel(II) complex with N and the O donor Schiff base towards Gram-positive and Gram-negative bacteria is explained by Overtone's concept and Tweedy's chelation theory [24].

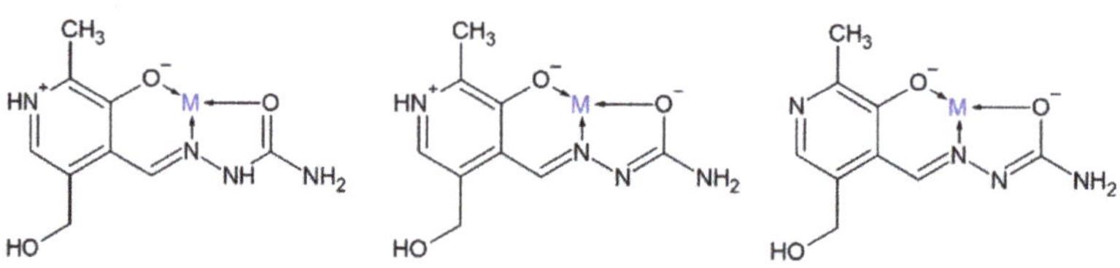

Figure 1. Different binding modes of PLSC ligand.

This paper presents results on the synthesis and structural analysis of a novel nickel(II)-PLSC complex with a coordinated sulfate anion and two molecules of water, $[Ni(PLSC)(SO_4)(H_2O)_2]$. The crystallographic structure was obtained by X-ray analysis. The most important stabilization interactions within the experimental structure were determined by Hirshfeld surface analysis. The structure optimization was performed by quantum chemical methods, and donor interactions within the coordination sphere were examined by the Quantum Theory of Atoms in Molecules (QTAIM). The antioxidant activity towards hydroxyl ($HO^{\bullet}$) and ascorbyl ($^{\bullet}Asc$) radicals was determined by EPR spectroscopy. The binding mechanisms towards the transport protein (HSA) and DNA were investigated by spectrofluorometric titration and molecular docking simulations.

2. Results and Discussion

2.1. Structural Analysis

Complex $[Ni(PLSC)(SO_4)(H_2O)_2]$ was obtained from the mixture of nickel(II) sulfate and the PLSC ligand in water. The crystal structure of the Ni(II)-PLSC included the PLSC ligand in neutral form, as verified by the presence of hydrogen atoms on pyridine and hydrazone nitrogen atoms (Figure 2a). This ligand contained phenolic hydroxyl oxygen, hydrazone nitrogen, and phenol oxygen as coordination sites. The experimental bond lengths and angles are presented in Tables S1 and S2. The geometry of the complex is a distorted octahedron with two water molecules and a sulfate group in coordination (Figure 2a). The position of the ligands around the central metal ion is influenced by the rigidity of the PLSC ligand and the relative flexibility of water molecules, along with the hydrogen bond between the water molecule and sulfate anion. The Ni-O_{water} bond lengths are 2.0411 and 2.0751 Å, while the distance between the sulfate oxygen atom and nickel(II) is 2.0555 Å. The shortest distance is found between the nitrogen atom and the central metal ion (2.0230 Å, Table S1).

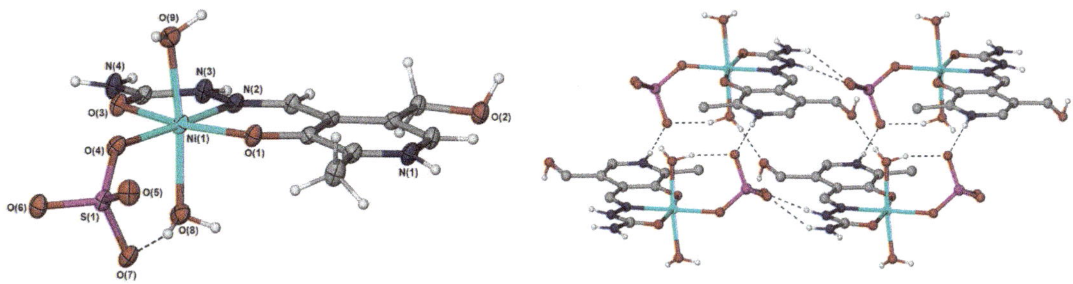

(a) (b)

Figure 2. (**a**) Molecular diagram of [Ni(PLSC)(SO_4)(H_2O)$_2$], with non-hydrogen atoms represented by 50% displacement ellipsoids and hydrogen atoms as spheres of arbitrary size. (**b**) The ball and stick representation shows part of the hydrogen bonding between the molecules. (Hydrogen-white, carbon-gray, nitrogen-blue, oxygen-red, sulfur-lilac, nickel-light blue).

This compound crystallizes in the monoclinic space group P2(1)/n. Additional details on the crystal structure are presented in the Methodology Section. Several hydrogen bonds are formed within unit cells. These interactions are listed in Table 1 and presented in Figures 2b and 3. Two intermolecular hydrogen bonds are formed between protonated nitrogen atoms in the aliphatic chain of the first unit and the sulfate anion of another (2.819(2) and 2.987(2) Å, Table 1). Due to steric hindrance, these bonds deviate from linearity (169 and 145°). Water molecules interact with the hydroxymethyl group with a bond distance of 2.662(2) Å and an angle of 165°. These interactions prove that co-crystalized solvent molecules are important for stabilizing crystal structure. Protonated hydrazine nitrogen forms hydrogen bonds with the sulfate anion (2.791 Å). Additionally, intramolecular interaction between the water molecule and sulfate anion stabilizes each unit. This interaction is characterized by the angle of 164°, which is limited by the overall geometry of the complex. The chains are propagating along the crystallographic *b* axes (Figure 3). Non-classical interactions in crystal structure are examined by the Hirshfeld surface analysis in the following section.

Table 1. Hydrogen bonds for [Ni(PLSC)(SO_4)(H_2O)$_2$ [Å and °].

D-H...A	d(D-H)	d(H...A)	d(D...A)	<(DHA)
O(2)-H(2)...O(6)#1	0.893(18)	1.772(18)	2.662(2)	174(3)
N(1)-H(1N)...O(7)#2	0.899(18)	1.912(19)	2.791(2)	165(3)
N(3)-H(3N)...O(5)#3	0.870(18)	1.961(19)	2.819(2)	169(3)
N(4)-H(4N)...O(6)#3	0.895(18)	2.21(2)	2.987(2)	145(2)
N(4)-H(5N)...O(4)#4	0.879(18)	2.24(2)	3.062(2)	155(3)

Table 1. *Cont.*

D-H...A	d(D-H)	d(H...A)	d(D...A)	<(DHA)
O(8)-H(8D)...O(7)	0.863(18)	1.85(2)	2.695(2)	164(3)
O(8)-H(8E)...O(2)#5	0.855(18)	1.925(19)	2.771(2)	170(3)
O(9)-H(9D)...O(5)#6	0.858(18)	1.84(2)	2.671(2)	162(4)
O(9)-H(9E)...O(1)#6	0.851(19)	2.00(2)	2.813(2)	160(4)

Symmetry transformations used to generate equivalent atoms: #1 x − 1/2, −y + 3/2, z − 1/2; #2 −x, −y + 1, −z + 1; #3 x, y + 1, z; #4 −x + 3/2, y + 1/2, −z + 3/2; #5 −x, −y + 2, −z + 1; #6 −x + 1, −y + 1, −z + 1.

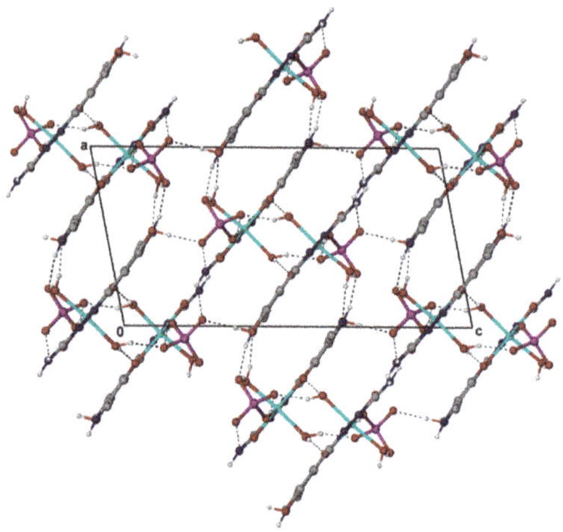

Figure 3. Cell packing is viewed down the b-axis, and the 3D hydrogen-bonded network is shown as comprising parallel layers of the Ni(PLSC) structural units.

2.2. Characterization

The FTIR spectra of the PLSC ligand and complex are presented in Figures S1 and S2. The FTIR spectrum of the PLSC ligand contains a sharp peak at 3461 cm^{-1} assigned to the O-H stretching vibration. This peak is absent from the spectrum of the ligand due to deprotonation. An intense band between 2800 and 3500 cm^{-1} contains vibrations of N-H and O-H in the ligand structure, and the formation of hydrogen bonds induces additional widening of the group [25,26]. The same band was found in the FTIR spectrum of the complex. The bands between 2950 and 2800 cm^{-1} are also assigned to the stretching vibration of the $^{+}$N-H group of the pyridine ring [1]. In the complex, the PLSC ligand is present in zwitterion form, as the migration of hydrogen atoms of the OH group to pyridine nitrogen occurred [27]. Another notable band in the FTIR spectrum of the ligand at 1680 cm^{-1} is assigned to C=O vibration, which coincides with the results presented in [1]. This band is absent from the FTIR spectrum of the complex. This supports the assumption that coordination occurred from this group to nickel(II) ion in imidazole form. The C=N vibration is located at 1570 cm^{-1} in the spectrum of the ligand. This band is shifted towards higher wavenumbers (1630 cm^{-1}) as this group is included in coordination with the central metal ion through the azomethine nitrogen atom [1,27]. The phenolic C-O group vibration is positioned at 1151 cm^{-1} in the ligand spectrum, as suggested in [9]. The position of this group is shifted to 1130 cm^{-1} due to the coordination of central metal ions. A weak band at 620 cm^{-1} is attributed to the sulfate ion vibration [9].

The UV-VIS spectrum of the complex was prepared in water, and it is depicted in Figure S3. The experimental spectrum contains two wide bands centered around 300 and 400 nm. Their structure is complex, suggesting that they are composed of two close position bands, which coincide well with the previously described spectrum of $Ni(PLSC)Cl_2 \cdot 3.5H_2O$ [6]. As explained in the previous reference, the band at 300 nm includes two types of transitions, the n→π* imine and n→π of both the imine and pyridine ring. These transitions can be seen in the spectrum of the PLSC ligand [28]. The other band at 400 nm is assigned to the LMCT transition between the central metal ion and azomethine nitrogen. Low-intensity d→d bands are usually located in this region, although the CT transitions probably mask them. The authors have also discussed that the octahedral geometry of the complex could be suggested from the position of this band [7], which, in the present case, was confirmed by X-ray crystallography.

The obtained value of the molar conductivity for the complex solution was 111 S cm^2, which falls within the typical range for 1:1 electrolytes (100–140 S cm^2) [29]. This could be explained by the partial replacement of the sulfate group by the solvent molecules, as discussed for similar complexes in [6]. The obtained value is much lower than that of PLSC complexes with Ni(II) and chloride/nitrate ions [6]. This exchange has significant implications, as the sulfate ion can be removed from the structure in the solution, changing the possible interactions with biomolecules, as presented in the following sections.

2.3. Hirshfeld Surface Analysis

Stabilization interactions are a crucial part of crystal structures, influencing the spatial distribution of units. In this contribution, Hirshfeld surface analysis was used for their identification and quantification. The fingerprint plots of the most important contacts are given in Figure S4.

The most important stabilization interactions were formed between oxygen and hydrogen atoms, denoted as O···H, which amounts to 52.9% of all contacts. The presence of a sulfate ion greatly increases the amount of these interactions within crystal structures compared to similar compounds [30,31]. The sulfate ion oxygen atoms act as hydrogen atom acceptors. On the other side, oxygen atoms of water molecules and the hydroxymethyl group are hydrogen atom donors in these interactions. Another important type of contact can be marked as H···H (29.7%). The abundance of hydrogen atoms in the structure of ligands and water molecules is responsible for compound stability through these interactions. Protonated pyridine, hydrazone, and amino nitrogen atoms are part of N···H contacts (4.5%). The interactions between carbon and hydrogen atoms amount to 11.8%. The PLSC core includes the aromatic ring and aliphatic chain, and the abundance of hydrogen atoms contributes to these interactions. Additionally, interactions between positively charged hydrogen atoms attached to electronegative atoms and negatively charged π aromatic cloud contribute to the mentioned contacts [32].

The interactions between electronegative atoms are characterized by much lower percentages, namely 0.6 (O···O) and 1.1% (O···N). Interaction, including carbon atoms and oxygen/nitrogen atoms, has a similar contribution to the total number of contacts, around 1.1%. It is important to emphasize that the central metal ion is not included in the contacts due to the octahedral geometry and bulkiness of the ligands. The same applies to the sulfur atom surrounded by four oxygen atoms.

2.4. Theoretical Structural Analysis

The crystallographic structure of the complex was further used for structural optimization by DFT methods and the B3LYP functional. For nonmetallic atoms, a 6-311++G(d,p) basis set was employed, while for the nickel(II) ion, two basis sets were checked, namely LanL2DZ and def2-TZVP, to determine the appropriate level of theory. These functional/basis set combinations were previously used for other nickel(II) complex compounds [33–35]. The comparison between experimental and theoretical structures included the calculation of the correlation coefficient (R) and the mean absolute error (MAE). The latter parameter determines the average value of the absolute difference between two data sets. The experimental and theoretical bond lengths and angles and R and MAE parameters are presented in Tables S1 and S2. The optimized structure at the B3LYP/6-311++G(d,p)(H,C,N,O,S)/LanL2DZ(Ni) level of theory is shown in Figure 4.

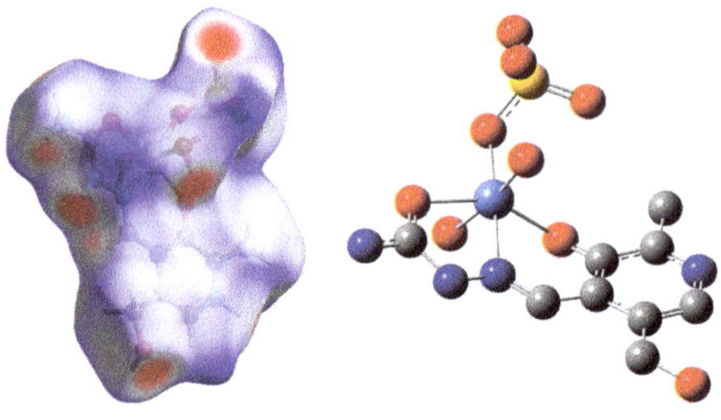

(**a**) (**b**)

Figure 4. (**a**) Hirshfeld surface and (**b**) optimized structure (hydrogen atoms are omitted for clarity) at the B3LYP/6-311++G(d,p)(H,C,N,O,S)/LanL2DZ(Ni) level of theory of [Ni(PLSC)(SO_4)(H_2O)$_2$]. (Hydrogen—white, carbon—gray, nitrogen—blue, oxygen—red, sulfur—yellow, nickel—light blue).

The correlation coefficients for the bond lengths are 0.988 (LanL2DZ) and 0.984 (def2-TZVP), with MAE values of 0.036 and 0.039 Å, respectively. These differences are comparable to the experimental error, showing that both levels of theory are applicable for examining the system. During the optimization, specific differences in bond lengths were observed. Where bonds including a central metal ion are concerned, the MAE values are 0.086 (LanL2DZ) and 0.098 Å (def2-TZVP). It is important to mention that an additional hydrogen bond was found between the sulfate ion and water molecule, leading to a slight change in Ni-O bond lengths with the water molecule and anion. Also, the differences in values of sulfate ion S-O bonds due to the interaction with water molecules were much more pronounced in the optimized structure, which disrupted the local symmetry of this anion.

Bond angles are much more prone to change during optimization, mainly if new interactions are formed between ligands. The correlation coefficients are still very high, 0.97 in both cases, although the MAE values are 3.45 (LanL2DZ) and 3.33° (def2-TZVP). The difference between the experimental and optimized structures in angle O4-Ni-O3 is around 10° for both theoretical structures, which is a direct consequence of the slight incline of the sulfate ion towards the water molecule. The angle that includes the water molecule (O1-Ni-O8) is also larger by 11° (94.03 vs. 105.41/104.88°). The strength of the formed hydrogen bond is responsible for the elongation of the O8-H bond to 1.515 Å, which is 0.6 Å more extended than the other O-H bond. These interactions are examined in more detail in the QTAIM analysis section. The octahedral geometry around the central metal ion is distorted, leading to the lower O1-Ni1-O3 value in the optimized structure. This value is 167.5° in the experimental and 150.8/151.7° in the theoretical structure. These changes result from the optimization of the isolated complexes in a vacuum, while other intramolecular interactions influence the stability of the crystal structure, as explained in the previous section. The angles within the PLSC ligand are not influenced by the optimization procedure, as this ligand contains extended delocalization and rigidity that prevents any conformational change. Therefore, the selected level of theory for the optimization and further structural analysis of the obtained complex was B3LYP/6-311++G(d,p)(H,C,N,O,S)/LanL2DZ(Ni), as the relative differences between the correlation coefficients/MAE values between two structures were in favor of this effective core potential. Similar results were found for other Ni(II) and Cu(II) complexes with PLSC ligands [9].

The stabilization interactions within the structure of [Ni(PLSC)(SO_4)(H_2O)$_2$] were examined by the QTAIM approach. Particular emphasis was put on the interactions between donor atoms and the central metal ion, which were quantified through parameters of Bond Critical Points (BCPs), such as the electron density (ρ(r)), Laplacian ($\nabla^2\rho$(r)), Lagrangian kinetic electron density (*G*(r)), potential electron density (*V*(r)), density of total electron energy ($H(r) = G(r) + V(r)$), and interatomic bond energy ($E_{bond} = V(r)/2$) [36]. Bader and Essen proposed a classification of interactions based on the values of electron density and the Laplacian. Shared (covalent) interactions have an electron density higher than 0.1 a.u., while closed-shell interactions (hydrogen bonds, ionic bonds, and van der Waals interactions) have a much lower electron density (usually around 0.01 a.u.) [37]. The mentioned parameters are shown in Table S3, along with the positions of BCPs in the structure of the obtained complex in Figure S5.

The Laplacian values of the bonds between donor atoms and the central metal ion are positive, between 0.150 and 0.320 a.u. The kinetic and potential energy values are included in Bianchi and coworkers' more detailed classification of interactions [38]. The shared shell region of covalent bonds has a –G(r)/V(r) ratio lower than 1; the intermediate region of dative and ionic bonds with a weak covalent degree has a ratio between 1 and 2, and closed-shell interactions (ionic bonds and van der Walls interactions) are characterized by a ratio higher than 2. The highest electron density values were obtained for the interactions between the central metal ion and sulfur oxygen and water oxygen atoms (0.068 and 0.075 a.u., Table S3). These interactions have –G(r)/V(r) values equal to 1, classifying them as ionic bonds with the highest interatomic bond energy of −142.3 and −156.3 kJ mol^{-1}. The second water molecule interacts with nickel(II) through a bond with a weak covalent character (–G(r)/V(r) = 1.1) and interatomic bond energy of −66.6 kJ mol^{-1}. Among the three donor atoms of PLSC, the highest electron density was found for interactions with a phenyl oxygen atom (0.54 a.u.). These interatomic bond energies were −92.2 (N2), −104.1 (O1), and −74.1 kJ mol^{-1} (O3). An aromatic ring and other electronegative atoms influence the interaction energy. These values follow results obtained for various nickel(II) complexes with similar ligands [39].

It is important to observe that electron densities and interatomic bond energies are significantly different for two molecules of water that interact with a central metal ion. The reason is additional interactions between ligand molecules. As presented in Table S4, there is a weak interaction between one of the oxygen atoms of the sulfate anion and the methyl group of PLSC, with an electron density of 0.017 a.u. and interatomic bond energy of −16.5 kJ mol^{-1}. This is a classic example of a weak hydrogen bond. Another hydrogen bond is found between the sulfate anion and the hydrogen atom attached to the O9 atom. This is a stronger bond with an interaction energy of −26.2 kJ mol^{-1}. A weak covalent character is proven by a –G(r)/V(r) equal to 1.1. In the crystallographic structure, one of the hydrogen bonds was particularly important for stabilization (O7···H-O8). This interaction was present in the optimized structure, and it has a high value of electron density (0.280 a.u.) and interatomic bond energy of −656.2 kJ mol^{-1} proving the formation of a covalent bond (–G(r)/V(r) = 0.2). The negative H(r) value of −1312.4 kJ mol^{-1} is additional proof of the covalent character. This result is expected due to the high electronegativity of the sulfate anion. As previously discussed, the deprotonation of water molecules leads to a much stronger interaction between O8 and nickel(II) ions. It should be emphasized that interactions between the central metal ion and water molecules are probably influenced by the optimization in a vacuum and that significant changes in strength and bond length can be expected in the solution.

2.5. Antioxidant Activity Determination

The antioxidant activity of the complex towards two biologically relevant radicals, $^{\bullet}$Asc and HO$^{\bullet}$, was examined by EPR spectroscopy. The first radical is a long-living species that can be directly monitored by EPR spectroscopy. On the other hand, for HO$^{\bullet}$, a spin trap was used. The hydroxyl radical was generated in the Fenton system, as described in the methodology section. Figure 5 presents the DPMPO-HO$^{\bullet}$ spectra without and with different amounts of complex. The low-field signals were used to determine antioxidant activity (Equation (1)), and the results are expressed as the amount of radical scavenged by the complex. It is important to outline that additional peaks were present in the spectra upon adding a complex, probably due to the use of DMSO as a solvent. The amount of reduced $HO^{\bullet}$ was 82.5, 87.0, 88.8, and 90.4% for the final concentration of the complex between 0.3 and 2 μM. The reduction process was controlled by the concentration of the complex. Similar activity (86.2%) for the 0.5 μM solution of $\{RuCl(\eta^6\text{-}p\text{-cymene})\}_2(\mu\text{-Cl})(\mu\text{-1-}N,N'\text{-naphthyl})]Cl$ was determined by the same methodology [40], which was comparable to the standard antioxidant, ascorbic acid. As the obtained complex does not contain any hydrogen-donating groups, it can be postulated that radical adduct formation is a dominant reduction mechanism [41]. The $^{\bullet}$Asc species has a much simpler spectrum, as shown in Figure 5. For this analysis, a nickel(II)-PLSC complex concentration range was taken between 0.5 and 2 μM. The reduction of radical was 53.0 (0.5 μM), 69.0 (1 μM), and 86.9 (2 μM). Again, the reduction process was concentration-dependent. These results show that the obtained complex acts as a radical scavenger and could be included in protecting cells from oxidative stress or influencing processes in cancer cells that occur through free radicals. Other nickel(II) complexes have also shown considerable radical-scavenging activity towards hydroxyl radicals and other reactive oxygen species [42–44].

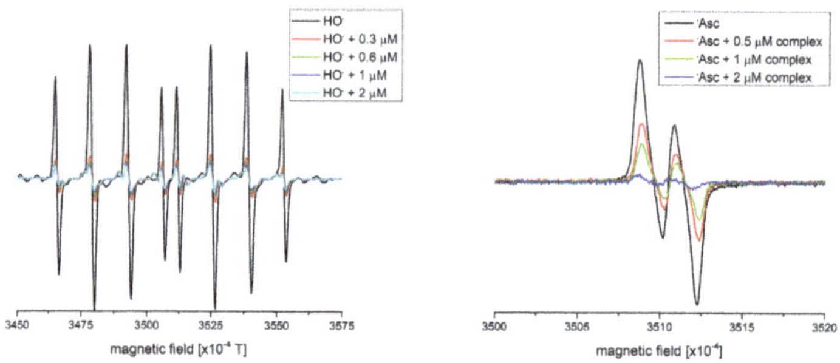

(a) (b)

Figure 5. The EPR spectra of the (**a**) DEPMPO-$HO^{\bullet}$ adduct and (**b**) ascorbyl radical in the absence (black line) and presence of different concentrations of $[Ni(PLSC)(SO_4)(H_2O)_2]$.

2.6. Experimental and Theoretical Protein-Binding Affinity

The distribution of compounds in the bloodstream usually occurs through their interactions with transport proteins such as HSA. Spectrofluorimetric titration was employed to examine the thermodynamics of the binding process. The HSA solution was irradiated by 280 nm, and the emission spectra were obtained between 300 and 500 nm, as shown in Figure 6. The fluorescence emission spectra were recorded upon adding the complex in equimolar amounts. The emission maxima value was dependent on the concentration of a quencher. The measurements were repeated at three temperatures in a range mimicking physiological conditions. The obtained fluorescence values were corrected because of the absorption of the complex at the excitation and emission wavelengths, as shown in Equation (2).

Upon the addition of the complex, the fluorescence intensity decreased. From this dependency, the Stern–Volmer constants of the process were calculated (Equation (3)). These constants and their dependency on temperature were further used to determine the fluorescence quenching mechanism. If static quenching occurs, a complex is formed between the fluorophore and quencher, and K_{SV} values decrease with temperature. On the other side, dynamic quenching includes the collision between the excited molecule and the quencher, increasing K_{SV} with temperature. The calculated KSV values are presented in Table 2. It should be noted that the intercept in these calculations was set to 1. As found, these values were in the narrow range between 1.95×10^4 (27 °C) and 1.84×10^4 M^{-1} (37 °C) (Figure S6). These values lead to the conclusion that strong interactions were formed by the complex and HSA molecules [25], and the quenching mechanism was denoted as static. Other nickel(II) complexes showed the same quenching mechanism towards HSA [45,46]. If the fluorophore lifetime was taken as 4.43 ns, as proposed by Hu and coworkers, the quenching constant is of the order of 10^{13} M^{-1} s^{-1}. This value is much higher than the collisional quenching constant of 2×10^{10} M^{-1} s^{-1} [47]. The double-log Stern–Volmer equation (Equation (4)) was used to determine the binding constant and the number of binding positions within the HSA molecule. These values are also given in Table 2.

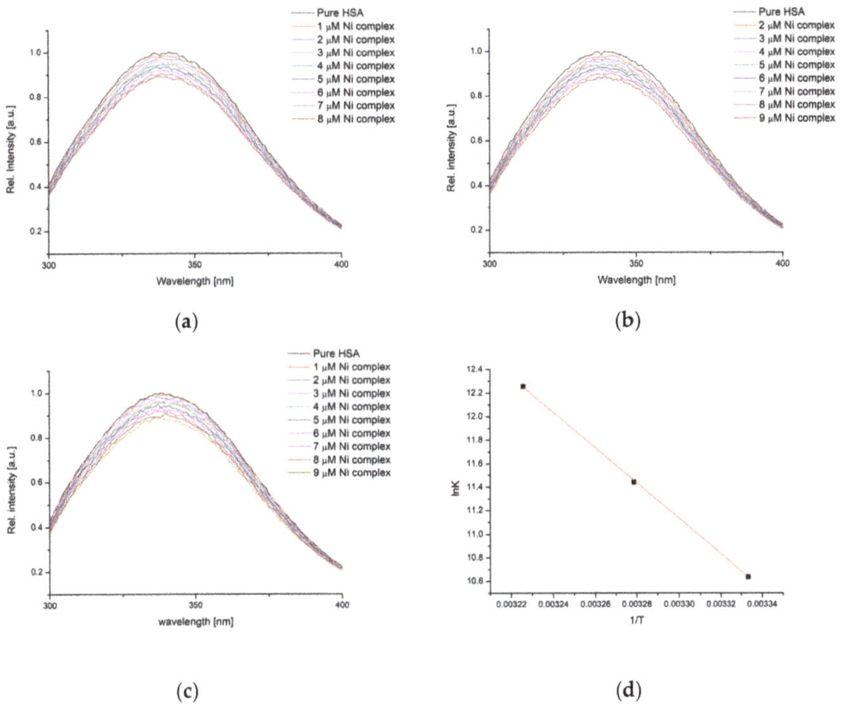

Figure 6. The fluorescence emission spectra of HSA for the titration with various concentrations of $[Ni(PLSC)(SO_4)(H_2O)_2]$ at (**a**) 27°, (**b**) 32°, and (**c**) 37 °C, and (**d**) the van 't Hoff plot for the binding process.

Table 2. The Stern–Volmer constant (K_{SV}), binding constant (K_b), Hill coefficient (n), and thermodynamic parameters of $[Ni(PLSC)(SO_4)(H_2O)_2]$ binding to HSA.

Compound	T [K]	K_{SV} [M^{-1}]	K_b [M^{-1}]	n	ΔH_b [kJ mol^{-1}]	ΔS_b [J mol^{-1} K^{-1}]	ΔG_b [kJ mol^{-1}]
1	300	1.95×10^4	3.87×10^4	1.08	170.5	655.9	−26.3
	305	1.87×10^4	1.08×10^5	1.18			−29.5
	310	1.84×10^4	3.51×10^5	1.28			−32.8

The change in the fluorescence intensity linearly decreased with the increase in the quencher concentration. The number of binding positions was between 1.08 and 1.28, which proved that one complex molecule was bound to one HSA molecule. The binding constants were 3.87×10^4 (27 °C), 1.08×10^5 (32 °C), and 3.51×10^5 (37 °C) (Figure S6). The binding constants were of the same order of magnitude as previously determined for similar nickel complexes with similar ligands [31,45]. The thermodynamic parameters of binding were calculated through Equation (4). The entropy and binding enthalpy changes were 170.5 kJ mol^{-1} and 655.9 J mol^{-1} K^{-1}. The positive values of these parameters are characteristic of the formation of non-specific interactions, mainly van der Waals and hydrogen bonds [48]. The changes in Gibbs free energy of binding were between −26.3 and

−32.8 kJ mol^{-1}, proving that the binding was spontaneous in the examined range. The spontaneity of the process is entropy-driven, as the entropic term (TΔS_b) is higher than the enthalpic. The binding mechanism was further investigated through molecular docking simulations. The negative value of the change in enthalpy (−11.026 kJ mol^{-1}) and positive value of the change in entropy (46.396 J mol^{-1} K^{-1}) were obtained for the octahedral nickel(II) complex with the N,N^0-dibenzylethane-1,2-diamine ligand in [49]. Much lower changes in Gibbs free energy of binding (between −24.9 and −25.5 kJ mol^{-1}) compared to the obtained complex result from the lower number of reactive groups in the ligand structure. A similar range in ΔG_b values (between −28.30 and −34.97 kJ mol^{-1}) was determined for the binding of the octahedral nickel(II) complex with 2-(2-nitrophenyl)imidazo [4,5-f]1,10-phenanthroline) in reference [50]. The electrostatic interactions between prepared nickel(II) complexes with nalidixic acid, 2,2′-bipyridine, and the 1,10-phenanthroline ligands and HSA were determined as the most appropriate based on the negative change in enthalpy and positive change in the entropy of binding, while the changes in Gibbs free energy were around −30 kJ mol^{-1} [51].

Several starting positions were examined for the binding of the nickel(II) complex to HSA. The best binding configuration at the FA9 binding site had a change in Gibbs free energy of −26.4 kJ mol^{-1}, comparable to the experimentally obtained value of −26.3 kJ mol^{-1}. The possible influence of hydrogen phosphate ($HPO_4{}^{2-}$) and dihydrogen phosphate ($H_2PO_4{}^{-}$) anions from the buffer solution on the binding affinity was examined in subsequent simulations. Separate investigations were undertaken to determine the interactions of each species ($HPO_4{}^{2-}$ and $H_2PO_4{}^{-}$) with the target molecule. Two anions had slightly lower binding affinities compared to the nickel(II) complex, namely −25.9 ($HPO_4{}^{2-}$) and −20.5 kJ mol^{-1} ($H_2PO_4{}^{-}$). The FA9 binding sight is the preferred location for all three ligands. The presence of the nickel(II) complex did not affect the anions' binding position or energy. However, both anions directed the complex towards the FA8 binding site. The presence of $H_2PO_4{}^{-}$ slightly decreased the binding energy of the Ni(II) complex (−25.8 kJ mol^{-1}), while $HPO_4{}^{2-}$ had virtually no effect on its binding energy (−26.5 kJ mol^{-1}), although the binding site of the complex was changed to FA8. This is a significant result, as the FA8 binding site is near the fluorescent amino acid Trp213 (Figure 7). Therefore, it can be expected that the interactions of the complex with surrounding amino acids change the chemical environment of Trp213, leading to a decrease in fluorescence emission, as experimentally observed. Also, the presence of additional charged species significantly influences the binding position.

Although bound to the FA8 binding site in the IIA subdomain, the nickel(II) complex interacts with amino acids from three subdomains and one interdomain region surrounding this binding site. These include the IB subdomain (Lys195), the IIA subdomain (Arg218, Arg222, and Asn295), the interdomain region between subdomains IIA and IIB (Val293 and Asn295), and the IIIA subdomain (Pro447 and Cys448) (Figure 8).

Figure 8 presents the most stable configuration. The obtained complex, rich in electronegative oxygen and nitrogen atoms, requires a specific environment for proper stabilization. This complex is bound near the polar (Lys195, Asn295, Pro447, and Cys 448) and positively charged amino acids (Arg218 and Arg222) (Figure 8). As a result, favorable intermolecular interactions include classic hydrogen bonds formed between the polar NH and OH groups of the complex and the polar (-NH_2 and -C=O) or charged (-$C(NH_2)^+$ and -$NH_3{}^+$) groups of nearby amino acids. Additionally, the charged group of amino acids participates in favorable electrostatic interactions with the $SO_4{}^{2-}$ group. However, due to the bulkiness of the amino acid residues, the positively charged -$C(NH_2)^{2+}$ group is located in the vicinity of the Ni^{2+} ion, leading to electrostatic repulsion or steric clashes between coordinated H_2O molecule and the -$NH_3{}^+$ group. Classic hydrogen bonds dominate over carbon–hydrogen bonds in this environment. Hydrophobic interactions are not observed, as the coordinated $SO_4{}^{2-}$ anion and two water molecules shield aromatic rings of the coordinated PLSC ligand from hydrophobic amino acid residues. Instead, the sulfate group forms two carbon–hydrogen bonds with the C-H group of Lys195 and Arg218. A third carbon–hydrogen bond is formed between the C=O group of Asn195 and the aromatic CH_2 group of the octahedral Ni(II) complex.

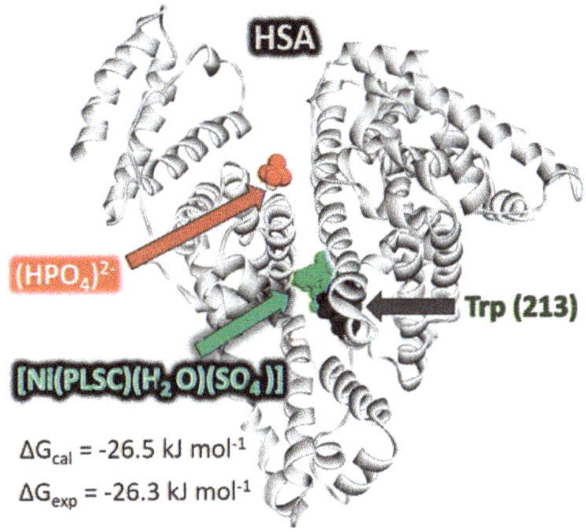

Figure 7. HSA molecule (PDB ID: 1AO6) with bound ligands: $[Ni(PLSC)(H_2O)_2(SO_4)]$ complex and HPO_4^{2-} anion, occupying FA9 and FA8 binding sites, respectively. Ligands and tryptophane are depicted using ball representation; each is colored distinctly. HPO_4^{2-} ion from buffer solution is colored by element, Ni(II) complex is shown in light green, and Trp213 is represented in dark grey.

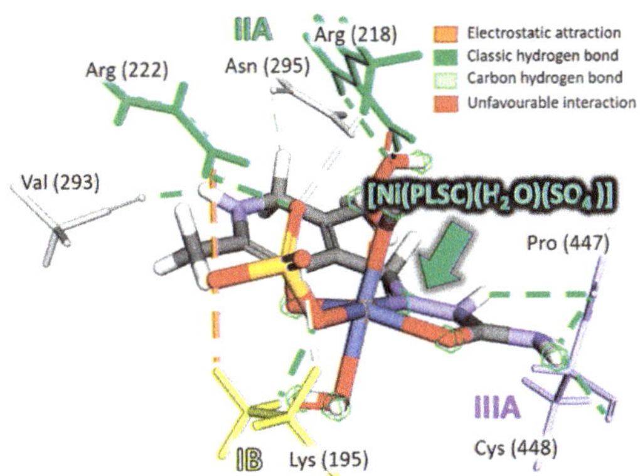

Figure 8. A 3D representation of the supramolecular interactions of $[Ni(PLSC)(H_2O)_2(SO_4)]$ located in the FA8 binding site. Only the interacting parts of the amino acids are shown, with colors corresponding to their respective regions of the HSA molecule: yellow for subdomain IB, green for subdomain IIA, interdomain region between subdomains IIA and IIB is light grey, and subdomain IA is violet. For the representation of nickel(II), complex sticks colored by the element were used. Supramolecular interactions are represented by dashed lines colored according to the type of interaction denoted in the figure's legend.

2.7. Experimental and Theoretical DNA Binding Affinity

2.7.1. Spectrofluorimetric Titration of $[Ni(PLSC)(H_2O)_2(SO_4)]$ by CT-DNA

The interactions between DNA and complexes are often examined as the first part of the biological assessment that leads to anticancer activity [52,53]. The complex showed relatively stable fluorescence upon irradiation at 280 nm. The emission spectra of the complex were recorded between 300 and 500 nm (Figure 9) upon the addition of CT-DNA, as explained in reference [54]. The intensity of fluorescence emission was low, although measurable with 10 nm slits. The measurements were repeated at three temperatures, as explained previously for HSA binding experiments. The K_{SV} values, binding constants, number of binding positions, and thermodynamic parameters of binding are presented in Table 3.

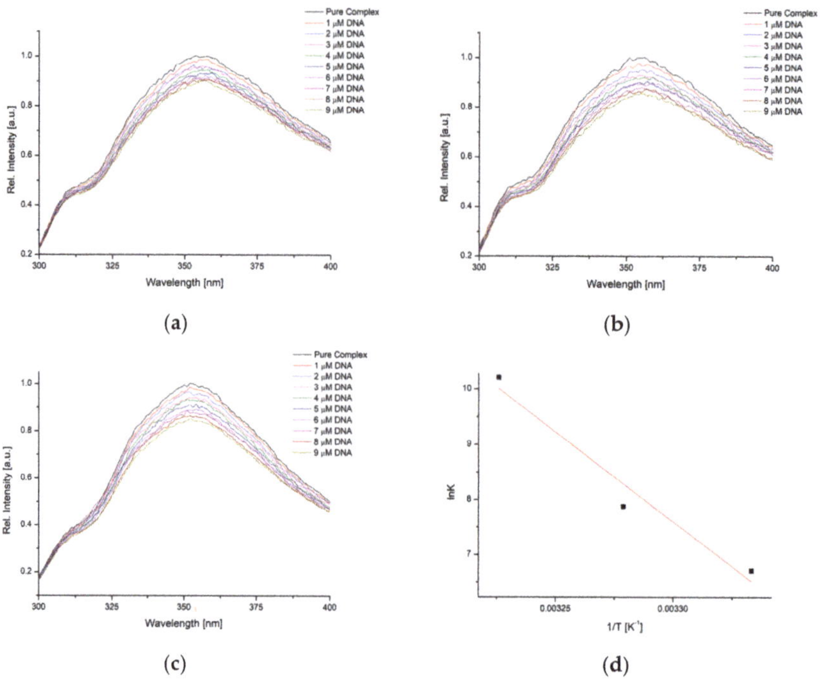

Figure 9. Fluorescence emission spectra of $[Ni(PLSC)(H_2O)_2(SO_4)]$ for the titration with various concentrations of CT-DNA at (**a**) 27°, (**b**) 32°, and (**c**) 37 °C, and (**d**) the van 't Hoff plot for the binding process.

Table 3. The Stern–Volmer constant (K_{SV}), binding constant (K_b), Hill coefficient (n), and thermodynamic parameters of $[Ni(PLSC)(SO_4)(H_2O)_2]$ binding to DNA.

Compound	T [K]	K_{SV} [M^{-1}]	K_b [M^{-1}]	n	ΔH_b [kJ mol^{-1}]	ΔS_b [J mol^{-1} K^{-1}]	ΔG_b [kJ mol^{-1}]
	300	1.71×10^4	8.23×10^2	0.77			−16.2
1	305	1.86×10^4	2.64×10^3	0.83	270.4	955.5	−21.0
	310	1.89×10^4	2.74×10^4	1.03			−25.8

The K_{SV} binding constants for the additions of CT-DNA to the solution of the complex are in a very narrow range, between 1.71 and 1.89×10^4 M^{-1}. The trend is reversed from the previous discussion, and the K_{SV} values increase with temperature, which characterizes the dynamic quenching of fluorescence [54]. The collision of particles leads to the nonradioactive relaxation of the complex. When the modified Stern–Volmer equation is applied, the binding constants are 8.23×10^2 (27 °C), 2.64×10^3 (32 °C), and 2.74×10^4 (32 °C). These changes in binding constants are additional proof that temperature significantly affects the interactions between CT-DNA and the complex. The number of binding positions is around one, showing that one complex molecule interacts with one CT-DNA. The changes in enthalpy and entropy of the process are both positive—270.4 kJ mol^{-1} and 955.5 J mol^{-1} K^{-1}, respectively. The formation of non-specific interactions, such as hydrophobic interactions, dominates the process, as explained in [54]. These values of thermodynamic parameters lead to a wide range of changes in Gibbs free energy of the process, between −16.2 and −25.8 kJ mol^{-1}. The binding process is spontaneous at all three temperatures, although the actual value of this parameter increases by more than 9 kJ mol^{-1}. Again, the entropic term is dominant, and the interactions between two species lead to the increase in order. Shahabadi and coworkers examined water-soluble nickel(II) complex with 2,20-bipyridine and 1,10-phenanthroline-5,6-dione ligands in a much wider temperature range (between 293 and 318 K) and the change in Gibbs free energy of binding changed between −29.11 and −21.01 kJ mol^{-1} [55]. The stability and voluminosity of ligands in the mentioned case were crucial for the interactions with CT-DNA. Positive changes in the enthalpy (63.75 kJ mol^{-1}) and entropy (271.04 J mol^{-1} K^{-1}) of binding were obtained for nickel(II) complexes with the drug mesalamine in [56], indicating that hydrophobic forces are dominant for the process, similar to the results for the nickel(II)-PLSC complex.

The possible binding mechanism was further investigated by the potassium iodide quenching experiments. This ion is a potent quencher of the intrinsic fluorescence of compounds, but it is repelled by the negatively charged DNA phosphate backbone. When the complex is intercalated into or positioned within a minor groove of the CT-DNA structure, the quenching constant should decrease compared to the measurements without CT-DNA. On the other hand, if the K_{SV} values are similar, external binding is more probable [57]. The fluorescence emission spectra of the complex in the absence and presence of CT-DNA and varying concentrations of KI are shown in Figure 10. As presented, adding KI led to a decrease in fluorescence emission intensity dependent on the amount of KI. The reduction in fluorescence was much faster in the presence of CT-DNA. The values of K_{SV} were 62.5 and 38.0 M^{-1} when CT-DNA was absent and present, respectively. This decrease in the K_{SV} value is characteristic of intercalation/minor groove binding of complexes, as previously shown for planar copper(II) complexes [58]. This is an interesting result, as octahedral complexes are usually too voluminous for this type of binding. This experimental result is further examined through molecular docking simulations to verify the obtained mode of interaction with DNA.

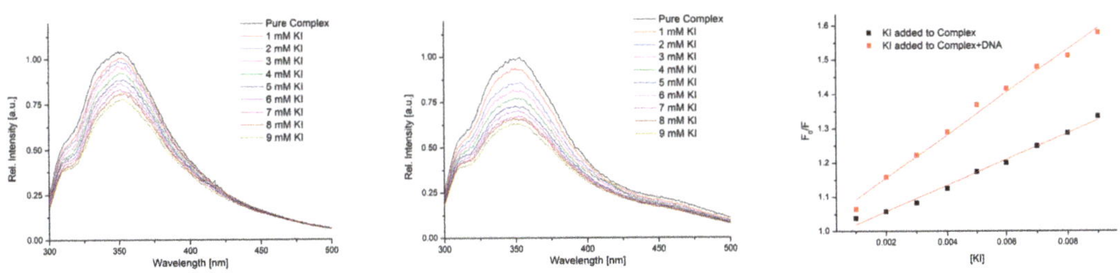

Figure 10. Fluorescence emission spectra of [Ni(PLSC)(H_2O)$_2$(SO_4)] without CT-DNA (**a**) and with CT-DNA (**b**) in the presence of different concentrations of KI, and (**c**) the Stern–Volmer plots for the complex fluorescence quenching by KI.

2.7.2. Ethidium Bromide Displacement Studies

Ethidium bromide (EB) displacement studies examine the binding mechanism of different compounds to DNA [59]. This method is based on removing EB and comparing the interaction strength. The intercalation of EB in the DNA structure, as a result, has a formation of a highly fluorescent species CT-DNA-EB with an excitation wavelength of 520 nm and emission maximum at 600 nm [60]. Upon the addition of the complex, the structure of CT-DNA-EB is decomposed, which is observed as the fluorescence emission intensity decreases. The unbound EB is easily quenched by the present species in the solution.

The emission spectra of CT-DNA-EB in the presence of various concentrations of the complex at 27 °C are shown in Figure 11. As expected, the emission maxima lowers as the concentration of the complex increases. The Stern–Volmer constant for this process is 1.14 × 10^4 M^{-1}. Previously investigated nickel(II) complexes with 3-amino-5-(4-

fluorophenyl) isoxazole derivatives showed K_{SV} values that were an order of magnitude lower than the one obtained in this experiment, which is a result of ligand stability through delocalization of the structure [61]. The K_{SV} value in the case of the nickel(II) complex with the Schiff base derived from 3–ethoxy salicylaldehyde and o–phenylenediamine was 6.01×10^4 M^{-1} [62], which is comparable to the obtained results in this study due to the existence of electronegative groups in the ligand structure. The binding constant obtained through the double log equation is 6.73×10^3 M^{-1}, with the value of n equal to 0.95 (Figure 11). These values show that one molecule of EB is replaced by one complex molecule. The binding affinity, calculated from the binding constant, is −22.0 kJ mol^{-1}.

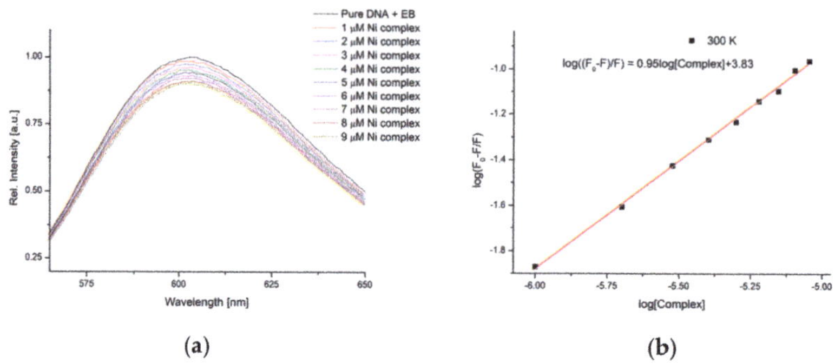

Figure 11. (**a**) Fluorescence emission spectra of CT-DNA-EB for the titration with the complex at 27 °C and (**b**) the double-log Stern–Volmer dependency of intensity on the concentration of $[Ni(PLSC)(H_2O)_2(SO_4)]$.

2.7.3. Molecular Docking Simulations of the Binding Mechanism to DNA

The binding mechanism of the title complex to DNA was elucidated through molecular docking simulations. After numerous attempts, the obtained binding energies of the octahedral complex were significantly higher than the experimental ones, and the binding position was not in line with the experimental data. Considering that the experimentally obtained binding mode included intercalation and minor groove binding, it was assumed that specific changes in the octahedral geometry might happen. Intercalation of the Ni(II) complex was only possible after the dissociation of two coordinated water molecules, resulting in the transformation from the octahedral geometry to the square planar Ni(II) complex, $[Ni(PLSC)(SO_4)]$. A similar change in geometry explained the minor groove binding of the Cr(III) complex with the (Z)-2-((pyridine-2-ylimino)methyl)phenol ligand through removal of two chlorine and one water ligand [63]. Upon binding to the DNA molecule, the square planar Ni(II) complex favored intercalation (ΔG_1 = −25.6 kJ mol^{-1}) over minor groove binding (ΔG_5 = −17.4 kJ mol^{-1}). The first four binding positions are located in the intercalation region, while the fifth binding position is in the minor groove (Figure 12). No binding to the major groove was observed. In spectrofluorometric experiments involving the replacement of ethidium bromide, the binding energy (−22.0 kJ mol^{-1}) was comparable to the calculated value (−22.5 kJ mol^{-1}). The binding energy from spectrofluorimetric measurement of DNA (ΔG_{exp} = −16.2 kJ mol^{-1}) closely matches the estimated value for minor groove binding (ΔG_{cal} = −17.4 kJ mol^{-1}).

Following the dissociation of coordinated water molecules from the octahedral Ni(II) complex, the π systems of nucleobases become available to the SO4 group and hydrogen atoms bound to electronegative or aromatic carbon atoms (Figure 13). This enables the square planar arrangement of the obtained complex to engage in additional interactions involving π electronic systems, such as π-sigma, π-sulfur, and π-alkyl interactions. The $SO_4{}^{2-}$ group forms the majority of these interactions, both in number and type, which is expected due to its prominent position and negative charge.

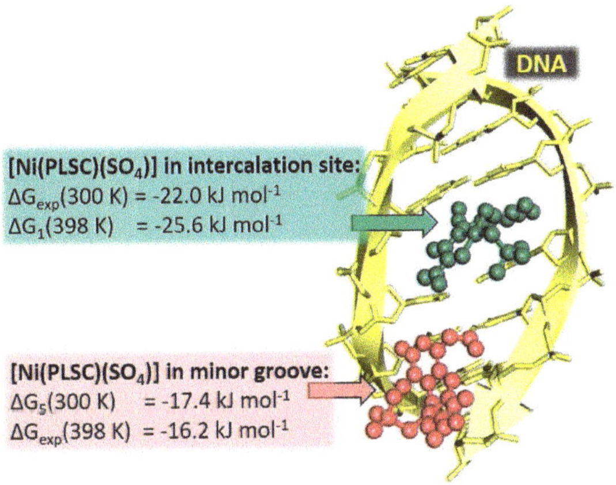

Figure 12. Binding of square planar Ni(II) complex with PLSC ligand at two distinct sites: intercalation site (depicted in dark green, ball-and-stick representation) and minor groove (shown in pink, ball-and-stick representation). DNA molecule is colored yellow. Experimentally determined binding energy (ΔG_{exp}), best-calculated binding energy ($\Delta G1$), and fifth calculated binding energy value (ΔG_5) are also indicated.

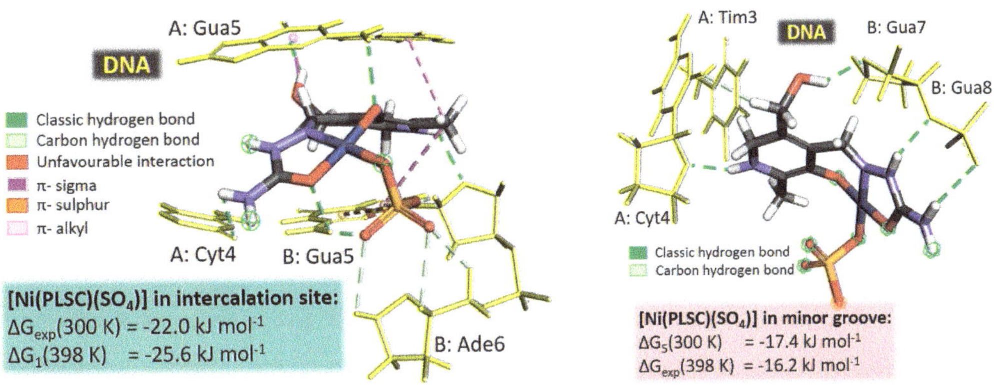

(a) (b)

Figure 13. The supramolecular interactions of the square planar Ni(II) complex in (**a**) the intercalation site and (**b**) the major groove. Only the interacting parts of the nucleobases are shown colored in yellow. For the representation of the square planar Ni(II) complex, sticks colored by element were used. Supramolecular interactions are represented by dashed lines colored according to the type of interaction denoted in the figure's legend. The experimentally determined binding energy (ΔG_{exp}) and the calculated binding energy values are also indicated.

In the minor groove, the supramolecular interactions are limited to hydrogen bonds, as depicted in Figure 13. In this binding position, the SO_4 group of the square planar Ni(II) complex is not involved. Instead, the interactions primarily involve hydrogen atoms bound to electronegative or carbon atoms. These results verify the experimental findings and assumption that certain changes in the geometry are possible in solution.

3. Materials and Methods

3.1. Chemicals

All of the chemicals were obtained from Merck (Darmstadt, Germany) and used without further purification.

3.2. Synthesis of PLSC

PLSC ligand was synthesized according to the previous procedure [64]. The obtained ligand was further examined by FTIR spectroscopy and elemental analysis. Anal. Calcd. for PLSC·2H_2O ($C_9H_{16}N_4O_5$, 260.25): C, 41.53; H, 6.19; N, 20.52. Found: C, 42.02; H, 6.82; N, 20.52. FTIR: 3461 (υ(O-H) stretching), 1570 (υ(C=N)), 1500 (υ(C-N)), 2956 cm^{-1} (υ($^+$NH)); broad band above 3000 cm^{-1} is attributed to hydrogen bonded groups. These positions match those of the PLSC ligand in reference [1].

3.3. Synthesis of [Ni(PLSC)(SO_4)(H_2O)$_2$]

A measured amount of 0.5 mmol (0.78 g) $NiSO_4$ and 0.5 mmol (0.13 g) of PLSC·2H_2O ligand were dissolved in 15 cm^3 of water without additional substances. The molar ratio of nickel salt and ligand was 1:1. With the addition of water, a gentle dissolution began. The dissolution was intensified upon heating at 100 °C. After ten minutes of boiling, the solid compounds were completely dissolved, leading to a clear solution of intense dark green color. The solution was left to crystalize at room temperature. After seven days, dark green crystals were separated. The obtained crystals were washed with cold water as the crystals were not highly soluble. This helped remove water-soluble impurities without dissolving the product. Yield: 0.120 g (76%) Anal. Calcd. for [Ni(PLSC)(SO_4)(H_2O)$_2$] ($C_9H_{16}NiNO_9S$, 415.03): C, 26.04; H, 3.86; N, 3.38; S, 7.72. Found: C, 26.14; H, 3.55; N, 3.42; S, 7.60.

3.4. Characterization

Ligand and complex FTIR spectra were recorded on a Thermo Nicolet-Avatar 370 FTIR spectrometer (Thermo Fisher Scientific, Waltham, MA, USA) between 4000 and 600 cm^{-1}. The KBr pellet technique was used to obtain spectra with a mass ratio of 3:150 mg (compound: KBr). The UV-VIS spectrum was prepared between 250 and 800 nm on a Thermo Scientific UV-VIS Spectrophotometer (Thermo Fisher Scientific, Waltham, MA, USA) with a resolution of 1 nm and integration time of 0.20 s. The UV-VIS spectrum was obtained for the 1.3×10^{-4} M solution of complex in water. The elemental analysis of air-dried samples was conducted on an Elementar Vario El III (Elementar Analysesysteme GmbH, Langenselbold, Germany). The mass percentages were calculated based on the intensity corresponding to different oxides. Molar conductivity of freshly prepared 1.3×10^{-4} M solution complex in water was recorded on WTW inoLab Cond 720 Conductometer (Thermo Fisher Scientific, Waltham, MA, USA).

3.5. X-Ray Crystallographic Analysis

A representative pale green plate-like crystal with dimensions 0.122 × 0.083 × 0.021 mm was selected and mounted on a nylon cryoloop. Diffraction data were collected at 123 K using CuKα radiation (λ = 1.54184 Å) on a Rigaku Synergy S diffractometer fitted with a HYPIX 6000 hybrid photon counting detector. Data were collected and processed, including an empirical (multi-scan) absorption correction, with CrysAlisPro software [65]. The structure was solved and refined by standard methods using the SHELX software suite in conjunction with the Olex2 graphical interface [66,67]. Non-hydrogen atoms were refined with anisotropic displacement ellipsoids, and hydrogen atoms attached to carbon were placed in calculated positions using a riding model. The positions of hydrogen atoms attached to oxygen and nitrogen were apparent in the difference Fourier map and were refined with restrained geometry (DFIX/DANG); d(N-H) = 0.91(2) or d(O-H) = 0.88(2) Å.

Crystallographic data were deposited in the Cambridge Crystallographic Data Centre (CCDC, 12 Union Road, Cambridge CB2 IEZ, UK; e-mail: depos-it@ccdc.cam.ac.uk) with the CCDC number 2348551 for [Ni(PLSC)(SO_4)(H_2O)$_2$]. The crystal data and structure refinement details are listed in Table 4.

Table 4. Crystal data of the newly obtained complex.

Empirical Formula	$[Ni(PLSC)(SO_4)(H_2O)_2]$ $C_9H_{16}NiNO_9S$
Formula weight	415.03
Temperature (K)	123 (2)
Crystal system	monoclinic
Space group	$P2_1/n$
Radiation/Wavelength [Å]	CuKα/1.54184 Å
Volume (Å^3)	1442.02(3)
Unit cell dimension (Å/°)	a = 8.96650(10) b = 9.28740(10) c = 17.6603(3) β = 101.327(2)
Z	4
Volume	1442.02(3) Å^3
Calculated density	1.912 g cm^{-3}
Goodness-of-fit on F^2	856
h, k, l_{max}	11, 11, 22
N_{ref}	3087
Bond precision: C-C	0.0030 Å
Data Completeness	0.978
Θ max [°]	80.075
R1 [$I > 2s(I)$], R1 (*all*)	0.0350
*w*R2 [$I > 2s(I)$], *w*R2 (*all*)	0.1009
CCDC no.	2348551

3.6. Hirshfeld Surface Analysis

The stabilization interactions in the crystallographic structures were examined in the CrystalExplorer [68] program through Hirshfeld surface analysis. These interactions are depicted through a graph characterized by two distances, one between the two nearest nuclei (de) and the other connecting nuclei with the external surface (d_i) [30,69,70]. If distances are shorter, equal, or longer than van der Waals separations between nuclei, they are colored red, white, and blue. The normalized distances are between −0.7224 (red) and 1.2925 (blue). The percentages of specific contacts are determined from the fingerprint plots shown in the Supplementary Information.

3.7. Quantum Chemical Analysis

The crystallographic structure of the title compound was optimized in the Gaussian 09 program package (Gaussian 09, revision C 01) [71] using B3LYP functional [72] in conjunction with 6-311++G(d,p) [73] basis set for H, C, N, O, and S atoms and LanL2DZ/def2-TZVP basis sets [74–76] for nickel atoms. The optimization was performed without any geometrical constraints, and the absence of imaginary frequencies was used as proof that the minimum on the potential energy surface was obtained. The stabilization interactions were quantified by the Quantum Theory of Atoms in Molecules (QTAIM), an approach proposed by Bader [77,78]. Particular emphasis was put on the donor atoms–central metal interactions. Electron density, Laplacian, and other parameters were determined in the Bond Critical Points (BCPs) [79]. The AIMAll program package [80] was used for these calculations, starting from the .wfx files from the Gaussian 09.

3.8. Antioxidant Activity

To investigate the scavenging activity of the samples towards $HO^\bullet$ radicals, a solution containing the samples and a Fenton reaction with the DEPMPO (5-diethoxyphosphoryl-5-methyl-1-pyrroline-N-oxide) spin-trap was used [81]. The reaction mixture, with a total volume of 29 μL, consisted of 2 μL H_2O_2 (final concentration 0.35 mM), 1 μL DEPMPO (final concentration 3.5 mM), and 1/2/3/6 μL of 10 μM sample solution (dissolved in water/DMSO mixture; final concentration 0.3/0.6/1/2 μM) and the rest was deionized water. Following this, 1 μL $FeSO_4$ (final concentration 0.15 mM) was added to the mixture and transferred to a gas-permeable Teflon tube, which was then placed in the EPR resonator (Bruker ELEXSYS-II E540 (Bruker, Billerica, MA, USA). The EPR signal of the DEPMPO/OH spin adduct was recorded after 2 min, using the following parameters: microwave power 10 mW, 9.85 GHz microwave frequency, 100 kHz modulation frequency, and 1 G modulation amplitude. Control measurements were conducted by replacing the sample with an equivalent volume of solvent. The antioxidant activity of the sorghum extracts (AA) was calculated using the formula

$$AA = 100 \cdot \frac{I_c - I_a}{I_c}\ (\%) \quad (1)$$

where I_c and I_a refer to the average intensity of the two most intense signals in the low field in the control spectrum and spectrum with added complex.

To evaluate the samples' activity against ascorbyl radicals, the EPR signal of $^{\bullet}$Asc in a DMSO solution was measured following a previously established method [82]. Briefly, 10 μL of EDTA (final concentration 250 μM) and 1 μL of $FeCl_3$ (final concentration 8 μM) were combined to form the Fe(III)-EDTA complex in 74 μL of DMSO. Subsequently, 5/10/20 μL of the sample (dissolved in water/DMSO mixture; final concentration 0.5/10/20 μM) was added. The $^{\bullet}$Asc radical was then generated by adding 10 μL of ascorbic acid (final concentration 250 μM) to the mixture, and 30 μL of it was transferred to a gas-permeable Teflon tube. The X-band EPR spectra of the $^{\bullet}$Asc radical were recorded 2 min later, under the following parameters: 10 mW microwave power, 9.85 GHz microwave frequency, 100 kHz modulation frequency, and 2 G modulation amplitude. Control measurements were performed by replacing the sample with solvent. The antiradical activity of the samples was calculated as described previously.

3.9. Spectrofluorimetric Determination of HSA Binding Affinity

The protein and DNA binding affinities of the obtained complex were examined by spectrofluorimetric titration on a Cary Eclipse MY2048CH03 spectrophotometer (Agilent Technologies, Santa Clara, CA, USA). The scan rate was set to 600 nm min^{-1} with both slits of 5 nm. The HSA fluorescence was excited at 280 nm, sufficient for activating tryptophan residues. The protein concentration was constant at 5×10^{-6} M in phosphate buffer saline at pH = 7.4 (concentrations of NaCl and KCl were 137 and 2.7 mM). The number concentration of complex changed between 1 and 10 μM. The fluorescence spectra were recorded between 300 and 500 nm two minutes after adding the quencher. The emission spectra maximum was located around 340 nm. The emission intensity was corrected for absorption of excitation wavelength and re-emission of emitted light by the following equation [83,84]:

$$F_c = F_m e^{\frac{A_1 + A_2}{2}} \quad (2)$$

In this equation, F_c and F_m are the corrected and measured fluorescence emissions, while A_1 and A_2 are the absorbances of the complex at the excitation and emission wavelengths. Throughout the manuscript, only the corrected fluorescence intensities are presented.

The quenching mechanism was determined by the Stern–Volmer equation through the dependence of the Stern–Volmer constant (K_{SV}) and temperature [85,86]. K_{SV} values were obtained in the following equation, in which F_0 and F denote the fluorescence intensity of pure compounds (HSA or complex) and compounds with added quenchers ([Q]).

$$\frac{F_0}{F} = 1 + K_{SV}[complex] = 1 + \tau_0 k_q[Q] \quad (3)$$

The double-log Stern–Volmer equation was employed to further examine the mechanism [87]. In this equation, F_0 and F have the same meaning, as previously discussed, while K_b is the binding constant and n is the cooperativity number [88].

$$\log\left(\frac{F_0 - F}{F}\right) = \log K_b + n\log[complex] \tag{4}$$

The thermodynamic parameters of binding were calculated through the dependency of the binding constant on temperature. The change in enthalpy and entropy were obtained from the slope and intercept, respectively:

$$\ln K_b = -\frac{\Delta H_b}{RT} + \frac{\Delta S_b}{R} \tag{5}$$

3.10. Spectrofluorimetric Determination of DNA Binding Affinity

The DNA binding studies were followed through complex fluorescence quenching by CT-DNA. For these experiments, calf thymus DNA was chosen. The concentration of CT-DNA solution was determined from the absorbance value at 260 nm (molar extinction coefficient 6600 $dm^3\ mol^{-1}$ cm). The absorbance ratio (A260/A280) was approximately 1.9, indicating that CT-DNA was sufficiently free from proteins [46]. These experiments were conducted using a Thermo Scientific Evolution 220 spectrophotometer (Thermo Fisher Scientific, Waltham, MA, USA). The concentration of the complex was kept constant at 10^{-5} M. The excitation wavelength (280 nm) was selected as it gave a measurable fluorescence. The emission spectra were recorded between 300 and 500 nm, with both slits set to 10 nm. The standard solution of CT-DNA (2.3×10^{-4} M) in phosphate buffer saline was added stepwise, and fluorescence spectra were recorded two minutes after the addition.

The iodide quenching experiment was performed by keeping the concentrations of complex and DNA constant at 10 and 20 μM. The concentration of KI ranged between 0 and 9 mM, with a step of 1 mM. The excitation wavelength was determined from the absorption spectra of the complex (280 nm), while the emission spectra were recorded between 300 and 500 nm.

The spectrofluorimetric titration was also used for the ethidium bromide (EB) displacement studies in the presence of the obtained complex. The concentrations of CT-DNA and EB were held constant at 50 and 5 μM in phosphate buffer saline pH = 7.4 (concentrations of NaCl and KCl were 137 and 2.7 mM). The concentration of the complex was changed from 0 to 9 μM in steps of 1 μM. The excitation wavelength for CT-DNA-EB was set to 520 nm, and the emission spectra were recorded between 540 and 650 nm. Both slits were set to 10 nm. The thermodynamic parameters of the displacement were determined from the binding constant [89].

3.11. Molecular Docking

The synthesized octahedral Ni(II) complex, $[Ni(PLSC)(H_2O)_2(SO_4)]$, contains an abundance of electronegative atoms, enabling a variety of potential interactions. However, the bulkiness of the complex introduces steric hindrance. Molecular docking calculations were employed to investigate the binding process to target molecules (BSA and DNA) at the molecular level. The crystal structure of the HSA molecule without any bound ligands (PDB ID: 1AO6 [90]) was selected as one target. Another target was the crystal structure of DNA with PDB ID:1XRW [91], which is suitable for comparison with spectrofluorimetric measurements, including ethidium bromide displacement e and Ni(II) complex binding to free DNA. The original crystal structures from the PDB were cleaned from water and ligand molecules before docking studies. Density Functional Theoretical (DFT) calculations using GAUSSIAN09 were performed at the B3LYP/6-311++G(d,p)(H,C,N,O,S)/LanL2DZ(Ni) level to optimize the crystal structure of the synthesized $[Ni(PLSC)(H_2O)_2(SO_4)]$ complex. All molecular docking calculations were conducted at ambient temperature (25 °C) and physiological pH (7.4) using AutoDock 4.2.6 within the AMDock program, version 1.5.2 [92].

4. Conclusions

The obtained complex [Ni(PLSC)(SO_4)(H_2O)$_2$] contains a PLSC ligand in neutral form, a sulfate ion directly coordinated with the central metal ion, and two molecules of water. The compound crystallized in the monoclinic space group P2(1)/n. The crystal structure was stabilized by several hydrogen bonds, of which the hydrogen bond between the sulfate ion and molecules of water directly affected the pseudo-octahedral geometry. The most numerous contacts were formed between oxygen and hydrogen atoms (52.9%) due to their high abundance, as determined by the Hirshfeld surface analysis. The interactions between hydrogen atoms accounted for 29.7% of all contacts. Other interactions between hydrogen and nitrogen/carbon atoms were present to a lesser extent. Optimization of a structure employing two pseudopotentials for nickel(II) ions led to structures very similar to the experimental one. Based on the comparison between bond lengths and angles, the B3LYP/6-311++G(d,p)(H,C,N,O,S)/LanL2DZ(Ni) level of theory was chosen for further theoretical analysis. The QTAIM parameters of the interactions between the donor atoms and nickel(II) ion showed the effects of the intramolecular hydrogen bond between the sulfate ion and water molecule. Additional stabilization interactions were identified within the structure of the obtained complex. The EPR measurements showed that the amount of reduced HO$^•$ was 82.5, 87.0, 88.8, and 90.4% for the final concentration of complex between 0.3 and 2 μM, while in the case of the ascorbyl radical, the reduction percentages were 53.0 (0.5 μM), 69.0 (1 μM), and 86.9 (2 μM). The binding of the complex to HSA was spontaneous in the temperature range between 27 and 37 °C. The changes in Gibbs free energy of binding were between −26.9 and −32.8 kJ mol^{-1}. The quenching of the HSA fluorescence was marked as static. The molecular docking results verified that the binding occurs at the FA8 binding site near the fluorescent amino acid Trp213 with the change in Gibbs free energy of −26.5 kJ mol^{-1}. The binding to DNA was also spontaneous in the same range, although dynamic. The potassium iodide experiments proved that minor groove/intercalation was the probable quenching mechanism. The binding affinities predicted by the molecular docking studies showed that the geometry of the complex in solution was changed to square planar through the removal of water molecules. The experimental binding energies were well reproduced. Further experimental studies on the cytotoxicity of the compound are advised.

Supplementary Materials: The following supporting information can be downloaded at: https://www.mdpi.com/article/10.3390/inorganics12110280/s1, Table S1: Experimental and theoretical (functionals for Ni given below in conjunction with B3LYP/6-311++G(d,p)(H,C,N,O,S) level of theory) bond lengths of [Ni(PLSC)(SO_4)(H_2O)$_2$] (in Å); Table S2: Experimental and theoretical (functionals for Ni given below in conjunction with B3LYP/6-311++G(d,p)(H,C,N,O,S) level of theory) bond lengths (in °) of [Ni(PLSC)(SO_4)(H_2O)$_2$]; Figure S1: FTIR spectrum of PLSC ligand; Figure S2: FTIR spectrum of [Ni(PLSC)(SO_4)(H_2O)$_2$]; Figure S3: UV-VIS spectrum of [Ni(PLSC)(SO_4)(H_2O)$_2$]; Figure S4: Fingerprint plots of the most numerous contacts within crystallographic structure of [Ni(PLSC)(SO_4)(H_2O)$_2$]; Table S3: The calculated Bond Critical Point (BCP) properties at the DFT/B3LYP-D3BJ/6-311+G(d,p)/LanL2DZ level of theory: the electron density (ρ(r)) and its Laplacian ($\nabla^2\rho(r)$); the Lagrangian kinetic electron density (G(r)) and the potential electron density (V(r)); the density of the total energy of electrons (H(r))—Cremer–Kraka electronic energy density; the interatomic bond energy, E_{bond}; Figure S5: Bond Critical Points within structure of Ni(PLSC)(SO_4)(H_2O)$_2$]; Figure S6: Stern-Volmer and double-log Stern-Volmer plots for the binding of [Ni(PLSC)(SO_4)(H_2O)$_2$] to HSA.

Author Contributions: Conceptualization, V.J., D.D., J.D.M. and A.R. (Aleksandra Rakić); methodology, L.G., O.A.O.A. and M.S.A.; software, T.Y.A.A., A.R. (Aleksandra Radulović) and A.R. (Aleksandra Rakić); validation, Đ.N., J.D.M., L.G. and O.A.O.A.; formal analysis, M.S.A., T.Y.A.A. and A.R. (Aleksandra Radulović); investigation, L.G., O.A.O.A., M.S.A. and T.Y.A.A.; resources, A.R. (Aleksandra Radulović), Đ.N. and A.R. (Aleksandra Rakić); data curation, L.G., O.A.O.A. and

M.S.A.; writing—original draft preparation, T.Y.A.A., A.R. (Aleksandra Radulović), Đ.N. and A.R. (Aleksandra Rakić); writing—review and editing, D.D., J.D.M. and V.J.; visualization, D.D.; supervision, J.D.M., D.D. and V.J.; project administration, V.J.; funding acquisition, V.J. All authors have read and agreed to the published version of the manuscript.

Funding: This research was funded by the Scientific Research Deanship at the University of Ha'il, Kingdom of Saudi Arabia, grant number RG-23080.

Data Availability Statement: Data are contained in this article. Further inquiries can be directed to the corresponding author.

Acknowledgments: The authors are thankful to the University of Ha'il, Kingdom of Saudi Arabia. This research was funded by the Scientific Research Deanship at the University of Ha'il, Kingdom of Saudi Arabia, through Project Number RG-23080.

Conflicts of Interest: The authors declare no conflicts of interest. The funders had no role in the design of the study; in the collection, analyses, or interpretation of data; in the writing of the manuscript; or in the decision to publish the results.

References

1. Knežević, N.Ž.; Leovac, V.M.; Jevtović, V.S.; Grgurić-Šipka, S.; Sabo, T.J. Platinum(IV) complex with pyridoxal semicarbazone. *Inorg. Chem. Commun.* **2003**, *6*, 561–564. [CrossRef]
2. Cordes, E.H.; Jencks, W.P. Semicarbazone Formation from Pyridoxal, Pyridoxal Phosphate, and Their Schiff Bases. *Biochemistry* **1962**, *1*, 773–778. [CrossRef] [PubMed]
3. Vidovic, D.; Radulovic, A.; Jevtovic, V. Synthesis, characterization and structural analysis of new copper(II) complexes incorporating a pyridoxal-semicarbazone ligand. *Polyhedron* **2011**, *30*, 16–21. [CrossRef]
4. Poleti, D.; Karanović, L.; Leovac, V.M.; Jevtović, V.S. Dibromo(pyridoxal semicarbazone-κ 3N1,O3,O3′)copper(II). *Acta Crystallogr. Sect. C Cryst. Struct. Commun.* **2003**, *59*, 73–75. [CrossRef] [PubMed]
5. Vojinović-Ješić, L.S.; Jovanović, L.S.; Leovac, V.M.; Radanović, M.M.; Rodić, M.V.; Barta Holló, B.; Mészásaros Szécsényi, K.; Ivković, S.A. Transition metal complexes with thiosemicarbazide-based ligands. Part 63. Syntheses, structures and physicochemical characterization of the first chromium(III) complexes with pyridoxal semi- and thiosemicarbazones. *Polyhedron* **2015**, *101*, 196–205. [CrossRef]
6. Leovac, V.M.; Jovanović, L.S.; Divjaković, V.; Pevec, A.; Leban, I.; Armbruster, T. Transition metal complexes with thiosemicarbazide-based ligands. Part LIV. Nickel(II) complexes with pyridoxal semi- (PLSC) and thiosemicarbazone (PLTSC). Crystal and molecular structure of $[Ni(PLSC)(H_2O)_3](NO_3)_2$ and $[Ni(PLTSC\text{-}H)py]NO_3$. *Polyhedron* **2007**, *26*, 49–58. [CrossRef]
7. Leovac, V.M.; Jovanović, L.S.; Jevtović, V.S.; Pelosi, G.; Bisceglie, F. Transition metal complexes with thiosemicarbazide-based ligand—Part LV: Synthesis and X-ray structural study of novel Ni(II) complexes with pyridoxal semicarbazone and pyridoxal thiosemicarbazone. *Polyhedron* **2007**, *26*, 2971–2978. [CrossRef]
8. Leovac, V.M.; Marković, S.; Divjaković, V.; Szécsényi, K.M.; Joksović, M.D.; Lebanc, I. Structural and DFT studies on molecular structure of Ni(II) chloride complex with pyridoxal semicarbazone (PLSC). Unusual coordination mode of PLSC. *Acta Chim. Slov.* **2008**, *55*, 850–860.
9. Jevtovic, V.; Alshammari, N.; Latif, S.; Alsukaibi, A.K.D.; Humaidi, J.; Alanazi, T.Y.A.; Abdulaziz, F.; Matalka, S.I.; Pantelić, N.Đ.; Marković, M.; et al. Synthesis, Crystal Structure, Theoretical Calculations, Antibacterial Activity, Electrochemical Behavior, and Molecular Docking of Ni(II) and Cu(II) Complexes with Pyridoxal-Semicarbazone. *Molecules* **2022**, *27*, 6322. [CrossRef]
10. Pharma, D.; Aduri, S.; Ramana, V.; Ch, R. Synthesis, Characterization, DNA binding, DNA Cleavage and Antibacterial Studies of Ni(II) and Cu(II) Complexes of Pyridoxal Semicarbazone. *Der Pharma Chem.* **2017**, *9*, 90–98.
11. Rufino-Felipe, E.; Valdés, H.; Morales-Morales, D. C-S Cross-Coupling Reactions Catalyzed by Well-Defined Copper and Nickel Complexes. *Eur. J. Org. Chem.* **2022**, *2022*, e202200654. [CrossRef]
12. Zhang, W.; Hong, J.; Zheng, J.; Huang, Z.; Zhou, J.; Xu, R. Nickel–Thiolate Complex Catalyst Assembled in One Step in Water for Solar H 2 Production. *J. Am. Chem. Soc.* **2011**, *133*, 20680–20683. [CrossRef]
13. Gan, L.; Groy, T.L.; Tarakeshwar, P.; Mazinani, S.K.S.; Shearer, J.; Mujica, V.; Jones, A.K. A Nickel Phosphine Complex as a Fast and Efficient Hydrogen Production Catalyst. *J. Am. Chem. Soc.* **2015**, *137*, 1109–1115. [CrossRef]
14. Meinhard, D.; Wegner, M.; Kipiani, G.; Hearley, A.; Reuter, P.; Fischer, S.; Marti, O.; Rieger, B. New Nickel(II) Diimine Complexes and the Control of Polyethylene Microstructure by Catalyst Design. *J. Am. Chem. Soc.* **2007**, *129*, 9182–9191. [CrossRef] [PubMed]
15. Derafa, W.; Aggoun, D.; Messasma, Z.; Houchi, S.; Bouacida, S.; Ourari, A. An unexpected single crystal structure of nickel(II) complex: Spectral, DFT, NLO, magnetic and molecular docking studies. *J. Mol. Struct.* **2022**, *1264*, 133190. [CrossRef]
16. Atasever Arslan, B.; Kaya, B.; Şahin, O.; Baday, S.; Saylan, C.C.; Ülküseven, B. The iron(III) and nickel(II) complexes with tetradentate thiosemicarbazones. Synthesis, experimental, theoretical characterization, and antiviral effect against SARS-CoV-2. *J. Mol. Struct.* **2021**, *1246*, 131166. [CrossRef] [PubMed]

17. Bhandarkar, S.E.; Pathare, P.P.; Khobragade, B.P. New Nickel (II), Copper (II) and Cobalt (II) Complexes Based Salicyaldehyde Schiff Base: Synthesis, Characterisation, and Antiviral Activity. *Mater. Today Proc.* **2023**, *92*, 807–816. [CrossRef]
18. Heng, M.P.; Sinniah, S.K.; Teoh, W.Y.; Sim, K.S.; Ng, S.W.; Cheah, Y.K.; Tan, K.W. Synthesis of a DNA-targeting nickel (II) complex with testosterone thiosemicarbazone which exhibits selective cytotoxicity towards human prostate cancer cells (LNCaP). *Spectrochim. Acta Part A Mol. Biomol. Spectrosc.* **2015**, *150*, 360–372. [CrossRef]
19. Ay, B.; Gönül, İ.; Demir, B.S.; Saygıdeğer, Y.; Kani, İ. Synthesis, structural characterization, and in vitro anticancer activity of two new nickel complexes bearing imine bonds. *Inorg. Chem. Commun.* **2020**, *114*, 107824. [CrossRef]
20. Bakr, E.A.; Al-Hefnawy, G.B.; Awad, M.K.; Abd-Elatty, H.H.; Youssef, M.S. New Ni(II), Pd(II) and Pt(II) complexes coordinated to azo pyrazolone ligand with a potent anti-tumor activity: Synthesis, characterization, DFT and DNA cleavage studies. *Appl. Organomet. Chem.* **2018**, *32*, e4104. [CrossRef]
21. Zhong, X.; Yi, J.; Sun, J.; Wei, H.-L.; Liu, W.-S.; Yu, K.-B. Synthesis and crystal structure of some transition metal complexes with a novel bis-Schiff base ligand and their antitumor activities. *Eur. J. Med. Chem.* **2006**, *41*, 1090–1092. [CrossRef] [PubMed]
22. Perontsis, S.; Hatzidimitriou, A.G.; Papadopoulos, A.N.; Psomas, G. Nickel-diflunisal complexes: Synthesis, characterization, in vitro antioxidant activity and interaction with DNA and albumins. *J. Inorg. Biochem.* **2016**, *162*, 9–21. [CrossRef] [PubMed]
23. Raza, A.; Bano, S.; Xu, X.; Zhang, R.X.; Khalid, H.; Iqbal, F.M.; Xia, C.; Tang, J.; Ouyang, Z. Rutin–Nickel Complex: Synthesis, Characterization, Antioxidant, DNA Binding, and DNA Cleavage Activities. *Biol. Trace Elem. Res.* **2017**, *178*, 160–169. [CrossRef] [PubMed]
24. Venkateswarlu, K.; Ganji, N.; Daravath, S.; Kanneboina, K.; Rangan, K. Shivaraj Crystal structure, DNA interactions, antioxidant and antitumor activity of thermally stable Cu(II), Ni(II) and Co(III) complexes of an N,O donor Schiff base ligand. *Polyhedron* **2019**, *171*, 86–97. [CrossRef]
25. Osman, U.M.; Silvarajoo, S.; Kamarudin, K.H.; Tahir, M.I.M.; Kwong, H.C. Ni(II) complex containing a thiosemicarbazone ligand: Synthesis, spectroscopy, single-crystal X-ray crystallographic and conductivity studies. *J. Mol. Struct.* **2021**, *1223*, 128994. [CrossRef]
26. Devi, J.; Yadav, M.; Jindal, D.K.; Kumar, D.; Poornachandra, Y. Synthesis, spectroscopic characterization, biological screening and in vitro cytotoxic studies of 4-methyl-3-thiosemicarbazone derived Schiff bases and their Co (II), Ni (II), Cu (II) and Zn (II) complexes. *Appl. Organomet. Chem.* **2019**, *33*, e5154. [CrossRef]
27. Ferrari Belicchi, M.; Fava Gasparri, G.; Leporati, E.; Pelizzi, C.; Tarasconi, P.; Tosi, G. Structure of Pyridoxal Thiosemicarbazone Trihydrate and Spectroscopic Properties of its Metal Complexes t. *J. Chem. Soc. Dalt. Trans.* **1986**, *3*, 2455–2461. [CrossRef]
28. Beraldo, H.; Nacif, W.F.; Teixeira, L.R.; Rebouças, J.S. Cobalt(II) and nickel(II) complexes of N(4′) substituted 3- and 4-acetylpyridine thiosemicarbazones. *Transit. Met. Chem.* **2002**, *27*, 85–88. [CrossRef]
29. Geary, W.J. The use of conductivity measurements in organic solvents for the characterisation of coordination compounds. *Coord. Chem. Rev.* **1971**, *7*, 81–122. [CrossRef]
30. Jevtovic, V.; Alhar, M.S.O.; Milenković, D.; Marković, Z.; Dimitrić Marković, J.; Dimić, D. Synthesis, Structural Characterization, Cytotoxicity, and Protein/DNA Binding Properties of Pyridoxylidene-Aminoguanidine-Metal (Fe, Co, Zn, Cu) Complexes. *Int. J. Mol. Sci.* **2023**, *24*, 14745. [CrossRef]
31. Jevtovic, V.; Alshamari, A.K.; Milenković, D.; Dimitrić Marković, J.; Marković, Z.; Dimić, D. The Effect of Metal Ions (Fe, Co, Ni, and Cu) on the Molecular-Structural, Protein Binding, and Cytotoxic Properties of Metal Pyridoxal-Thiosemicarbazone Complexes. *Int. J. Mol. Sci.* **2023**, *24*, 1910. [CrossRef] [PubMed]
32. Gak Simić, K.; Đorđević, I.; Lazić, A.; Radovanović, L.; Petković-Benazzouz, M.; Rogan, J.; Trišović, N.; Janjić, G. On the supramolecular outcomes of fluorination of cyclohexane-5-spirohydantoin derivatives. *CrystEngComm* **2021**, *23*, 2606–2622. [CrossRef]
33. Sciortino, G.; Lihi, N.; Czine, T.; Maréchal, J.; Lledós, A.; Garribba, E. Accurate prediction of vertical electronic transitions of Ni(II) coordination compounds via time dependent density functional theory. *Int. J. Quantum Chem.* **2018**, *118*, e25655. [CrossRef]
34. Sağlam, E.G.; Zeyrek, C.T.; Dal, H.; Ünver, H.; Ebinç, A. New {bis-pyridine-bis-[3-methyl-1-butoxy-(p-methoxyphenyl) phosphonodithioato]}nickel(II) complex: Synthesis, characterization, single crystal structure and theoretical studies. *J. Mol. Struct.* **2020**, *1201*, 127185. [CrossRef]
35. Chai, L.; Xu, L.; Zhang, X.; Li, Y. Two dinuclear copper (II) and nickel (II) complexes based on 4-(diethylamino)salicylaldehyde: X-ray structures, spectroscopic, electrochemical, antibacterial, Hirshfeld surfaces analyses, and time-dependent density functional theory calculations. *Appl. Organomet. Chem.* **2021**, *35*, e6068. [CrossRef]
36. Kasalović, M.P.; Jelača, S.; Milanović, Ž.; Maksimović-Ivanić, D.; Mijatović, S.; Lađarević, J.; Božić, B.; Marković, Z.; Dunđerović, D.; Rüffer, T.; et al. Novel triphenyltin(iv) compounds with carboxylato *N*-functionalized 2-quinolones as promising potential anticancer drug candidates: In vitro and in vivo evaluation. *Dalt. Trans.* **2024**, *53*, 8298–8314. [CrossRef]
37. Soliman, S.M.; Albering, J.; Abu-Youssef, M.A.M. Structural analyses of two new highly distorted octahedral copper(II) complexes with quinoline-type ligands; Hirshfeld, AIM and NBO studies. *Polyhedron* **2017**, *127*, 36–50. [CrossRef]
38. Bianchi, R.; Gervasio, G.; Marabello, D. Experimental Electron Density Analysis of $Mn_2(CO)_{10}$: Metal-Metal and Metal-Ligand Bond Characterization. *Inorg. Chem.* **2000**, *39*, 2360–2366. [CrossRef] [PubMed]
39. Jevtovic, V.; Rakić, A.; Alshammari, O.A.O.; Alhar, M.S.; Alenezi, T.; Rakic, V.; Dimić, D. Theoretical Study of the Effects of Different Coordination Atoms (O/S/N) on Crystal Structure, Stability, and Protein/DNA Binding of Ni(II) Complexes with Pyridoxal-Semi, Thiosemi, and Isothiosemicarbazone Ligand Systems. *Inorganics* **2024**, *12*, 251. [CrossRef]

40. Eichhorn, T.; Kolbe, F.; Mišić, S.; Dimić, D.; Morgan, I.; Saoud, M.; Milenković, D.; Marković, Z.; Rüffer, T.; Dimitrić Marković, J.; et al. Synthesis, Crystallographic Structure, Theoretical Analysis, Molecular Docking Studies, and Biological Activity Evaluation of Binuclear Ru(II)-1-Naphthylhydrazine Complex. *Int. J. Mol. Sci.* **2023**, *24*, 689. [CrossRef]
41. Galano, A.; Mazzone, G.; Alvarez-Diduk, R.; Marino, T.; Alvarez-Idaboy, J.R.; Russo, N. Food Antioxidants: Chemical Insights at the Molecular Level. *Annu. Rev. Food Sci. Technol.* **2016**, *7*, 335–352. [CrossRef] [PubMed]
42. Sathyadevi, P.; Krishnamoorthy, P.; Butorac, R.R.; Cowley, A.H.; Bhuvanesh, N.S.P.; Dharmaraj, N. Effect of substitution and planarity of the ligand on DNA/BSA interaction, free radical scavenging and cytotoxicity of diamagnetic Ni(ii) complexes: A systematic investigation. *Dalt. Trans.* **2011**, *40*, 9690. [CrossRef] [PubMed]
43. El-Gammal, O.A.; Abd Al-Gader, I.M.; El-Asmy, A.A. Synthesis, characterization, biological activity of binuclear Co(II), Cu(II) and mononuclear Ni(II) complexes of bulky multi-dentate thiosemicarbazide. *Spectrochim. Acta Part A Mol. Biomol. Spectrosc.* **2014**, *128*, 759–772. [CrossRef] [PubMed]
44. Patel, R.N.; Singh, Y.; Singh, Y.P.; Butcher, R.J.; Kamal, A.; Tripathi, I.P. Copper(II) and nickel(II) complexes with N′-[(Z)-phenyl(pyridin-2-yl)methylidene]acetohydrazide: Synthesis, crystal structures, DFT calculations and antioxidant effects. *Polyhedron* **2016**, *117*, 20–34. [CrossRef]
45. Barma, A.; Ghosh, D.; Karmakar, P.; Roy, P. Synthesis and characterization of a mononuclear nickel(II) complex with N,O-donor ligand: Its DNA/HSA protein binding properties and tumor suppressive function. *J. Mol. Struct.* **2022**, *1250*, 131687. [CrossRef]
46. Nithya, P.; Helena, S.; Simpson, J.; Ilanchelian, M.; Muthusankar, A.; Govindarajan, S. New cobalt(II) and nickel(II) complexes of benzyl carbazate Schiff bases: Syntheses, crystal structures, in vitro DNA and HSA binding studies. *J. Photochem. Photobiol. B Biol.* **2016**, *165*, 220–231. [CrossRef]
47. Hu, K.; Liu, C.; Li, J.; Liang, F. Copper(II) complexes based on quinoline-derived Schiff-base ligands: Synthesis, characterization, HSA/DNA binding ability, and anticancer activity. *Medchemcomm* **2018**, *9*, 1663–1672. [CrossRef]
48. Sun, J.; Huang, Y.; Zheng, C.; Zhou, Y.; Liu, Y.; Liu, J. Ruthenium (II) Complexes Interact with Human Serum Albumin and Induce Apoptosis of Tumor Cells. *Biol. Trace Elem. Res.* **2015**, *163*, 266–274. [CrossRef]
49. Wu, S.-S.; Yuan, W.-B.; Wang, H.-Y.; Zhang, Q.; Liu, M.; Yu, K.-B. Synthesis, crystal structure and interaction with DNA and HSA of (N,N′-dibenzylethane-1,2-diamine) transition metal complexes. *J. Inorg. Biochem.* **2008**, *102*, 2026–2034. [CrossRef]
50. Ouyang, Y.; Zhao, X.-F.; Qiao, X.; Zhang, R.; Bao, W.-G.; Xie, C.-Z.; Xu, J.-Y. A mononuclear nickel(II) complex based on polypyridyl ligand: Synthesis, characterization, and biological activity. *J. Coord. Chem.* **2016**, *69*, 475–484. [CrossRef]
51. Kumar, M.; Lal, N.; Luthra, P.M.; Masram, D.T. Exploring the binding and cleavage activities of nickel II complexes towards DNA and proteins. *New J. Chem.* **2021**, *45*, 6693–6708. [CrossRef]
52. Abdalla, E.M.; Al-Sulami, A.I.; Aly, S.A.; Abd-Allah, M.T.; Nasr, G.M.; Albohy, S.A.H.; Hosny, S. Synthesis, characterization, DNA binding, DFT, anticancer, antibacterial, and the effect of gamma irradiation of novel Co(II), Ag (I), and Cd (II) complexes with hydrazone derivatives. *J. Saudi Chem. Soc.* **2023**, *27*, 101770. [CrossRef]
53. Palchaudhuri, R.; Hergenrother, P.J. DNA as a target for anticancer compounds: Methods to determine the mode of binding and the mechanism of action. *Curr. Opin. Biotechnol.* **2007**, *18*, 497–503. [CrossRef]
54. Shahabadi, N.; Kashanian, S.; Mahdavi, M.; Sourinejad, N. DNA Interaction and DNA Cleavage Studies of a New Platinum(II) Complex Containing Aliphatic and Aromatic Dinitrogen Ligands. *Bioinorg. Chem. Appl.* **2011**, *2011*, 525794. [CrossRef] [PubMed]
55. Shahabadi, N.; Kashanian, S.; Khosravi, M.; Mahdavi, M. Multispectroscopic DNA interaction studies of a water-soluble nickel(II) complex containing different dinitrogen aromatic ligands. *Transit. Met. Chem.* **2010**, *35*, 699–705. [CrossRef]
56. Shahabadi, N.; Moradi Fili, S.; Shahlaei, M. Synthesis, characterization and comparative DNA interaction studies of new copper(II) and nickel(II) complexes containing mesalamine drug using molecular modeling and multispectroscopic methods. *J. Coord. Chem.* **2015**, *68*, 3667–3684. [CrossRef]
57. Indumathy, R.; Radhika, S.; Kanthimathi, M.; Weyhermuller, T.; Unni Nair, B. Cobalt complexes of terpyridine ligand: Crystal structure and photocleavage of DNA. *J. Inorg. Biochem.* **2007**, *101*, 434–443. [CrossRef]
58. Kashanian, S.; Khodaei, M.M.; Roshanfekr, H.; Shahabadi, N.; Mansouri, G. DNA binding, DNA cleavage and cytotoxicity studies of a new water soluble copper(II) complex: The effect of ligand shape on the mode of binding. *Spectrochim. Acta Part A Mol. Biomol. Spectrosc.* **2012**, *86*, 351–359. [CrossRef]
59. Parsekar, S.U.; Haldar, P.; Antharjanam, P.K.S.; Kumar, M.; Koley, A.P. Synthesis, characterization, crystal structure, DNA and human serum albumin interactions, as well as antiproliferative activity of a Cu(II) complex containing a Schiff base ligand formed in situ from the Cu(II)-induced cyclization of 1,5-bis(salicylidene). *Appl. Organomet. Chem.* **2021**, *35*, e6152. [CrossRef]
60. Psomas, G. Mononuclear metal complexes with ciprofloxacin: Synthesis, characterization and DNA-binding properties. *J. Inorg. Biochem.* **2008**, *102*, 1798–1811. [CrossRef]
61. Ramesh, G.; Daravath, S.; Ganji, N.; Rambabu, A.; Venkateswarlu, K. Shivaraj Facile synthesis, structural characterization, DNA binding, incision evaluation, antioxidant and antimicrobial activity studies of Cobalt(II), Nickle(II) and Copper(II) complexes of 3-amino-5-(4-fluorophenyl)isoxazole derivatives. *J. Mol. Struct.* **2020**, *1202*, 127338. [CrossRef]
62. Zaki, M.; Hairat, S.; Kamaal, S.; Aljarba, N.H.; AL–Johani, N.S.; Alkahtani, S. Synthesis, crystal structure elucidation and DNA/HSA binding profile of Ni(II) complex of Schiff base derived from 3–ethoxy salicylaldehyde and o–phenylenediamine. *J. Mol. Struct.* **2022**, *1265*, 133351. [CrossRef]

63. Abdel-Rahman, L.H.; Basha, M.T.; Al-Farhan, B.S.; Shehata, M.R.; Abdalla, E.M. Synthesis, characterization, potential antimicrobial, antioxidant, anticancer, DNA binding, and molecular docking activities and DFT on novel Co(II), Ni(II), VO(II), Cr(III), and La(III) Schiff base complexes. *Appl. Organomet. Chem.* **2022**, *36*, e6484. [CrossRef]
64. Jevtovic, V. *Cu, Fe, Ni and V complexes with Pyridoxal Semicarbazones, Synthesis, Physical and Chemical Properties, Structural Analyses and Biological Activities*; Lambert Academic Publishing: Saarbrücken, Germany, 2010.
65. Oxford Diffraction/Agilent Technologies UK Ltd. *CrysAlisPRO*; Oxford Diffraction/Agilent Technologies UK Ltd.: Yarnton, UK, 2017.
66. Sheldrick, G.M. IUCr Crystal structure refinement with *SHELXL*. *Acta Crystallogr. Sect. C Struct. Chem.* **2015**, *71*, 3–8. [CrossRef] [PubMed]
67. Dolomanov, O.V.; Bourhis, L.J.; Gildea, R.J.; Howard, J.A.K.; Puschmann, H. OLEX2: A complete structure solution, refinement and analysis program. *J. Appl. Crystallogr.* **2009**, *42*, 339–341. [CrossRef]
68. Turner, M.J.; McKinnon, J.J.; Wolff, S.K.; Grimwood, D.J.; Spackman, P.R.; Jayatilaka, D.; Spackman, M.A. *CrystalExplorer17 2017*; University of Western Australia: Perth, Australia, 2017.
69. Spackman, M.A.; Byrom, P.G. A novel definition of a molecule in a crystal. *Chem. Phys. Lett.* **1997**, *267*, 215–220. [CrossRef]
70. Spackman, M.A.; Jayatilaka, D. Hirshfeld surface analysis. *CrystEngComm* **2009**, *11*, 19–32. [CrossRef]
71. Frisch, M.J.; Trucks, G.W.; Schlegel, H.B.; Scuseria, G.E.; Robb, M.A.; Cheeseman, J.R.; Scalmani, G.; Barone, V.; Mennucci, B.; Petersson, G.A.; et al. *Gaussian 09, Revision C.01*; Gaussian, Inc.: Wallingford, CT, USA, 2009.
72. Becke, A.D. Density-functional thermochemistry. III. The role of exact exchange. *J. Chem. Phys.* **1993**, *98*, 5648. [CrossRef]
73. Dunning, T.H. Gaussian basis sets for use in correlated molecular calculations. I. The atoms boron through neon and hydrogen. *J. Chem. Phys.* **1989**, *90*, 1007. [CrossRef]
74. Hay, P.J.; Wadt, W.R. Ab initio effective core potentials for molecular calculations. Potentials for the transition metal atoms Sc to Hg. *J. Chem. Phys.* **1985**, *82*, 270–283. [CrossRef]
75. Hay, P.J.; Wadt, W.R. Ab initio effective core potentials for molecular calculations. Potentials for K to Au including the outermost core orbitale. *J. Chem. Phys.* **1985**, *82*, 299–310. [CrossRef]
76. Weigend, F. Accurate Coulomb-fitting basis sets for H to Rn. *Phys. Chem. Chem. Phys.* **2006**, *8*, 1057. [CrossRef] [PubMed]
77. Bader, R.F.W. Atoms in molecules. *Acc. Chem. Res.* **1985**, *18*, 9–15. [CrossRef]
78. Bader, R.F.W. *Atoms in Molecules: A Quantum Theory*; Oxford University Press: Oxford, UK, 1990.
79. Caramori, G.F.; Parreira, R.L.T.; Ferreira, A.M.D.C. Isatin-Schiff base copper(II) complexes—A DFT study of the metal-ligand bonding situation. *Int. J. Quantum Chem.* **2012**, *112*, 625–646. [CrossRef]
80. Todd, A. *AIMAll, Version 19.10.12*; Keith, T.K., Ed.; Gristmill Software: Overland Park, KS, USA, 2019.
81. Savić, A.G.; Mojović, M. Free Radicals Identification from the Complex EPR Signals by Applying Higher Order Statistics. *Anal. Chem.* **2012**, *84*, 3398–3402. [CrossRef] [PubMed]
82. Nakarada, Đ.; Pejin, B.; Tommonaro, G.; Mojović, M. Liposomal integration method for assessing antioxidative activity of water insoluble compounds towards biologically relevant free radicals: Example of avarol. *J. Liposome Res.* **2020**, *30*, 218–226. [CrossRef]
83. Yan, J.; Zhang, G.; Pan, J.; Wang, Y. α-Glucosidase inhibition by luteolin: Kinetics, interaction and molecular docking. *Int. J. Biol. Macromol.* **2014**, *64*, 213–223. [CrossRef]
84. Suo, Z.; Sun, Q.; Yang, H.; Tang, P.; Gan, R.; Xiong, X.; Li, H. Combined spectroscopy methods and molecular simulations for the binding properties of trametinib to human serum albumin. *RSC Adv.* **2018**, *8*, 4742–4749. [CrossRef]
85. Zazeri, G.; Povinelli, A.P.R.; Lima, M.d.F.; Cornélio, M.L. Detailed Characterization of the Cooperative Binding of Piperine with Heat Shock Protein 70 by Molecular Biophysical Approaches. *Biomedicines* **2020**, *8*, 629. [CrossRef]
86. Zazeri, G.; Povinelli, A.; Lima, M.; Cornélio, M. Experimental Approaches and Computational Modeling of Rat Serum Albumin and Its Interaction with Piperine. *Int. J. Mol. Sci.* **2019**, *20*, 2856. [CrossRef]
87. Wu, D.; Liu, D.; Zhang, Y.; Zhang, Z.; Li, H. Unravelling the binding mechanism of benproperine with human serum albumin: A docking, fluorometric, and thermodynamic approach. *Eur. J. Med. Chem.* **2018**, *146*, 245–250. [CrossRef] [PubMed]
88. Bakar, K.A.; Feroz, S.R. A critical view on the analysis of fluorescence quenching data for determining ligand–protein binding affinity. *Spectrochim. Acta Part A Mol. Biomol. Spectrosc.* **2019**, *223*, 117337. [CrossRef] [PubMed]
89. Jevtovic, V.; Golubović, L.; Alshammari, O.A.O.; Alhar, M.S.; Alanazi, T.Y.A.; Rakic, V.; Ganguly, R.; Dimitrić Marković, J.; Rakić, A.; Dimić, D. The Counterion (SO_4^{2-} and NO_3^-) Effect on Crystallographic, Quantum-Chemical, Protein-, and DNA-Binding Properties of Two Novel Copper(II)–Pyridoxal-Aminoguanidine Complexes. *Crystals* **2024**, *14*, 814. [CrossRef]
90. Sugio, S.; Kashima, A.; Mochizuki, S.; Noda, M.; Kobayashi, K. Crystal structure of human serum albumin at 2.5 Å resolution. *Protein Eng. Des. Sel.* **1999**, *12*, 439–446. [CrossRef]
91. Baruah, H.; Wright, M.W.; Bierbach, U. Solution Structural Study of a DNA Duplex Containing the Guanine-N7 Adduct Formed by a Cytotoxic Platinum-Acridine Hybrid Agent. *Biochemistry* **2005**, *44*, 6059–6070. [CrossRef]
92. Valdés-Tresanco, M.S.; Valdés-Tresanco, M.E.; Valiente, P.A.; Moreno, E. AMDock: A versatile graphical tool for assisting molecular docking with Autodock Vina and Autodock4. *Biol. Direct* **2020**, *15*, 12. [CrossRef]

inorganics

MDPI

Article

In Vitro Evaluation of Silver-NHC Complexes Against a Clinical Isolate of *Acanthamoeba castellanii*: Time- and Dose-Dependent Effects

Zübeyda Akın-Polat [1], Neslihan Şahin [2,*], Shaima Hkiri [3], Bui Minh Thu Ly [3], İsmail Özdemir [4] and David Sémeril [3,*]

[1] Departments of Medical Parasitology, Cumhuriyet University School of Medicine, 58140 Sivas, Türkiye
[2] Department of Science Education, Faculty of Education, Cumhuriyet University, 58040 Sivas, Türkiye
[3] Synthèse Organométallique et Catalyse, UMR-CNRS 7177, Strasbourg University, 67008 Strasbourg, France
[4] Drug Application and Research Center, İnönü University, 44280 Malatya, Turkey
* Correspondence: neslihan@cumhuriyet.edu.tr (N.Ş.); dsemeril@unistra.fr (D.S.)

Abstract: The synthesis of a series of six chloro[*N*-alkyl-*N*-cinnamyl-benzimidazol-2-yliden]silver(I) complexes was successfully achieved, wherein allyl (**3a**), methoxymethyl (**3b**), benzyl (**3c**), 3-fluorobenzyl (**3d**), 4-fluorobenzyl (**3e**) and 4-methyl-benzyl (**3f**) substituents were grafted on the benzimidazole ring. The isolated silver *N*-heterocyclic carbene (NHC) complexes were identified by microanalyses and mass spectrometry and characterized by FT-IR and NMR spectroscopic techniques. Conclusive evidence for the structures of complexes **3c** and **3d** was provided by single-crystal X-ray crystallography. The *in vitro* inhibitory activity of the six Ag-NHC complexes was tested against trophozoites and cysts of the pathogenic *Acanthamoeba castellanii* strain and the efficacy sequence is as follows: **3d** > **3c** > **3f** > **3a** > **3b** > **3e**. At a concentration of 100 μM in complexes **3c**, **3d** and **3f** and after 72 h of incubation, 5.3, 3.2 and 6.3% *A. castellanii* trophozoite viabilities were observed, respectively. The utilization of elevated silver(I) drug concentrations, 1000 μM, resulted in the near-total eradication of pathogenic protozoa.

Keywords: silver *N*-heterocyclic carbene complexes; spectroscopic techniques; X-ray crystal structures; *Acanthamoeba castellanii*; *in vitro* evaluation

1. Introduction

Acanthamoeba spp. are free-living amoebae that are widely distributed in soil, freshwater, and the air [1,2]. Some species have been identified as opportunistic human pathogens. These protozoa can cause severe infections, including *Acanthamoeba* keratitis (AK), which is a painful corneal disease and granulomatous amoebic encephalitis, which is a typically fatal infection of the central nervous system occurring in immunocompromised individuals [3–5]. *Acanthamoeba* species exhibit a biphasic life cycle, characterized by the alternating phase between a metabolically active trophozoite stage and a dormant cyst stage. The latter stage is distinguished by its resilience to environmental stress and therapeutic agents. Furthermore, it has been demonstrated that both stages are capable of initiating infection [2,4].

AK has been identified as the most prevalent disease caused by *Acanthamoeba*. It primarily affects contact lens wearers, particularly those with corneal microtrauma or poor lens care practices, such as inadequate cleaning or exposure to contaminated water during activities like swimming or showering. The clinical manifestations include severe ocular pain, conjunctival hyperaemia, blurred vision, photophobia, and a foreign body sensation in the eye. Given the non-specific nature of these symptoms, they frequently masquerade as other forms of keratitis. This underscores the critical importance of a prompt and accurate diagnosis to avert severe consequences. In the absence of intervention, AK has the potential to escalate to a state of corneal ulceration, which may consequently lead to irreversible visual impairment or even total blindness [6–9].

Presently, there is an absence of an FDA-approved standard treatment protocol for AK. Treatment regimens generally consist of prolonged therapeutic interventions that incorporate topical antiseptics, such as polyhexamethylene biguanide and chlorhexidine, in combination with antimicrobials, including diamidines and azoles. Notwithstanding these therapeutic regimens, the degree of therapeutic success remains limited, and recurrence is not uncommon. Moreover, the majority of commercial contact lens disinfectants demonstrate an absence of amoebicidal activity against *Acanthamoeba*, thereby increasing the risk of infection for lens wearers [10–12]. These challenges underscore the pressing need to develop more effective, targeted therapeutic strategies for AK [13,14].

The exploration of organometallic compounds as potential drug candidates has emerged as a promising field within the bioinorganic chemistry [15,16]. Among these compounds, metal-*N*-heterocyclic carbene (NHC) complexes have attracted considerable attention in synthetic chemistry [17–19]. The introduction of NHCs into organometallic chemistry is attributed to Fischer in 1964. However, it was not until 1991 that a free NHC was successfully isolated, despite the existence of early reports of NHC metal complexes dating back to 1968 [20–24]. NHCs are distinguished by their robust stability, which is attributed to the coordination of two nitrogen atoms in close proximity, enabling them to form strong bonds with transition metals [25]. The bonding interactions between NHC ligands and metal centers are shaped by a combination of steric and electronic effects [26]. Initially, it was hypothesized that NHC complexes function exclusively as σ-donors. However, subsequent research has revealed a more intricate nature of these interactions. The inherent stability of NHCs, in conjunction with their facile synthesis, has resulted in their extensive utilization in the domain of coordination chemistry. Among metal-NHC complexes, Ag-NHC complexes are notable for their biological properties [27–30].

While biguanides, diamidines, and azoles have demonstrated efficacy in the treatment of *Acanthamoeba*, limitations such as prolonged treatment durations, toxicity to human corneal cells, and incomplete cyst eradication have prompted the exploration of novel agents [10,12]. Silver-based compounds, including silver nitrate, silver sulfadiazine, and silver acetate, have been recognized for their antimicrobial properties for a considerable duration. These compounds have been employed as effective microbiocidal agents for millennia [31–34]. Furthermore, studies have demonstrated the efficacy of silver nanoparticles in combating *Acanthamoeba* [35–37]. Notably, silver compounds exhibit minimal toxicity toward human cells, particularly mammalian cell membranes.

To date, and to the best of our knowledge, there has been no report of research on the use of silver-*N*-heterocyclic carbene complexes against *A. castellanii*. In the present article, we report on the synthesis of six chloro[*N*-alkyl-*N*-cinnamyl-benzimidazol-2-yliden]silver(I) complexes, **3a–f** (Figure 1), and their *in vitro* tests against *A. castellanii*.

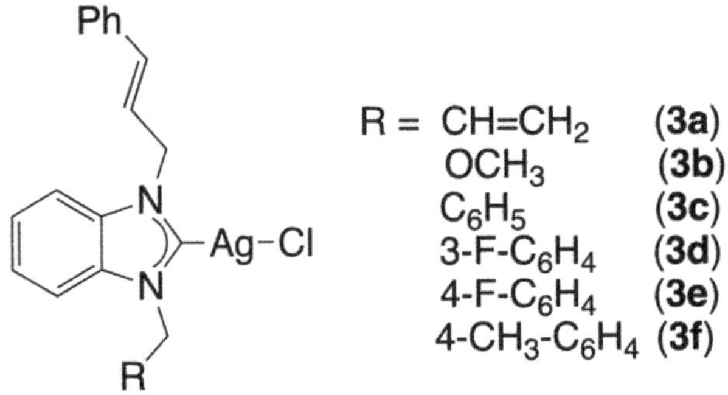

Figure 1. Silver complexes **3a–f** tested against *Acanthamoeba castellanii* in the present study.

2. Results and Discussion

2.1. Synthesis of Silver Complexes

Six silver(I) complexes (**3a-f**), namely chloro[1-cinnamyl-3-allyl-benzimidazol-2-yliden]silver(I) (**3a**), chloro[1-methoxymethyl-3-cinnamyl-benzimidazol-2-yliden]silver(I) (**3b**), chloro[1-benzyl-3-cinnamyl-benzimidazol-2-yliden]silver(I) (**3c**), chloro[1-(3-fluorobenzyl)-3-cinnamyl-benzimidazol-2-yliden]silver(I) (**3d**), chloro[1-(4-fluorobenzyl)-3-cinnamyl-benzimidazol-2-yliden]silver(I) (**3e**) and chloro[1-(4-methyl-benzyl)-3-cinnamyl-benzimidazol-2-ylidene]silver(I) (**3f**), were obtained in satisfactory yields (33–86%) from a reaction of Ag_2O with the benzimidazolium salt (**1a–f**) in dichloromethane for 24 h at room temperature, with the reaction protected from light by aluminum foil (Scheme 1). The benzimidazolium salts **1a–f** were previously prepared by alkylation of 1-cinnamyl-benzimidazole (**2**) and aryl chloride. Note that the alkyl substituents grafted on the benzimidazole only slightly modify the σ-donor properties of the *N*-heterocyclic carbenes. Indeed, the $^1J_{CH}$ constants demonstrate minimal variability within the range of 220–225 Hz (Table 1). This finding indicates that these ligands exhibit weak σ-donor properties, which are comparable to those of 1,3-dicyclohexyl-imidazolium salt [38].

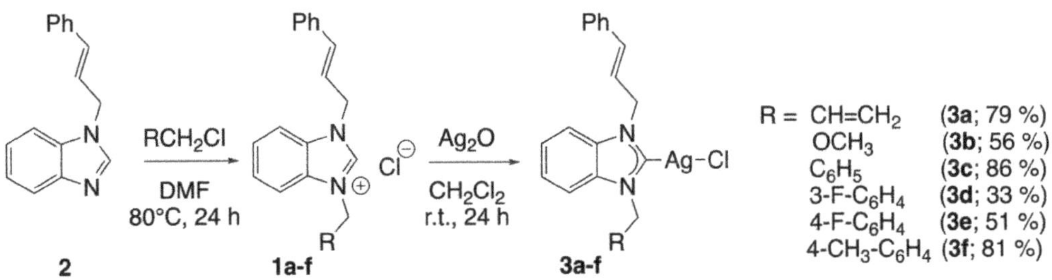

Scheme 1. Synthesis of silver complexes **3a–f**.

The silver(I) complexes demonstrated stability in the presence of air and moisture; however, they exhibited sensitivity to light. In the dark, these complexes remained stable in solution of dimethyl sulfoxide (DMSO) for at least three days. The characterization of the inorganic samples was performed using Fourier transform infrared spectroscopy (FT-IR), nuclear magnetic resonance spectroscopy (^1H, $^{13}C\{^1H\}$, and $^{19}F\{^1H\}$ NMR) and mass spectrometry (see the experimental section, Supplementary Materials and Table 1).

Table 1. Spectroscopic data of silver(I) complexes **3a–f** and their corresponding benzimidazolium salts **1a–f**.

Benzimidazolium Salts					**Silver Complexes**		
		FT-IR ν**(CN) (cm**$^{-1}$**)**	^{1}**H NMR NCHN (CDCl**$_3$**; ppm)**	^{13}**C{**^{1}**H} NMR NCHN (CDCl**$_3$**; ppm)**		**FT-IR** ν**(CN) (cm**$^{-1}$**)**	^{13}**C{**^{1}**H} NMR NC(Ag)N (DMSO-**d_6**; ppm)**
1a		1556	11.78 ($^{1}J_{CH}$ = 220 Hz)	143.88	**3a**	1387	188.49
1b		1556	11.94 ($^{1}J_{CH}$ = 223 Hz)	144.14	**3b**	1379	Not observed
1c	[39]	1560	12.02 ($^{1}J_{CH}$ = 221 Hz)	144.03	**3c**	1394	Not observed
1d	[39]	1562	12.15 ($^{1}J_{CH}$ = 220 Hz)	144.26	**3d**	1385	189.04
1e	[39]	1562	10.07 ($^{1}J_{CH}$ = 220 Hz)	142.71	**3e**	1391	188.64
1f		1556	12.17 ($^{1}J_{CH}$ = 225 Hz)	144.09	**3f**	1387	Not observed

The benzimidazolium salts **1a–f** exhibited a distinctive ν(CN) band within the FT-IR spectra, with a frequency range of 1556–1562 cm^{-1}. Following the formation of silver complexes, no peaks were detected in this region, and a downshift was observed for the ν(CN) bands (1379–1394 cm^{-1}). These values are in agreement with those previously observed by Panda, Ghosh and co-workers in the chloro[1-benzyl-3-*tert*-butylimidazol-2-yliden]silver(I) complex (ν(CN) at 1375 cm^{-1}) [40]. The observed change in stretching frequency can be attributed to a weakening of the CN bond, which becomes a partial double bond after coordination. This assertion is supported by the X-ray structures of two complexes (*vide infra*), in which the distances of carbon-nitrogen bonds are longer than that of a C=N bond (Table 1).

A thorough analysis of the ^{1}H NMR spectra of the silver-NHC complexes (**3a–f**) revealed the absence of the downfield-shifted resonance (δ = 10.07–12.17 ppm) of the acid proton (NC*H*N) from the spectra of the *N*-heterocyclic carbene precursor salts, indicative of deprotonation of the benzimidazolium carbene. In the ^{13}C{^{1}H} NMR spectra of these salts, the signals from the carbon atom between the two nitrogen atoms (NCHN), present in the 142.71–144.26 ppm range, undergo significant deshielding after coordination to the silver atom. The analysis of the spectra of the complexes reveals the presence of NC(Ag)N carbon peaks in the 188.49–189.04 ppm range for **3a, 3d** and **3e**. As is the case with numerous silver(I) complexes, the signal in question could not be observed in the case of complexes **3b**, **3c** and **3f** [41]. During the coordination step, the *trans* conformation of the double bond of the cinnamyl substituent remains unchanged, the coupling constant between the two protons of the double bond is approximately 16 Hz (Table 1).

A detailed analysis of the mass spectra of the six silver(I) complexes confirmed the successful formation of the desired complexes. Each organometallic compound displays a peak that corresponds to the cations [M − Cl]$^+$ at m/z = 481.05, 385.05, 431.06 and 445.08 for complexes **3a**, **3b**, **3c** and **3f**, respectively and/or [M − Cl + CH_3CN]$^+$ at m/z = 422.07, 472.08, 490.08, 490.10 and 486.11 for complexes **3a**, **3c**, **3d**, **3e** and **3f**, respectively, with the expected isotopic profiles.

2.2. Single-Crystal X-Ray Diffraction Studies

The formation of a silver(I) complex of the (NHC)Ag-Cl type was confirmed through two single-crystal X-ray diffraction studies of complexes **3c** and **3d**. The complexes exhibit centrosymmetric crystallization in the $P\overline{1}$ triclinic space group, with the center of inversion situated at the midpoint of the [Ag_2Cl_2] bridge core.

The formation of dimers is achieved through the establishment of an additional Ag–Cl bond (Ag1–Cl1a of 2.968(3) and 2.9320(6) Å for complexes **3c** and **3d**, respectively), which are substantially longer than the direct Ag1–Cl1 bond (Ag1–Cl1 of 2.3854(13) and 2.3972(5) Å for complexes **3c** and **3d**, respectively) (Figures 2 and 3). Dimer formation is frequently observed during the crystallization of (NHC)Ag-Cl complexes, as evidenced in chloro[1-(9-acridinyl)-3-methylimidazol-2-yliden]silver(I) [42] or chloro[1-(*N*,*N*-diethylcarbamoylmethyl)-3-mesitylimidazol-2-ylidene]silver(I) [43] complexes.

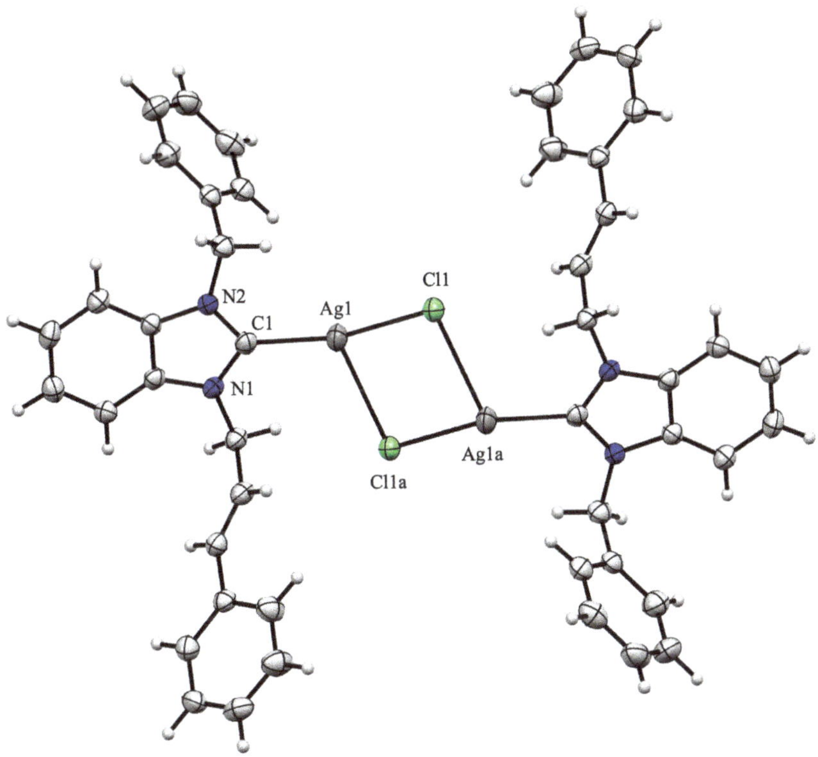

Figure 2. Molecular structure of silver(I) complex **3c**. The ORTEP drawing, with 50% probability thermal ellipsoid, shows the atom labelling. Important bond lengths (Å) and angles (°): Ag1-Cl1 2.3854(13), Ag1-Cl1a 2.968(3), Ag1-C1 2.096(3), N1-C1 1.350(4), N1-C2, 1.396(3), C1-Ag1-Cl1 163.55(7), Cl1-Ag1-Cl1a 86.49(6), Cl1a-Ag1-C1 109.24(9), N1-C1-Ag1 128.69(19), Ag1-C1-N2 125.1(2) and N2-C1-N1 105.7(2).

The $C_{carbene}$-Ag-Cl alignment is angled (C1-Ag1-Cl1 163.55(7) and 162.33(6)° for complexes **3c** and **3d**, respectively), and due to dimer formation, the Ag_2Cl_2 moiety is out of the plane of the benzimidazole ring, resulting in dihedral angles of 12.5 and 14.6° for complexes **3c** and **3d**, respectively. The Ag1-Cl1 (2.3854(13) and 2.3972(5) Å for complexes **3c** and **3d**, respectively) and Ag1-C1 distances (2.096(3) and 2.102(2) for complexes **3c** and **3d**, respectively) are within the usual range [44,45]. The two aromatic rings, which correspond to the cinnamyl and benzyl/3-fluorobenzyl substituents, are inclined toward the benzimidazole moiety, forming dihedral angles of 79.3/64.1° and 64.6/79.8°, respectively.

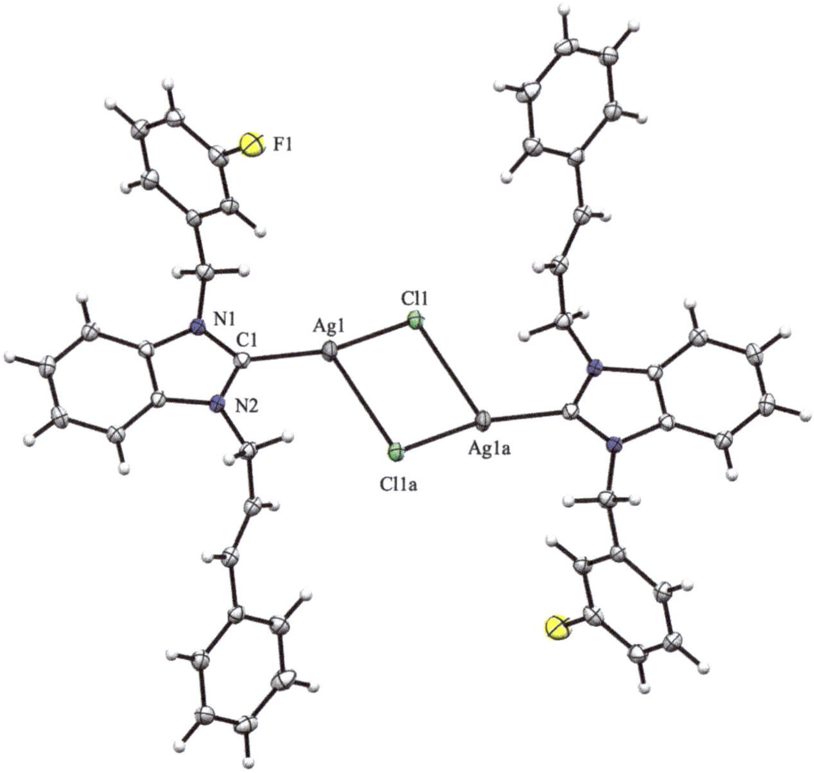

Figure 3. Molecular structure of silver(I) complex **3d**. The ORTEP drawing, with 50% probability thermal ellipsoid, shows the atom labelling. Important bond lengths (Å) and angles (°): Ag1-Cl1 2.3972(5), Ag1-Cl1a 2.9320(6), Ag1-C1 2.102(2), N1-C1 1.364(3), N1-C2, 1.355(3), C1-Ag1-Cl1 162.33(6), Cl1-Ag1-Cl1a 86.06(2), Cl1a-Ag1-C1 110.52(6), N1-C1-Ag1 125.43(15), Ag1-C1-N2 128.48(15) and N2-C1-N1 105.58(18).

2.3. *In Vitro Evaluation of Acanthamoeba castellanii*

2.3.1. Effect on *A. castellanii* Trophozoites

The antiamoebic activities of the six silver(I) complexes (**3a–f**), the six benzimidazolium salts (**1a–f**), $AgNO_3$, and Ag_2O were evaluated against *A. castellanii* trophozoites after 24, 48 or 72 h of incubation at concentrations of 10, 100 or 1000 µM (Figures 4 and 5).

A significant decrease in trophozoite viability was observed to be time- and concentration-dependent. The silver(I) complexes **3a–f** exhibited significantly higher antiamoebic activities than their corresponding benzimidazolium salts (**1a–f**). At the highest concentration (1000 µM), all complexes demonstrated strong trophozoite-killing effects within 24 h. Complexes **3a**, **3c**, **3d** and **3f** completely eliminated trophozoite viability within 72 h. The residual viability in the **3b** and **3e** treatments was 3.4 and 2.2%, respectively ($p < 0.05$). In contrast, the corresponding carbene precursor salts exhibited only minimal inhibition at this concentration, with no statistically significant reduction in viability compared with the control group.

Overall, efficacy was reduced at 100 µM, but a similar trend was observed. Complexes **3c**, **3d** and **3f** exhibited significant activity, with 5.3, 3.2 and 6.3% trophozoite viability, respectively, recorded after 72 h ($p < 0.05$). Ligands at the same concentration did not demonstrate any inhibitory effects. At the lowest tested concentration (10 µM), the complexes displayed minimal activities. All ligands yielded viability rates comparable to those of the untreated control group.

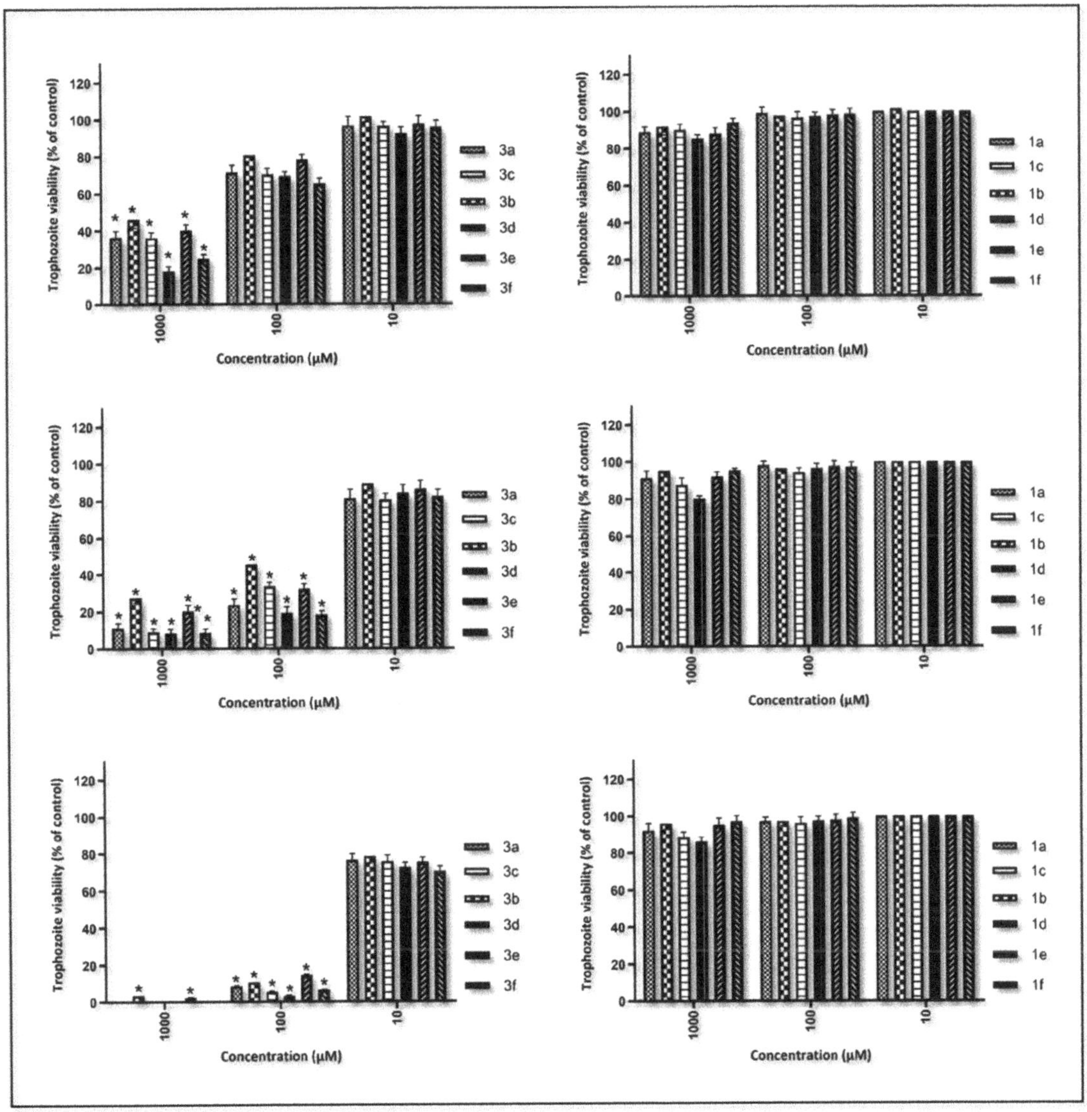

Figure 4. The viability of trophozoites at concentrations of 1000, 100 and 10 μM against the silver(I) complex (**3a–f**) and the benzimidazolium salts (**1a–f**) was assessed after 24, 48 and 72 h against the pathogenic *A. castellanii*. Data are expressed as the mean ± SD of four independent experiments (* $p < 0.05$ vs. control).

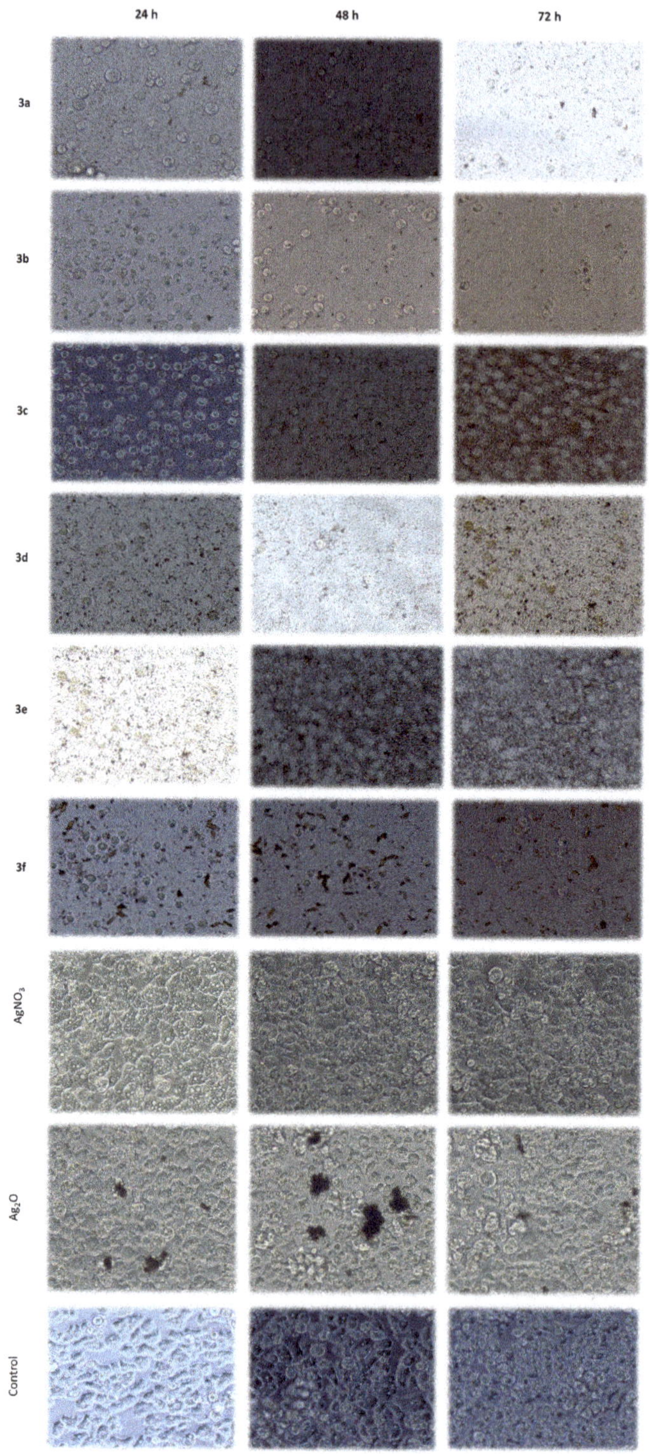

Figure 5. Inverted microscope images showing the viability of *A. castellanii* trophozoites treated with 1000 µM concentrations of silver(I) complexes (**3a–f**), $AgNO_3$ and Ag_2O after 24, 48 and 72 h (magnification: 200×).

As illustrated in Figure 4, the efficacy of silver(I) complexes **3a–f** against *A. castellanii* exhibits variation in the following sequence: **3d** > **3c** > **3f** > **3a** > **3b** > **3e**. This sequence provides compelling evidence that the lipophilicity [46,47] of the silver(I) complexes does not significantly impact the efficacy of these drugs. In fact, the two complexes with the lowest calculated logP [48] (complexes **3a** and **3b** with logP of 3.99 and 3.42, respectively; Table 2) are the least effective against *A. castellanii*. This observation underscores the pivotal function of the arylmethyl substituent grafted on to the *N*-heterocyclic carbene. The nature (H, F or CH_3) and position (*meta* or *para*) of the substituent carried on this aromatic cycle are also of crucial importance. Indeed, the **3d** complex, which contains a fluorine atom in the *meta* position of the benzylic substituent, exhibits higher reactivity compared with its less polar **3e** counterpart, which possesses the fluorine atom in *para* position on the aromatic ring.

Table 2. Calculated logP [48] for silver(I) complexes **3a–f**.

Silver Complex	Calculated logP
3a	3.99
3b	3.42
3c	4.60
3d	4.82
3e	4.86
3f	5.01

A study was conducted to ascertain the impact of $AgNO_3$ and Ag_2O on trophozoite viability. The study encompassed all concentrations and exposure times to ensure a comprehensive investigation. The findings indicate that neither compound exhibited a statistically significant reduction in trophozoite viability compared with untreated controls. Morphological evaluations substantiated these findings, disclosing the absence of cytopathic effects or aberrations from the conventional structure and motility of viable trophozoites (see Figure 5). The findings suggest that the silver salts examined were ineffective against *Acanthamoeba* trophozoites under these conditions, thereby emphasizing the enhanced efficacy and selectivity of the silver(I) complexes described in the present study.

2.3.2. Effect on *A. castellanii* Cysts

Following a 72 h observation period, no excystation or trophozoite formation was detected in cysts treated with complexes **3a**, **3c**, **3d** or **3f** at the highest tested concentration (1000 µM). This finding suggests the complete cysticidal activity of the complexes. Conversely, exposure to complexes **3b** and **3e** resulted in limited and delayed growth in cysts. Furthermore, minimal excystation was observed after 72 h at a concentration of 100 µM, and after 48 h at a concentration of 1000 µM, in cysts treated with complexes **3a**, **3c**, **3d** and **3f**. These results are of particular significance in light of the well-documented resistance of *A. castellanii* cysts to conventional antimicrobials [13]. The absence of any detectable activity by a ligand alone further emphasizes the importance of complexing with silver to achieve anti-amoebic potency.

At the lowest tested concentration (10 µM), none of the silver(I) complexes exhibited detectable cysticidal activity. In a similar manner, the ligands (**1a–f**) did not affect cyst viability at any time point (24, 48 or 72 h), and trophozoite outgrowth occurred at levels comparable to the untreated control. These results indicate that the ligands, when administered individually, were not efficacious against *A. castellanii* cysts under the conditions of the study. The present findings suggest that the silver(I) **3a–f** complexes may offer an effective alternative or adjunctive therapeutic strategy by targeting both the trophozoite and cyst stages.

3. Materials and Methods

3.1. General

The synthesis of silver(I) complexes was carried out under an inert atmosphere of argon with dried solvents. ^{1}H, ^{13}C{^{1}H} and ^{19}F{^{1}H} NMR spectra were recorded with Bruker Avance III spectrometer (300 or 500 M Hz). The spectra were calibrated according to the residual protonated solvent in $CDCl_3$ (δ = 7.26 and 77.16 ppm for ^{1}H and ^{13}C, respectively) or in DMSO-d_6 (δ = 2.50 and 39.52 ppm for ^{1}H and ^{13}C, respectively). ^{19}F NMR spectra are given relative to external CCl_3F. Chemical shifts and coupling constants are reported in ppm and Hz, respectively. ESI-TOF spectra were recorded on a Bruker MicroTOF spectrometer. Infrared spectra were recorded on a Bruker FT-IR Alpha-P spectrometer. 1-Cinnamyl-benzimidazole (**2**) [49], 1-benzyl-3-cinnamyl-benzimidazolium chloride (**1c**) [39], 1-(3-fluorobenzyl)-3-cinnamyl-benzimidazolium chloride (**1d**) [39], and 1-(4-fluorobenzyl)-3-cinnamyl-benzimidazolium chloride (**1e**) [39] were prepared according to literature procedures.

3.2. General Procedure for the Synthesis of Benzimidazolium Salts

The 1-cinnamyl-benzimidazole (**2**) (234 mg, 1 mmol) was first dissolved in DMF (5 mL) before the addition of aryl chloride (1 mmol). The resulting solution was then subjected to a heating process at a temperature of 80 °C for a duration of 24 h. Subsequent to a period of cooling to room temperature, the salt underwent precipitation by the addition of Et_2O (100 mL). The white solid was filtered and washed twice with Et_2O (2 × 10 mL) prior to being dried under vacuum.

1-Cinnamyl-3-allyl-benzimidazolium chloride (**1a**): Yield: 36%; FT-IR: ν(CN) 1556 cm^{-1}; ^{1}H NMR (300 M Hz, $CDCl_3$): δ = 11.78 (s, 1H, NCHN), 7.79–7.74 (m, 1H, arom CH), 7.72–7.67 (m, 1H, arom CH), 7.64–7.58 (m, 2H, arom CH), 7.41–7.37 (m, 2H, arom CH), 7.34–7.28 (m, 3H, arom CH), 6.92 (d, 1H, CH=*CH*Ph, $^3J_{HH}$ = 15.9 Hz), 6.39 (dt, 1H, *CH*=CHPh, $^3J_{HH}$ = 15.9 Hz, $^3J_{HH}$ = 6.7 Hz), 6.18–6.05 (m, 1H, *CH*=CH_2), 5.52–5.46 (m, 4H, *CH*$_2$CH=CH and CH=*CH*$_2$), 5.31 (d, 2H, *CH*$_2$CH=CH_2, $^3J_{HH}$ = 6.0 Hz); ^{13}C{^{1}H} NMR (126 M Hz, $CDCl_3$): δ = 143.88 (s, NCHN), 137.18 (s, CH=CHPh), 135.12, 131.67, 131.64, 128.99, 128.88, 127.30, 127.24, 127.08, 113.81, 113.78 (10 s, arom Cs), 129.71 (s, CH=CH_2), 122.02 (s, CH=CH_2), 120.32 (s, CH=CHPh), 50.37 (s, CH_2CH=CHPh), 50.21 (s, CH_2C=CH_2) ppm. Elemental analysis (%): calcd for $C_{19}H_{19}N_2Cl$ (310.82): C: 73.42; H: 6.16; N 9.01; found C: 73.34; H: 6.21; N: 8.97.

1-Methoxymethyl-3-cinnamyl-benzimidazolium chloride (**1b**): Yield: 38%; FT-IR: ν(CN) 1556 cm^{-1}; ^{1}H NMR (500 M Hz, $CDCl_3$): δ = 11.94 (s, 1H, NCHN), 7.81 (d, 1H, arom CH, $^3J_{HH}$ = 9.0 Hz), 7.72 (d, 1H, arom CH, $^3J_{HH}$ = 9.0 Hz), 7.60–7.55 (m, 2H, arom CH), 7.34 (d, 2H, arom CH, $^3J_{HH}$ = 8.5 Hz), 7.26–7.19 (m, 3H, arom CH), 6.86 (d, 1H, CH=*CH*Ph, $^3J_{HH}$ = 15.5 Hz), 6.39 (d br, 1H, *CH*=CHPh, $^3J_{HH}$ = 15.5 Hz), 6.03 (s, 2H, *CH*$_2$OCH$_3$), 5.42 (d, 2H, *CH*$_2$CH=CH, $^3J_{HH}$ = 2.0 Hz), 3.46 (s, 3H, CH_2O*CH*$_3$); ^{13}C{^{1}H} NMR (126 M Hz, $CDCl_3$): δ = 144.14 (s, NCHN), 137.35 (s, CH=CHPh), 135.03, 131.73, 131.16, 129.03, 128.89, 127.66, 127.59, 127.09, 114.42, 113.73 (10 s, arom Cs), 120.07 (s, CH=CHPh), 79.15 (s, CH_2OCH_3), 57.96 (s, CH_2OCH_3), 50.43 (s, CH_2CH=CHPh) ppm. Elemental analysis (%): calcd for $C_{18}H_{19}N_2OCl$ (314.81): C: 68.67; H: 6.08; N 8.90; found C: 68.64; H: 6.01; N: 8.86.

1-(4-Methyl-benzyl)-3-cinnamyl-benzimidazolium chloride (**1f**): Yield: 90%; FT-IR: ν(CN) 1556 cm^{-1}; ^{1}H NMR (300 M Hz, $CDCl_3$): δ = 12.17 (s, 1H, NCHN), 7.73 (d, 1H, arom CH, $^3J_{HH}$ = 9.9 Hz), 7.60–7.50 (m, 3H, arom CH), 7.42–7.39 (m, 4H, arom CH), 7.36–7.28 (m, 3H, arom CH), 7.18 (d, 2H, arom CH, $^3J_{HH}$ = 8.1 Hz), 6.92 (d, 1H, CH=*CH*Ph, $^3J_{HH}$ = 15.6 Hz), 6.44 (dt, 1H, *CH*=CHPh, $^3J_{HH}$ = 15.6 Hz, $^3J_{HH}$ = 6.6 Hz), 5.80 (s, 2H, *CH*$_2$C$_6$H$_4$CH$_3$), 5.50 (d, 2H, *CH*$_2$CH=CH, $^3J_{HH}$ = 6.6 Hz), 2.32 (s, 3H, C_6H_4*CH*$_3$); ^{13}C{^{1}H} NMR (126 M Hz, $CDCl_3$): δ = 144.09 (s, NCHN), 139.55, 137.25, 131.73, 131.54, 130.23, 129.70, 129.06, 128.93, 128.50, 127.21, 127.11, 120.27, 113.92, 113.75 (14 s, arom Cs), 135.10 (s, CH=CHPh), 127.23 (s,

CH=CHPh), 51.61 (s, $CH_2C_6H_4CH_3$), 50.30 (s, CH_2CH=CHPh), 21.34 (s, $C_6H_4CH_3$) ppm. Elemental analysis (%): calcd for $C_{24}H_{23}N_2Cl$ (374.90): C: 76.89; H: 6.18; N 7.47; found C: 77.02 H: 6.25, N: 7.39.

3.3. General Procedure for the Preparation of Silver Complexes

In a Schlenk tube, a solution of benzimidazolium salt (1.0 mmol) in CH_2Cl_2 (50 mL) was prepared under argon atmosphere. Subsequently, Ag_2O (2.2 mmol) was added to the mixture, which was stirred in a dark environment at room temperature for a period of 24 h. Following this, the reaction mixture was filtered through a celite bed, and the volume of the resulting solution was reduced to approximately 5 mL. The silver complex was precipitated with the addition of Et_2O (60 mL). The mixture was subsequently filtered, washed with Et_2O, and then dried under vacuum.

Chloro[1-cinnamyl-3-allyl-benzimidazol-2-yliden]silver(I) (**3a**): Yield: 79%; FT-IR: ν(CN) 1387 cm^{-1}; ^{1}H NMR 300 M Hz, DMSO-d_6): δ = 7.86–7.83 (m, 1H, arom CH), 7.78–7.74 (m, 1H, arom CH), 7.46–7.40 (m, 4H, arom CH), 7.30–7.21 (m, 3H, arom CH), 6.74 (d, 1H, CH=C*H*Ph, $^3J_{HH}$ = 15.9 Hz), 6.55 (dt, 1H, C*H*=CHPh, $^3J_{HH}$ = 15.9 Hz, $^3J_{HH}$ = 6.1 Hz), 6.14–6.02 (m, C*H*=CH_2), 5.28 (d, 2H, CH_2CH=CH, $^3J_{HH}$ = 6.1 Hz), 5.25 (dd, 1H, CH=CH_2, $^3J_{HH}$ = 9.9 Hz, $^3J_{HH}$ = 1.2 Hz), 5.20 (dd, 1H, CH=CH_2, $^3J_{HH}$ = 17.1 Hz, $^3J_{HH}$ = 1.2 Hz), 5.13 (d, 2H, CH_2CH=CH_2, $^3J_{HH}$ = 5,4 Hz), ^{13}C{^{1}H} NMR (126 M Hz, DMSO-d_6): δ = 188.49 (s, NCN), 135.66, 133.37, 133.34, 128.57, 128.01, 126.52, 124.58, 112.28, 112.25 (9 s, arom Cs), 133.36 (s, CH=CHPh), 124.04 (s, CH=CHPh), 124.02 (s, CH=CH_2), 118.29 (s, CH=CH_2), 50.87 (s, CH_2CH=CHPh), 50.59 (s, CH_2C=CH_2) ppm. MS (ESI-TOF): m/z = 381.05 $[M - Cl]^+$ and 422.07 $[M - Cl + CH_3CN]^+$ (expected isotopic profiles). Elemental analysis (%): calcd for $C_{19}H_{18}N_2AgCl$ (417.68): C: 54.64; H: 4.34; N 6.71; found C: 54.72; H: 4.39; N: 6.62.

Chloro[1-methoxymethyl-3-cinnamyl-benzimidazol-2-yliden]silver(I) (**3b**): Yield: 56%; FT-IR: ν(CN) 1379 cm^{-1}; ^{1}H NMR (500 M Hz, DMSO-d_6): δ = 7.88–7.83 (m, 2H, arom CH), 7.49–7.47 (m, 2H, arom CH), 7.42 (d, 2H, arom CH, $^3J_{HH}$ = 7.0 Hz), 7.28 (t, 2H, arom CH, $^3J_{HH}$ = 7.2 Hz), 7.22 (t, 1H, arom CH, $^3J_{HH}$ = 7.2 Hz), 6.75 (d, 1H, CH=C*H*Ph, $^3J_{HH}$ = 16.0 Hz), 6.55 (dt, 1H, C*H*=CHPh, $^3J_{HH}$ = 16.0 Hz, $^3J_{HH}$ = 6.0 Hz), 5.83 (s, 2H, CH_2OCH_3), 5.31 (d, 2H, CH_2CH=CH, $^3J_{HH}$ = 6.0 Hz), 3.31 (s, 3H, CH_2OCH_3); ^{13}C{^{1}H} NMR (126 M Hz, DMSO-d$_6$): δ = 135.57, 133.71, 133.12, 128.62, 128.08, 126.57, 124.49, 112.50, 112.48 (9 s, arom Cs), 133.40 (s, CH=CHPh), 124.46 (s, CH=CHPh), 80.01 (s, CH_2OCH_3), 56.19 (s, CH_2OCH_3), 50.73 (s, CH_2CH=CHPh) ppm. MS (ESI-TOF): m/z = 385.05 $[M - Cl]^+$ (expected isotopic profiles). Elemental analysis (%): calcd for $C_{18}H_{18}N_2OAgCl$ (421.67): C: 51.27; H: 4.30; N 6.64; found C: 51.18; H: 4.26; N: 6.57.

Chloro[1-benzyl-3-cinnamyl-benzimidazol-2-yliden]silver(I) (**3c**). Yield: 86%; FT-IR: ν(CN) 1394 cm^{-1}; ^{1}H NMR (300 M Hz, DMSO-d_6): δ = 6.82 (d, 1H, arom CH, $^3J_{HH}$ = 9.9 Hz), 6.70 (d, 1H, arom CH, $^3J_{HH}$ = 9.9 Hz), 6.43–6.23 (m, 12H, arom CH), 5.71 (d, 1H, CH=C*H*Ph, $^3J_{HH}$ = 15.9 Hz), 5.52 (dt, 1H, C*H*=CHPh, $^3J_{HH}$ = 15.9 Hz, $^3J_{HH}$ = 6.0 Hz), 4.70 (s, 2H, CH_2Ph), 4.26 (d, 2H, CH_2CH=CH, $^3J_{HH}$ = 6.0 Hz); ^{13}C{^{1}H} NMR (126 M Hz, DMSO-d_6): δ = 136.43, 135.78, 133.69, 129.00, 128.79, 128.26, 128.24, 127.54, 126.71, 124.36, 124.32, 112.52, 112.51 (13 s, arom Cs), 133.48 (s, CH=CHPh), 124.57 (s, CH=CHPh), 52.17 (s, CH_2Ph), 50.78 (s, CH_2CH=CHPh) ppm. MS (ESI-TOF): m/z = 431.06 $[M - Cl]^+$ and 472.08 $[M - Cl + CH_3CN]^+$ (expected isotopic profiles). Elemental analysis (%): calcd for $C_{23}H_{20}N_2AgCl$ (467.74): C: 59.06; H: 4.31; N 5.99; found C: 58.98; H: 4.35; N: 5.87.

Chloro[1-(3-fluorobenzyl)-3-cinnamyl-benzimidazol-2-yliden]silver(I) (**3d**): Yield: 33%; FT-IR: ν(CN) 1385 cm^{-1}; ^{1}H NMR (500 M Hz, DMSO-d_6): δ = 7.85 (dd, 1H, arom CH, $^3J_{HH}$ = 7.5 Hz, $^4J_{HH}$ = 1.5 Hz), 7.72 (dd, 1H, arom CH, $^3J_{HH}$ = 7.0 Hz, $^4J_{HH}$ = 1.5 Hz), 7.45–7.35 (m, 5H, arom CH), 7.29–7.19 (m, 5H, arom CH), 7.14 (td, 1H, arom CH, $^3J_{HH}$ = 8.7 Hz, $^4J_{HH}$ = 2.5 Hz), 6.74 (d, 1H, CH=C*H*Ph, $^3J_{HH}$ = 16.0 Hz), 6.55 (dt, 1H, CH=CHPh, $^3J_{HH}$ = 16.0 Hz, $^3J_{HH}$

= 6.0 Hz), 5.76 (s, 2H, $CH_2C_6H_4F$), 5.30 (d, 2H, CH_2CH=CH, $^3J_{HH}$ = 6.0 Hz); $^{13}C\{^1H\}$ NMR (126 M Hz, DMSO-d_6): δ = 189.04 (s, NCN), 162.19 (d, CF, $^1J_{CF}$ = 244.9 Hz), 135.66, 133.54, 133.29, 128.58, 128.02, 126.53, 124.21, 124.17, 112.41, 112.28 (10 s, arom Cs), 139.11 (d, arom C_{quat}, $^3J_{CF}$ = 7.3 Hz), 133.29 (s, CH=CHPh), 130.87 (d, arom CH, $^3J_{CF}$ = 8.3 Hz), 124.57 (s, CH=CHPh), 123.36 (d, arom CH, $^4J_{CF}$ = 2.9 Hz), 114.94 (d, arom CH, $^2J_{CF}$ = 20.8 Hz), 114.30 (d, arom CH, $^2J_{CF}$ = 21.9 Hz), 51.32 (s, $CH_2C_6H_4F$), 50.68 (s, CH_2CH=CHPh); $^{19}F\{^1H\}$ NMR (282 M Hz, DMSO-d_6): δ = -112.42 (s, C_6H_4F) ppm. MS (ESI-TOF): m/z = 490.08 [M − Cl + CH_3CN]$^+$ (expected isotopic profiles). Elemental analysis (%): calcd for $C_{23}H_{19}N_2FAgCl$ (485.73): C: 56.87; H: 3.94; N 5.77; found C: 56.85; H: 3.88; N: 5.74.

Chloro[1-(4-fluorobenzyl)-3-cinnamyl-benzimidazol-2-yliden]silver(I) (**3e**): Yield: 51%; FT-IR: ν(CN) 1391 cm^{-1}; ^1H NMR (500 M Hz, DMSO-d_6): δ = 7.84 (d, 1H, arom CH, $^3J_{HH}$ = 9.0 Hz), 7.73 (d, 1H, arom CH, $^3J_{HH}$ = 9.0 Hz), 7.46–7.39 (m, 6H, arom CH), 7.28 (t, 2H, arom CH, $^3J_{HH}$ = 7.5 Hz), 7.22 (t, 1H, arom CH, $^3J_{HH}$ = 7.2 Hz), 7.17 (t, 2H, arom CH, $^3J_{HH}$ = 9.0 Hz), 6.74 (d, 1H, CH=C*H*Ph, $^3J_{HH}$ = 16.0 Hz), 6.54 (dt, 1H, C*H*=CHPh, $^3J_{HH}$ = 16.0 Hz, $^3J_{HH}$ = 6.0 Hz), 5.72 (s, 2H, $CH_2C_6H_4F$), 5.29 (d, 2H, CH_2CH=CH, $^3J_{HH}$ = 6.0 Hz); $^{13}C\{^1H\}$ NMR (126 M Hz, DMSO-d_6): δ = 188.64 (s, NCN), 161.76 (d, CF, $^1J_{CF}$ = 244.8 Hz), 135.67, 133.55, 133.25, 128.59, 128.04, 126.55, 124.16, 124.13, 112.40, 112.34 (10 s, arom Cs), 133.31 (s, CH=CHPh), 132.54 (d, arom C_{quat}, $^4J_{CF}$ = 3.1 Hz), 129.59 (d, arom CH, $^3J_{CF}$ = 8.4 Hz), 124.57 (s, CH=CHPh), 115.64 (d, arom CH, $^2J_{CF}$ = 21.7 Hz), 51.20 (s, $CH_2C_6H_4F$), 50.68 (s, CH_2CH=CHPh); $^{19}F\{^1H\}$ NMR (282 M Hz, DMSO-d_6): δ = −114.15 (s, C_6H_4F) ppm. MS (ESI-TOF): m/z = 490.10 [M − Cl + CH_3CN]$^+$ (expected isotopic profiles). Elemental analysis (%): calcd for $C_{23}H_{19}N_2FAgCl$ (485.73): C: 56.87; H: 3.94; N 5.77; found C: 56.92; H: 4.11; N: 5.72.

Chloro[1-(4-methyl-benzyl)-3-cinnamyl-benzimidazol-2-ylidene]silver(I) (**3f**): Yield: 81%; FT-IR: ν(CN) 1387 cm^{-1}; ^1H NMR (300 M Hz, DMSO-d_6): δ = 7.83 (d, 1H, arom CH, $^3J_{HH}$ = 9.0 Hz), 7.73 (d, 1H, arom CH, $^3J_{HH}$ = 8.7 Hz), 7.45–7.39 (m, 4H, arom CH), 7.30–7.20 (m, 5H, arom CH), 7.12 (d, 2H, arom CH, $^3J_{HH}$ = 7.8 Hz), 6.72 (d, 1H, CH=C*H*Ph, $^3J_{HH}$ = 15.9 Hz), 6.53 (dt, 1H, C*H*=CHPh, $^3J_{HH}$ = 15.9 Hz, $^3J_{HH}$ = 6.0 Hz), 5.67 (S, 2H, $CH_2C_6H_4CH_3$), 5.27 (d, 2H, CH_2CH=CH, $^3J_{HH}$ = 6.0 Hz), 2.23 (s, 3H, $C_6H_4CH_3$); $^{13}C\{^1H\}$ NMR (126 M Hz, DMSO-d_6): δ = 137.35, 135.65 133.31, 133.29, 129.33, 128.58, 128.02, 127.43, 126.54, 124.09, 124.06, 112.41, 112.33 (13 s, arom Cs), 133.52 (s, CH=CHPh), 124.57 (s, CH=CHPh), 51.79 (s, $CH_2C_6H_4CH_3$), 50.61 (s, CH_2CH=CHPh), 20.64 (s, $C_6H_4CH_3$) ppm. MS (ESI-TOF): m/z = 445.08 [M − Cl]$^+$ and 486.11 [M − Cl + CH_3CN]$^+$ (expected isotopic profiles). Elemental analysis (%): calcd for $C_{24}H_{22}N_2ClAg$ (481.77): C: 59.83; H: 4.60; N 5.81; found C: 59.97 H: 4.49; N: 5.74.

3.4. X-Ray Crystal Structure Analysis

Single crystals of silver(I) complexes **3c** and **3d**, suitable for X-ray analysis, were obtained through the slow diffusion of Et_2O into a CH_2Cl_2 solution of the complex. The crystal structures of complexes **3c** and **3d** were determined using, respectively, a Bruker APEX-II CCD and a Bruker PHOTON-III CPAD with Mo-Kα radiation (λ = 0.71073 Å). The final structure was solved with SHELXT-2018/2 [50], which revealed the non-hydrogen atoms of the molecule. Following anisotropic refinement, the location of all hydrogen atoms was ascertained through the use of a Fourier difference map. The structure was refined with SHELXL-2019/3 [51] using the full-matrix least-square technique (use of F square magnitude; x, y, z, and βij for C, Cl, N, F and Ag atoms; x, y, and z in riding mode for H atoms) (Table 3). The Cambridge Crystallographic Data Centre (CCDC) contains the supplementary crystallographic data for the structures. The data can be obtained free of charge via www.ccdc.cam.ac.uk/structures.

Table 3. Selected crystallographic data for complexes **3c** and **3d**.

	3c	**3d**
CCDC depository	2453683	2451735
Chemical formula	$Ag_2Cl_2C_{46}H_{40}N_4$	$Ag_2Cl_2C_{46}H_{38}F_2N_4$
Molar mass (g.mol^{-1})	935.46	971.44
Temperature (K)	173(2)	120(2)
Crystal system	Triclinic	Triclinic
Space group	$P\overline{1}$	$P\overline{1}$
a (Å)	8.725(6)	8.6142(3)
b (Å)	10.626(9)	10.6406(4)
c (Å)	10.982(9)	11.1741(4)
α (°)	106.338(10)	107.7810(10)
β (°)	91.58(2)	91.7150(10)
γ (°)	100.33(2)	101.2310(10)
Volume (Å^3)	957.9(13)	952.22(6)
Z	1	1
$\rho_{calc.}$ (g·cm^{-3})	1.622	1.694
μ (mm^{-1})	1.201	1.219
$F000$	472	488
Crystral size (mm)	0.180 × 0.140 × 0.120	0.140 × 0.120 × 0.080
Radiation (λ/Å)	0.71073 Mo Kα	0.71073 Mo Kα
θ range for data collection (°)	$1.939 \leq \theta \leq 27.947$	$2.42 \leq \theta \leq 27.88$
Reflections collected	37155	35981
Rint	0.0508	0.0468
Goodness-of-fit on F^2	1.021	1.008
R_1, wR_2 [$I > 2\sigma(I)$]	$R_1 = 0.0334$, $wR_2 = 0.0738$	$R_1 = 0.0264$, $wR_2 = 0.0559$
R_1, wR_2 (all data)	$R_1 = 0.0487$, $wR_2 = 0.0816$	$R_1 = 0.0340$, $wR_2 = 0.0594$

3.5. *In Vitro Effects on A. castellanii*

3.5.1. *A. castellanii* Strain and Culture

The *A. castellanii* PAT06 strain (sequence type T4, GenBank accession no. EF429131), which was previously isolated from a patient with severe AK, was utilized in the study. The strain was cultivated axenically in proteose peptone-yeast extract-glucose (PYG) medium. The cultures were maintained in 25 cm^2 tissue culture flasks (Corning, USA) and incubated at 25 °C [52].

Trophozoites in the exponential growth phase (72–96 h) were harvested by centrifugation at 500× *g* for 10 min. Subsequently, the cell pellets were washed twice with a sterile Neff's saline solution (1.2 g NaCl, 0.4 g $MgSO_4 \cdot H_2O$, 0.4 g $CaCl_2 \cdot 2H_2O$, 1.42 g Na_2HPO_4, and 1.36 g KH_2PO_4 dissolved in 100 mL of distilled water). The cell concentrations were determined using a Neubauer hemocytometer, and the suspension was adjusted to a final concentration of 1 × 10^4 trophozoites/mL in PYG medium with ≥95% viability. The prepared suspension was utilized immediately for viability assays.

Mature cysts were obtained by culturing trophozoites on non-nutrient agar (NNA) plates presmeared with *Escherichia coli*, followed by incubation for six weeks at 25 °C. The resulting cysts were harvested by washing with a sterile Neff's saline solution, with the concentration subsequently adjusted to 1 × 10^4 cysts/mL for the ensuing assays.

3.5.2. *A. castellanii* Trophozoites

Stock solutions of the silver(I) complexes and their respective ligands were prepared in 100% dimethyl sulfoxide (DMSO). Working solutions at final concentrations of 10, 100, and 1000 μM were prepared by diluting the stock solutions with PYG medium. The final DMSO concentration in all experimental groups, including the highest tested concentration, was standardized to 10% (*v*/*v*). $AgNO_3$ and Ag_2O were prepared in PYG medium at final concentrations of 10, 100, and 1000 μM.

A. castellanii trophozoites were seeded into sterile 24-well tissue culture plates at a density of 10^4 trophozoites per mL per well. Subsequently, the cultures were subjected to incubation with the test compounds at the designated concentrations for a duration of 24, 48 and 72 h at a temperature of 25 °C. Following the incubation period, the adherent trophozoites were meticulously detached using a sterile cell scraper and resuspended by gentle pipetting. The resulting cell suspensions were stained with 0.3% methylene blue for a period of 10 min. The viability of the trophozoites was determined by staining them with methylene blue, a method that distinguishes between viable (unstained) and non-viable (blue-stained) cells. The cell count was performed using a Neubauer hemocytometer under a light microscope. The percentage of viable trophozoites was subsequently calculated relative to the untreated control.

The control group consisted of trophozoites cultured in PYG medium alone. The solvent control group was maintained in PYG medium containing 10% (*v*/*v*) DMSO. All experiments were performed in quadruplicate.

3.5.3. *A. castellanii* Cysts

In order to assess the impact of silver(I) complexes and ligands on *A. castellanii* cysts, the cysts were prepared at a final concentration of 1×10^4 cysts/mL. Subsequently, the samples were subjected to incubation with the test compounds at the indicated concentrations at 25 °C for 24, 48 and 72 h.

Following the incubation period, the cyst suspensions were centrifuged and washed three times with sterile Neff's saline. The treated cysts were then inoculated onto non-nutrient agar (NNA) plates that had been smeared with *Escherichia coli* and incubated at 25 °C for 14 days.

The viability of the cyst was monitored daily using phase-contrast microscopy. Cysts were considered viable if excystation occurred and motile trophozoites were observed. In the absence of trophozoite emergence, the cysts were considered non-viable.

3.6. *Statistical Analysis*

The experiments described herein were conducted in four independent biological replicates, with each replicate comprising technical triplicates. The results are expressed as the mean ± standard deviation (SD). The statistical analysis was carried out using GraphPad Prism version 9.0. To assess the statistical significance of the observed variations among the multiple treatment groups, we employed a one-way analysis of variance (ANOVA), which was then followed by a Tukey's multiple comparison test. For the purpose of performing for-a-value comparisons, a two-tailed Student's *t*-test was utilized. In the context of statistical analysis, a *p*-value of less than 0.05 was determined to be statistically significant.

4. Conclusions

In summary, the present study described the synthesis of six new chloro[*N*-alkyl-*N*-cinnamyl-benzimidazol-2-yliden]silver(I) complexes, with isolated yields ranging from 33 to 86%. These complexes were thoroughly characterized by a combination of analytical techniques, including FT-IR and NMR spectroscopic analyses, as well as microanalysis and mass spectrometry. Subsequently, a substantial *in vitro* inhibitory activity against trophozoites and cysts of the *A. castellanii* strain was exhibited, signifying a superior efficacy in comparison to the ligands and commercially available silver complexes as Ag_2O and $AgNO_3$. At a concentration of 100 μM, complexes bearing benzyl, 3-fluorobenzyl and 4-methyl-benzyl substituents on the benzimidazole ring exhibited significant activity, with 5.3,

3.2 and 6.3% trophozoite viability, respectively, recorded after 72 h. These findings provide substantial support for the preclinical development of silver-based NHC compounds as novel anti-*Acanthamoeba* agents, particularly for the treatment of AK. Subsequent studies will concentrate on the optimization of the substituents of the benzimidazole ligand in silver complexes and the investigation of their mode of action.

Supplementary Materials: The following supporting information can be downloaded at https://www.mdpi.com/article/10.3390/inorganics13060204/s1, characterizing data of 1-cinnamyl-3-allyl-benzimidazolium chloride (**1a**) with Figure S1. FT-IR spectrum, Figure S2. ^{1}H NMR spectrum and Figure S3. ^{13}C{^{1}H} NMR spectrum; characterizing data of 1-methoxymethyl-3-cinnamyl-benzimidazolium chloride (**1b**) with Figure S4. FT-IR spectrum, Figure S5. ^{1}H NMR spectrum and Figure S6. ^{13}C{^{1}H} NMR spectrum; characterizing data of 1-(4-methyl-benzyl)-3-cinnamyl-benzimidazolium chloride (**1f**) with Figure S7. FT-IR spectrum, Figure S8. ^{1}H NMR spectrum and Figure S9. ^{13}C{^{1}H} NMR spectrum; characterizing data of chloro[1-cinnamyl-3-allyl-benzimidazol-2-yliden]silver(I) (**3a**) with Figure S10. FT-IR spectrum, Figure S11. ^{1}H NMR spectrum, Figure S12. ^{13}C{^{1}H} NMR spectrum and Figures S13 and S14. Mass spectrum (ESI-TOF); characterizing data of chloro[1-methoxymethyl-3-cinnamyl-benzimidazol-2-yliden]silver(I) (**3b**) with Figure S15. FT-IR spectrum, Figure S16. ^{1}H NMR spectrum, Figure S17. ^{13}C{^{1}H} NMR spectrum and Figure S18. Mass spectrum (ESI-TOF); characterizing data of chloro[1-benzyl-3-cinnamyl-benzimidazol-2-yliden]silver(I) (**3c**) with Figure S19. FT-IR spectrum, Figure S20. ^{1}H NMR spectrum, Figure S21. ^{13}C{^{1}H} NMR spectrum and Figure S22. Mass spectrum (ESI-TOF), characterizing data of chloro[1-(3-fluorobenzyl)-3-cinnamyl-benzimidazol-2-yliden]silver(I) (**3d**) with Figure S23. FT-IR spectrum, Figure S24. ^{1}H NMR spectrum, Figure S25. ^{13}C{^{1}H} NMR spectrum, Figure S26. ^{19}F{^{1}H} NMR spectrum and Figure S27. Mass spectrum (ESI-TOF); characterizing data of chloro[1-(4-fluorobenzyl)-3-cinnamyl-benzimidazol-2-yliden]silver(I) (**3e**) with Figure S28. FT-IR spectrum, Figure S29. ^{1}H NMR spectrum, Figure S30. ^{13}C{^{1}H} NMR spectrum, Figure S31. ^{19}F{^{1}H} NMR spectrum and Figure S32. Mass spectrum (ESI-TOF); characterizing data of chloro[1-(4-methyl-benzyl)-3-cinnamyl-benzimidazol-2-ylidene]silver(I) (**3f**) with Figure S33. FT-IR spectrum, Figure S34. ^{1}H NMR spectrum, Figure S35. ^{13}C{^{1}H} NMR spectrum and Figures S36 and S37. Mass spectrum (ESI-TOF).

Author Contributions: Conceptualization, N.Ş. and D.S.; methodology, Z.A.-P., N.Ş. and D.S.; validation, Z.A.-P., N.Ş. and D.S.; formal analysis, Z.A.-P., N.Ş., S.H. and D.S.; investigation, Z.A.-P., N.Ş., S.H. and B.M.T.L.; resources, Z.A.-P. and D.S.; data curation, Z.A.-P., N.Ş. and D.S.; writing—original draft preparation, Z.A.-P., N.Ş. and S.H.; writing—review and editing, İ.Ö. and D.S.; visualization, Z.A.-P.; supervision, İ.Ö. and D.S. All authors have read and agreed to the published version of the manuscript.

Funding: This research received no external funding.

Institutional Review Board Statement: Not applicable.

Informed Consent Statement: Not applicable.

Data Availability Statement: The original contributions presented in the study are included in the article/Supplementary Material, further inquiries can be directed to the corresponding authors.

Acknowledgments: We would like to thank Julia Walochnik from the Department of Medical Parasitology at the Clinical Institute of Hygiene at the University of Vienna, Vienna, Austria, for providing the Acanthamoeba strain.

Conflicts of Interest: The authors declare no conflicts of interest.

References

1. Geisen, S.; Fiore-Donno, A.M.; Walochnik, J.; Bonkowski, M. *Acanthamoeba* everywhere: High diversity of *Acanthamoeba* in soils. *Parasitol. Res.* **2014**, *113*, 3151–3158. [CrossRef] [PubMed]
2. Schustera, F.L.; Visvesvara, G.S. Free-living amoebae as opportunistic and non-opportunistic pathogens of humans and animals. *Int. J. Parasitol.* **2004**, *34*, 1001–1027. [CrossRef]

3. Marciano-Cabral, F.; Cabral, G. *Acanthamoeba* spp. as agents of disease in humans. *Clin. Microbiol. Rev.* **2003**, *16*, 273–307. [CrossRef] [PubMed]
4. Siddiqui, R.; Khan, N.A. Biology and pathogenesis of *Acanthamoeba*. *Parasites Vectors* **2012**, *5*, 6. [CrossRef]
5. Vyas, S.; Jain, V.; Goyal, M.K.; Radotra, B.D.; Khandelwal, N. Granulomatous amoebic meningoencephalitis. *Neurol. India* **2013**, *61*, 530–531. [CrossRef]
6. Lorenzo-Morales, J.; Khan, N.A.; Walochnik, J. An update on *Acanthamoeba* keratitis: Diagnosis, pathogenesis and treatment. *Parasite* **2015**, *22*, 10. [CrossRef]
7. Niederkorn, J.Y. The biology of *Acanthamoeba* keratitis. *Exp. Eye Res.* **2021**, *202*, 108365. [CrossRef]
8. Zhang, Y.; Xu, X.; Wei, Z.; Cao, K.; Zhang, Z.; Liang, Q. The global epidemiology and clinical diagnosis of *Acanthamoeba* keratitis. *J. Infect. Public Health* **2023**, *16*, 841–852. [CrossRef] [PubMed]
9. Varacalli, G.; Di Zazzo, A.; Mori, T.; Dohlman, T.H.; Spelta, S.; Coassin, M.; Bonini, S. Challenges in *Acanthamoeba* keratitis: A review. *J. Clin. Med.* **2021**, *10*, 942. [CrossRef]
10. Petrillo, F.; Tortori, A.; Vallino, V.; Galdiero, M.; Fea, A.M.; De Sanctis, U.; Reibaldi, M. Understanding *Acanthamoeba* keratitis: An in-depth review of a sight-threatening eye infection. *Microorganisms* **2024**, *12*, 758. [CrossRef]
11. Büchele, M.L.C.; Nunes, B.F.; Filippin-Monteiro, F.B.; Caumo, K.S. Diagnosis and treatment of *Acanthamoeba* keratitis: A scoping review demonstrating unfavorable outcomes. *Contact Lens Anterior Eye* **2023**, *46*, 101844. [CrossRef]
12. Fanselow, N.; Sirajuddin, N.; Yin, X.-T.; Huang, A.J.W.; Stuart, P.M. *Acanthamoeba* keratitis, pathology, diagnosis and treatment. *Pathogens* **2021**, *10*, 323. [CrossRef] [PubMed]
13. Ahmed, U.; Anwar, A.; Ong, S.-K.; Anwar, A.; Khan, N.A. Applications of medicinal chemistry for drug discovery against *Acanthamoeba* infections. *Med. Res. Rev.* **2022**, *42*, 462–512. [CrossRef] [PubMed]
14. Reyes-Batlle, M.; Sifaoui, I.; Rodríguez-Expósito, R.L.; Piñero, J.E.; Lorenzo-Morales, J. New insights in *Acanthamoeba*. *Pathogens* **2022**, *11*, 609. [CrossRef] [PubMed]
15. Gasser, G.; Ott, S.; Metzler-Nolte, N. Organometallic anticancer compounds. *J. Med. Chem.* **2011**, *54*, 3–25. [CrossRef]
16. Noffke, A.L.; Habtemariam, A.; Pizarro, A.M.; Sadler, P.J. Designing organometallic compounds for catalysis and therapy. *Chem. Commun.* **2012**, *48*, 5219–5246. [CrossRef]
17. Middya, P.; Chattopadhyay, S. An overview of synthesis, structure and biological application of *N*-heterocyclic carbene complexes of silver. *J. Mol. Struct.* **2025**, *1324*, 140842. [CrossRef]
18. Ceramella, J.; Catalano, A.; Mariconda, A.; D'Amato, A.; Aquila, S.; Saturnino, C.; Rosano, C.; Sinicropi, M.S.; Longo, P. Silver *N*-heterocyclic carbene (NHC) complexes as antimicrobial and/or anticancer agents. *Pharmaceuticals* **2025**, *18*, 9. [CrossRef]
19. Şahin, N.; Mosrati, M.A.; Merghni, A.; Özdemir, İ.; Sellami, H.; Bedchiche, K.; Krayiem, S.; Aifa, S.; Abdelmalek, D.; Sémeril, D. Synthesis, antimicrobial and antibiofilm activities of silver(I) complexes with *N*-alkylbenzimidazole derivatives and their protein interaction modelling study. *J. Mol. Struct.* **2025**, *1322*, 140440. [CrossRef]
20. Fischer, E.O.; Maasböl, A. On the existence of a tungsten carbonyl carbene complex. *Angew. Chem. Int. Ed.* **1964**, *3*, 580–581. [CrossRef]
21. Herrmann, W.A.; Köcher, C. *N*-Heterocyclic carbenes. *Angew. Chem. Int. Ed.* **2017**, *36*, 2162–2187. [CrossRef]
22. Wanzlick, H.-W.; Schönherr, H.-J. Direct synthesis of a mercury salt-carbene complex. *Angew. Chem. Int. Ed.* **1968**, *7*, 141–142. [CrossRef]
23. Öfele, K. 1,3-Dimethyl-4-Imidazolinyliden-(2)-Pentacarbonylchrom ein neuer Übergangsmetall-Carben-Komplex. *J. Organomet. Chem.* **1968**, *12*, 42–43. [CrossRef]
24. Arduengo, A.J., III; Kline, M.; Calabrese, J.C.; Davidson, F. Synthesis of a reverse ylide from a nucleophilic carbene. *J. Am. Chem. Soc.* **1991**, *113*, 9704–9705. [CrossRef]
25. Herrmann, W.A. *N*-Heterocyclic carbenes: A new concept in organometallic catalysis. *Angew. Chem. Int. Ed.* **2002**, *41*, 1290–1309. [CrossRef]
26. Cavallo, L.; Correa, A.; Costabile, C.; Jacobsen, H. Steric and electronic effects in the bonding of *N*-heterocyclic ligands to transition metals. *J. Organomet. Chem.* **2005**, *690*, 5407–5413. [CrossRef]
27. Lasmari, S.; Ikhlef, S.; Boulcina, R.; Mokrani, E.H.; Bensouici, C.; Gürbüz, N.; Dündar, M.; Karcı, H.; Özdemir, İ.; Koç, A.; et al. New silver *N*-heterocyclic carbenes complexes: Synthesis, molecular docking study and biological activities evaluation as cholinesterase inhibitors and antimicrobials. *J. Mol. Struct.* **2021**, *1238*, 130399. [CrossRef]
28. Ashraf, R.; Bhatti, H.N.; Iqbal, M.A.; Jamil, Y. Synthesis of aryl linked binuclear silver *N*-heterocyclic carbene complexes, DNA interaction study and biological potentials. *Inorg. Chem. Commun.* **2020**, *119*, 108077. [CrossRef]
29. Hussaini, S.Y.; Haque, R.A.; Razali, M.R. Recent progress in silver(I)-, gold(I)/(III) -and palladium(II)-*N*-heterocyclic carbene complexes: A review towards biological perspectives. *J. Organomet. Chem.* **2019**, *882*, 96–111. [CrossRef]
30. Sánchez, O.; González, S.; Higuera-Padilla, Á.R.; León, Y.; Coll, D.; Fernández, M.; Taylor, P.; Urdanibia, I.; Rangel, H.R.; Ortega, J.T.; et al. Remarkable *in vitro* anti-HIV activity of new silver(I)- and gold(I)-*N*-heterocyclic carbene complexes. Synthesis, DNA binding and biological evaluation. *Polyhedron* **2016**, *110*, 14–23. [CrossRef]

31. Brown, M.R.W.; Anderson, R.A. The bactericidal effect of silver ions on *Pseudomonas aeruginosa*. *J. Pharm. Pharmacol.* **1968**, *20*, 1S–3S. [CrossRef] [PubMed]
32. Melaiye, A.; Youngs, W.J. Silver and its application as an antimicrobial agent. *Expert Opin. Ther. Pat.* **2005**, *15*, 125–130. [CrossRef]
33. Banti, C.N.; Hadjikakou, S.K. Anti-proliferative and anti-tumor activity of silver(I) compounds. *Metallomics* **2013**, *5*, 569–596. [CrossRef] [PubMed]
34. Medici, S.; Peana, M.; Nurchi, V.M.; Zoroddu, M.A. Medical uses of silver: History, myths, and scientic evidence. *J. Med. Chem.* **2019**, *62*, 5923–5943. [CrossRef]
35. Hendiger, E.B.; Padzik, M.; Sifaoui, I.; Reyes-Batlle, M.; López-Arencibia, A.; Rizo-Liendo, A.; Bethencourt-Estrella, C.J.; San Nicolás-Hernández, D.; Chiboub, O.; Rodríguez-Expósito, R.L.; et al. Silver nanoparticles as a novel potential preventive agent against *Acanthamoeba* keratitis. *Pathogens* **2020**, *9*, 350. [CrossRef]
36. Niyyati, M.; Sasani, R.; Mohebali, M.; Ghazikhansari, M.; Kargar, F.; Hajialilo, E.; Rezaeian, M. Anti-*Acanthamoeba* effects of silver and gold nanoparticles and contact lenses disinfection solutions. *Iran J. Parasitol.* **2018**, *13*, 180–185.
37. Padzik, M.; Hendiger, E.B.; Chomicz, L.; Grodzik, M.; Szmidt, M.; Grobelny, J.; Lorenzo-Morales, J. Tannic acid-modified silver nanoparticles as a novel therapeutic agent against *Acanthamoeba*. *Parasitol. Res.* **2018**, *117*, 3519–3525. [CrossRef]
38. Meng, G.; Kakalis, L.; Nolan, S.P.; Szostak, M. A simple ^{1}H NMR method for determining the σ-donor properties of *N*-heterocyclic carbenes. *Tetrahedron Lett.* **2019**, *60*, 378–381. [CrossRef]
39. Şahin, N.; Zengin, S.; Özdemir, İ.; Sémeril, D. C-H activation of furanyl and thiofuranyl substrates catalyzed by trans-dichloro [1-cinnamyl-3-arylmethyl-benzimidazol-2-yliden]pyridine palladium(II) complexes. *Polyhedron* **2024**, *261*, 117144. [CrossRef]
40. Ray, S.; Mohan, R.; Singh, J.K.; Samantaray, M.K.; Shaikh, M.M.; Panda, D.; Ghosh, P. Anticancer and antimicrobial metallopharmaceutical agents based on palladium, gold, and silver *N*-heterocyclic carbene complexes. *J. Am. Chem. Soc.* **2007**, *129*, 15042–15053. [CrossRef]
41. Garrison, J.C.; Youngs, W.J. Ag(I) *N*-heterocyclic carbene complexes: Synthesis, structure, and application. *Chem. Rev.* **2005**, *105*, 3978–4008. [CrossRef] [PubMed]
42. Gimeno, M.C.; Laguna, A.; Visbal, R. *N*-Heterocyclic carbene coinage metal complexes as intense blue-green emitters. *Organometallics* **2012**, *31*, 7146–7157. [CrossRef]
43. Tulloch, A.A.D.; Danopoulos, A.A.; Winston, S.; Kleinhenz, S.; Eastham, G. *N*-Functionalised heterocyclic carbene complexes of silver. *J. Chem. Soc. Dalton Trans.* **2000**, 4499–4506. [CrossRef]
44. Laidlaw, G.; Wood, S.H.; Kennedy, A.R.; Nelson, D.J. An *N*-heterocyclic carbene with a saturated backbone and spatially-defined steric impact. *Z. Anorg. Allg. Chem.* **2019**, *645*, 105–112. [CrossRef]
45. Tutar, U.; Çelik, C.; Üstün, E.; Özdemir, N.; Sahin, N.; Sémeril, D.; Gürbüz, N.; Özdemir, İ. Benzimidazol-2-ylidene silver complexes: Synthesis, characterization, antimicrobial and antibiofilm activities, molecular docking and theoretical investigations. *Inorganics* **2023**, *11*, 385. [CrossRef]
46. Asekunowo, P.O.; Haque, R.A.; Razali, M.R. A comparative insight into the bioactivity of mono- and binuclear silver(I)-N-heterocyclic carbene complexes: Synthesis, lipophilicity and substituent effect. *Rev. Inorg. Chem.* **2017**, *37*, 29–50. [CrossRef]
47. Ronga, L.; Varcamonti, M.; Tesauro, D. Structure-activity relationships in NHC-silver complexes as antimicrobial agents. *Molecules* **2023**, *28*, 4435. [CrossRef]
48. Admetlab. Available online: https://admetlab3.scbdd.com/ (accessed on 22 May 2025).
49. Tan, K.L.; Vasudevan, A.; Bergman, R.G.; Ellman, J.A.; Souers, A.J. Microwave-assisted C-H bond activation: A rapid entry into functionalized heterocycles. *Org. Lett.* **2003**, *5*, 2131–2134. [CrossRef]
50. Sheldrick, G.M. SHELXT-Integrated space-group and crystal-structure determination. *Acta Crystallogr. Sect. A* **2015**, *71*, 3–8. [CrossRef]
51. Sheldrick, G.M. Crystal structure refinement with SHELXL. *Acta Crystallogr. Sect. C* **2015**, *71*, 3–8.
52. Schuster, F.L. Cultivation of pathogenic and opportunistic free-living amebas. *Clin. Microbiol. Rev.* **2002**, *15*, 342–354. [CrossRef] [PubMed]

inorganics

MDPI

Article

Synthesis, Characterization, and Cytotoxicity Research of Sulfur-Containing Metal Complexes

Yanting Yang, Danqin Li and Mei Luo *

Department of Chemistry and Chemical Engineering, Hefei University of Technology, Hefei 230002, China; yangyanting0626@163.com (Y.Y.); 18325610866@163.com (D.L.)
* Correspondence: luomei@pku.edu.cn

Abstract: In this experiment, the excellent coordination ability of sulfur-containing ligands was utilized. Diphenylacetyl disulfide and 3,3′-diaminodiphenyl sulfone were selected as ligands, and $Cu(NO_3)_2{\cdot}3H_2O$, $Ni(NO_3)_2{\cdot}6H_2O$ and $ZnCl_2$ were reacted under one-pot conditions to synthesize three mononuclear complexes: $[C_4H_{18}CuO_{12}S_2]$(I), $[C_{12}H_{18}N_4NiO_{11}S]$(II) and $[C_{24}H_{24}Cl_2N_4O_4S_2Zn]$(III). Complex (I) belongs to the orthorhombic crystal system with space group Pbca, while complexes (II) and (III) belong to the monoclinic crystal system with space groups P21/n and P2/n. The crystal structure of the complex was determined using X-ray diffraction (XRD). The structure of the complex was analyzed using infrared Fourier transform infrared spectroscopy (FT-IR), ultraviolet–visible spectroscopy (UV–Vis), nuclear magnetic resonance (NMR), and electrospray mass spectrometry (ESI-MS), and the thermal stability and composition of the complex were detected via thermogravimetry (TGA). In terms of application, the biological activity of complexes (I)–(III) in human cancer cell lines (lung cancer A549, liver cancer SMMC-7721, breast cancer MDA-MB-231, and colon cancer SW480) was tested using the MTS method. The results showed that complex (II) had a good inhibitory effect on breast cancer MDA-MB-231.

Keywords: one-pot method; complex; crystal structure; cytotoxicity

1. Introduction

Metal–organic complexes are a type of compound formed by coordination bonds between metals and organic ligands [1]. They can be used to synthesize novel metal–organic complexes with different structures and functions through the selection of metal centers and ligands [2–4], and accordingly, they are widely used in fields such as catalysis [5], materials [6], drug design [7], and sensing technology [8]. Organosulfur metal complexes represent an important branch of metal–organic complexes and are characterized using sulfur-containing compounds as ligands to form stable coordination bonds with metal centers. Sulfur-containing compounds are an excellent class of ligands with multiple coordination modes (such as monodentate, bidentate, and bridging) [9], playing an important role in the synthesis of metal–organic complexes. Organic sulfur compounds are abundant in sources and exist in many plants in nature, with special chemical properties and biological activities [10–12]. Studies have shown that organic sulfur compounds can affect the expression of related proteins in cancer cells, thereby inhibiting their proliferation. This finding also proves that organic sulfur compounds have the potential to become anticancer drugs [13–16]. Some achievements have recently been made in the treatment of cancer with sulfur-containing compounds. In 2018, Soltani et al. [17] isolated five new sulfur-containing compounds from the broad-leaved Aweigen and conducted cytotoxicity experiments on the human cancer cell lines A2780, A549, HeLa, and HCT116. Some sulfur-containing compounds have moderate cytotoxicity [17]. In 2019, Balakrishnan et al. [18] synthesized two complex crystals by directly reacting potassium morpholine dithiocarbamate ($K^+C_5H_8NOS_2$) ligands with nickel and copper metals and studied the inhibitory effects of the two complexes on Gram-positive and Gram-negative bacteria. The experimental results revealed that the synthesized complexes had good antibacterial activity [18]. In 2020, Yekke-Ghasemi et al. [19] successfully synthesized three dithiocarbamate metal complexes and tested their cytotoxicity against HeLa and MCF-7 cancer cell lines. The results showed that the three novel complexes were more cytotoxic than cisplatin [19]. In 2023, Singh et al. [20] successfully prepared 2-((2-(benzylthio) phenyl) imino) methyl)-4-chlorophenol cobalt, nickel, copper, and zinc metal complexes, and the zinc complexes showed excellent anti-inflammatory and antidiabetic activities. In 2024, Czylkowska et al. [21] synthesized a series of metal complexes using 5-(1-methylpyrrole-2-yl) methyl)-4-(2-chlorophenyl)-1,2,4-triazoline-3-thione ligands. Under in vitro conditions, the metal complexes showed good cytotoxicity against A549 and HT-29 cancer cell lines, with the Cu(II) complex having the highest activity ($IC_{50} = 247.90 \pm 2.30$) [21]. Although many studies have been conducted on organic sulfur metal complexes, there is still much research space. In this work, two sulfur-containing ligands, namely, bis(phenylacetyl) disulfide and 3,3-diaminodiphenyl sulfone, were used to synthesize several novel metal complexes with metal salts under heating conditions using a one-pot method [22,23]. Their structures were characterized using the EA, XRD, FT-IR, UV–Vis, NMR, ESI-MS and TGA methods. The cytotoxicity of the complexes was preliminarily screened via the MTS method.

2. Results and Discussion

2.1. Crystal Structure Analyses of Complexes (I)–(III)

The Figure 1 shows the crystal structures of complexes (I)–(III) and S_8. Complexes (I) (Figure 1a) and S_8 (Figure 1b) have an orthorhombic crystal system with space groups Pbca and Fddd, respectively, whereas complexes (II) (Figure 1c) and (III) (Figure 1d) are both monoclinic with space groups $P2_1/n$ and P2/n, respectively.

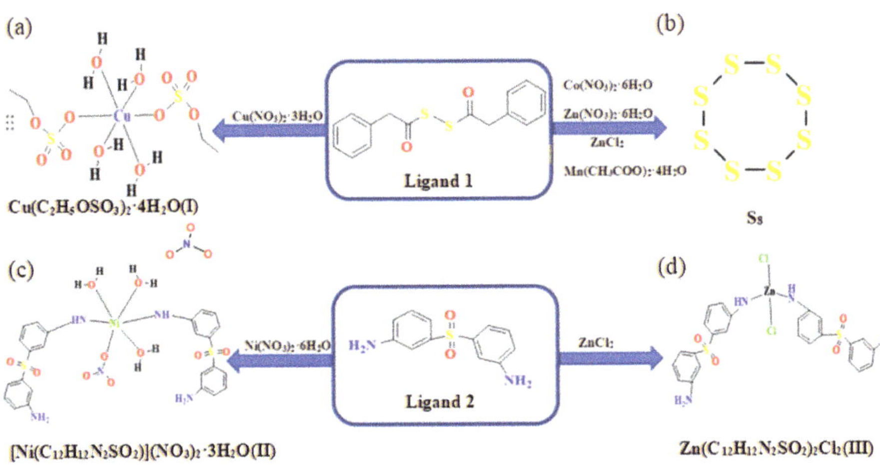

Figure 1. Synthesis protocol of complexes (I)–(III). (**a**) for (I), (**b**) for S_8, (**c**) for (II) and (**d**) for (III).

Complex (I) is not formed by direct chelation of metal ions and ligands but by ligand bond cleavage and recombination followed by coordination with metal salts; this may be due to the breaking of S–S bonds under the effect of Lewis acids, which further react with oxygen atoms in air and ethoxy groups in ethanol to form complexes similar to copper sulfate salts. The metal center Cu^{2+} of complex (I) is coordinated via six coordination units and has two symmetrical structural units. Metal Cu(II) is connected to six oxygen atoms, forming a slightly distorted octahedral configuration, where two O atoms (O1, $O1^{i}$) are derived from O in the ethyl sulfate group, and the other four O atoms (O5, $O5^{i}$, O6, $O6^{i}$) originate from four H_2O molecules. The longest bond length between Cu1-O1 and Cu1-$O1^{i}$ is 2.4052 (12) Å, whereas the shortest bond length between Cu1-O5 and Cu1-$O5^{i}$ is 1.9520 (9) Å. The cis angle and trans angle around the Cu1 atom are 97.15 (13)° and 150.61 (12)°, respectively. Molecules form hydrogen bonds within and between molecules through the O atoms on sulfate ions and the H atoms on water molecules, ultimately forming a three-dimensional network structure. The crystal stacking diagram is shown in the Figure 2. The S_8 crystal is a "crown-shaped" structure formed by eight S atoms, which are divided into upper and lower layers. The bond angles of the S–S bonds are all approximately 108°. The structure of the S_8 crystal has already been reported in great detail and will not be repeated here.

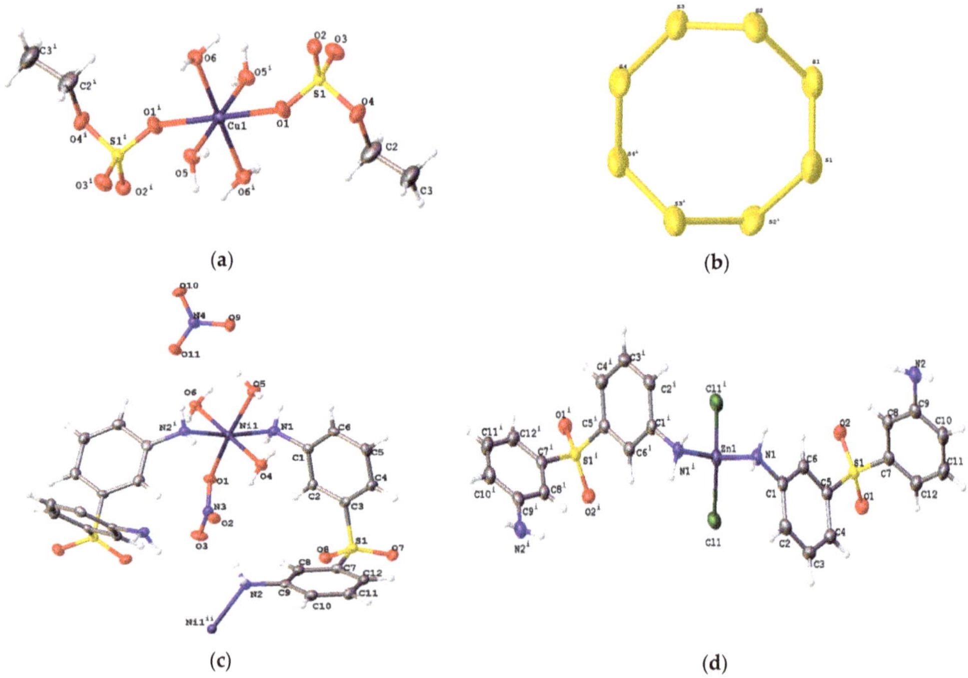

Figure 2. ORTEP molecular structures of complexes, (**a**) for (I), (**b**) for S_8, (**c**) for (II) and (**d**) for (III), with a 30% thermal ellipsoid probability.

Complex (II) is a twisted octahedral configuration centered on Ni(II), with a coordination number of six. The mononuclear crystal complex (II) contains one 3,3′-diaminodiphenyl sulfone ligand, three water molecules, and two NO_3^- ions (one of which is in a free state). From the crystal structure diagram below, it can be observed that the two molecules are bridged to Ni(II) through the N atom on the ligand NH_2. Molecules also communicate with each other through C(8)-H(8···O(2)#1, N(1)-H(1A)···O(9)#3, O(4)-H(4A)···O(3)#2, and other hydrogen bonds (see Table 1), which are stacked together and connected along the b-axis direction to form a three-dimensional chain structure.

Table 1. Partial hydrogen bond data for complexes (I)–(III).

	D-H···A	d(D-H)/Å	d(H···A)/Å	d(D···A)/Å	∠(DHA)/°
Complex (I)	O(5)-H(5B)···O(2)#2	0.87	1.91	2.7145(14)	152.6
	O(6)-H(6B)···O(3)#4	0.87	1.92	2.7392(15)	156.5
Complex (II)	C(8)-H(8)···O(2)#1	0.95	2.55	3.144(7)	120.9
	N(1)-H(1A)···O(9)#3	0.91	2.16	3.048(7)	166.5
	O(4)-H(4A)···O(3)#2	0.87	2.13	2.926(6)	152.4
Complex (III)	N(1)-H(1A)···N(2)#2	0.91	2.12	3.025(3)	176.0
	N(2)-H(2A)···O(2)#3	0.88	2.27	3.068(2)	150.4
	N(1)-H(1B)···Cl(1)#4	0.91	2.57	3.4355(18)	159.2
	C(2)-H(2)···Cl(1)#4	0.95	3.22	3.986(2)	139.3
	C(3)-H(3)···O(1)#5	0.95	2.66	3.254(3)	121.5
	C(6)-H(6)···N(2)#2	0.95	3.05	3.630(3)	120.5

Complex (III) is a slightly distorted tetrahedral configuration formed by the 4-coordination of Zn(II). The central ion Zn^{2+} is directly connected to two Cl^- and two N donors (N1, $N1^i$) on 3,3′-diaminodiphenyl sulfone ligands, with main bond lengths and bond angles of $d_{N1\text{-}Zn1} = d_{N1^i\text{-}Zn1}$ = 2.0530 (17) Å and $d_{Cl1\text{-}Zn1} = d_{Cl1^i\text{-}Zn1}$ = 2.2393 (6) Å, respectively, and ∠N1-Zn1-Cl1^i = 112.36(5)°, ∠Cl1-Zn1-Cl1^i = 109.43(3)°, and ∠N1-Zn1-N1^i = 105.00(9)°. Molecules interact with each other through four types of hydrogen bonds, namely, N-H···N, N-H···O, C-H···N, and C-H···O, forming a regular three-dimensional network structure.

2.2. IR Analyses of Complexes (I)–(III)

Figure 3 shows all the infrared absorption peaks of complexes (I)–(III) in the range of 4000–500 cm^{-1}. The stretching vibration absorption peak of OH in water molecules is located at 3491 cm^{-1}, whereas the absorption peak of free OH is generally at approximately 3700–3500 cm^{-1} [24]. However, due to the presence of intramolecular and intermolecular hydrogen bonds in complex (I) molecules (see Table 1), the absorption peak redshifts [25]. In the figure, two absorption peaks were observed at 2990 cm^{-1} and 2899 cm^{-1}, indicating the presence of both CH_3 and CH_2 in the molecular structure of complex (I). The in-plane bending vibration absorption peak of C-H on the methyl group was observed at 1380 cm^{-1}, and the stretching vibration frequency of the C–C bond on the alkyl group was observed near 1235 cm^{-1}. There are two characteristic absorption peaks of sulfate ions: one peak, located at 1200–1000 cm^{-1}, is the symmetric stretching vibration peak of sulfate ions [26], which is commonly used to determine whether sulfate ions are present in the molecule, and the other peak, located near 1300–1100 cm^{-1}, is the asymmetric stretching vibration peak of sulfate ions. The corresponding absorption peaks in the figure are at 1065 cm^{-1} and 1020 cm^{-1}.

Complexes (II) and (III) are two types of complex crystals synthesized from the same ligand with different metal salts. They share certain similarities in composition and structure, which are particularly evident in infrared images. We observe that complex (III) has two distinct absorption peaks at 3585 cm^{-1} and 3515 cm^{-1}, corresponding to the stretching vibration peaks of NH_2 [25], whereas complex (II) has only one absorption peak at 3512 cm^{-1}; this may be due to the simultaneous presence of -OH and -NH_2 groups in complex (II) and the similar range of their characteristic infrared absorption peaks. Therefore, the absorption peaks of the two may partially overlap, resulting in only one absorption peak of a primary amine being visible from the graph. Figure 3 shows that complex (II) indeed has a broad peak at 3400 cm^{-1}, which is consistent with the characteristic absorption of OH. Both complexes (II) and (III) have benzene ring structures, and the benzene ring generally has several characteristic absorption peaks. First, there are three weak absorption peaks at 3100–3000 cm^{-1}, which are caused by the C-H stretching vibration of the benzene ring. In the figure, they are specifically shown as complexes (II) (3180 cm^{-1}, 3095 cm^{-1}, and 2925 cm^{-1}) and (III) (3130 cm^{-1}, 3065 cm^{-1}, and 3010 cm^{-1}). Second, complexes (II) and (III) have two strong infrared absorption peaks at 1600 cm^{-1} and 1435 cm^{-1} and 1605 cm^{-1} and 1480 cm^{-1}, respectively, which correspond to the antisymmetric and symmetric stretching vibrations of the carbon skeleton of the benzene ring structure. In addition, because both complexes (II) and (III) are *meta*-disubstituted on the benzene ring, complex (II) absorbs out-of-plane bending vibrations of the C–H bond on the aromatic ring at 872 cm^{-1}, 795 cm^{-1}, and 715 cm^{-1} and complex (III) at 855 cm^{-1}, 790 cm^{-1}, and 712 cm^{-1}. As well as the benzene ring structure, there is an important sulfone group in the molecules of complexes (II) and (III), and an infrared absorption peak can be observed at 1297 cm^{-1} and 1300 cm^{-1}, respectively, corresponding to the asymmetric stretching vibration absorption peak of the S=O bond. At positions 1142 cm^{-1} and 1150 cm^{-1}, there are symmetric stretching vibrations of S=O in the two complex molecules, respectively. A strong absorption peak can be observed at 1045 cm^{-1} and 1092 cm^{-1}, which is the S=O stretching vibration characteristic infrared absorption peak of complexes (II) and (III) [27,28]. Although there are many similarities in the structures of complexes (II) and (III), there are also some differences. For example, the anions in complex molecules are not the same. Complex (II) contains NO_3^- ions in the molecule, and the strong stretching vibration peak range of the N–O bond of nitrate ions is generally between 1390 and 1370 cm^{-1}, as shown in the figure. Complex (II) shows a strong absorption peak at 1380 cm^{-1}, which proves that NO_3^- ions are indeed present in the molecule [29]. M-N exhibited a stretching vibration absorption peak in the range of 660–600 cm^{-1}, whereas M-O exhibited a stretching vibration absorption peak in the range of 600–500 cm^{-1}, indicating chelation between the metal and the ligand [30,31].

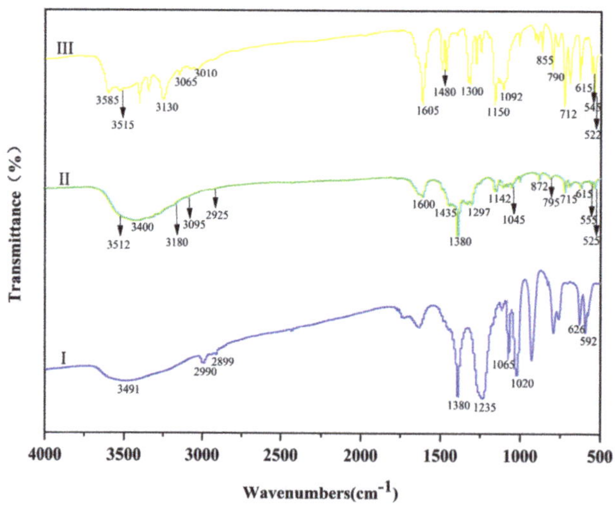

Figure 3. IR spectra of complexes (I) to (III) in the 4000–500 cm^{-1} region.

2.3. UV–Vis Spectral Analyses of Complexes (I)–(III)

The UV–visible spectra of complexes (I) and (III) at 200–800 nm are shown in Figure 4a,b shows the UV–visible spectra of complexes (I) and (II) at 300–1000 nm. Complex (I) has a weak absorption band at 260 nm, which is caused by the n → π* transition of the -S=O bond and belongs to the R absorption band. In addition, a wide absorption band at 810 nm can be observed in Figure 4b, which is caused by the d-d transition of Cu^{2+} with strong absorbance.

At 405 nm, a narrow absorption peak can be observed, which is caused by the d-d transition of Ni^{2+} with a lighter color and weaker absorbance [32]. The benzene rings in complexes (II) and (III) are both connected to the chromophore NH_2 and undergo n–π conjugation, causing both the E-band and the B-band of the benzene ring to redshift. Interestingly, the benzene rings in these two complexes are also replaced by the chromophore S=O. At this time, the double bond is conjugated with the benzene ring, and a K-band appears at 200–250 nm. The figure shows that complexes (II) and (III) have absorption bands at 228 nm and 230 nm, respectively. At the same time, the B-band of the benzene ring undergoes a significant redshift; that is, the B-band shifts from 255 nm to 315 nm.

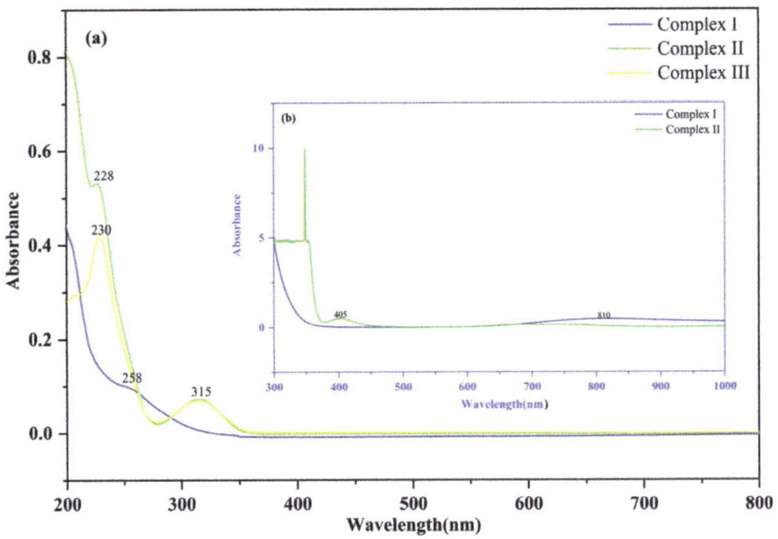

Figure 4. UV–Vis spectra of complexes (I)–(III) in the range of 200–1000 nm. ((**a**) shows the UV absorption spectra of complexes I–III at 200–800nm, while (**b**) shows the UV absorption spectra of complexes I–II at 300–1000 nm).

2.4. NMR Spectroscopy of Complex (III)

We also conducted ^{1}H NMR and ^{13}C NMR tests on complex (III), and Figure 5 shows the NMR spectra of this complex. The chemical shift of the hydrogen on the benzene ring is generally approximately 7.26 ppm [33]. In the figure, nuclear magnetic resonance signals can be observed at 7.08 ppm, 7.02 ppm, and 6.91 ppm, but the chemical shift is less than 7.26 ppm; this is most likely due to the substitution of the benzene ring with $-NH_2$ and $-SO_2$, which reduces the electron density on benzene carbons. The sulfone group is an electron-withdrawing group and has a deshielding effect on the hydrogen nucleus, resulting in an increase in the chemical shift value. The amino group is a strong electron-donating group, which increases the shielding effect when connected to the benzene ring, ultimately leading to a decrease in the chemical shift value. Due to the connection of two amino groups and one sulfone group in the molecule, the shielding effect is preserved. The nuclear magnetic resonance signal at δ = 5.20 ppm corresponds to the hydrogen atom on $-NH_2$. The chemical shift of amino hydrogen is influenced by neighboring functional groups, such as the conjugation and induction effects of benzene rings, as well as metal ions with low electronegativity. The ^{13}C NMR spectrum of complex (III) shows the presence of various types of carbon atoms in the complex, with chemical shift values of 112 ppm, 114 ppm, 118 ppm, 129 ppm, 142 ppm, and 149 ppm. The reason for the different chemical shift values is the same as that for the hydrogen spectrum and will not be elaborated on here. The position on the benzene ring where the substituent is substituted has the largest change in chemical shift values, followed by the *ortho-* and *para*-positions, with the *meta*-carbon having the smallest impact [34].

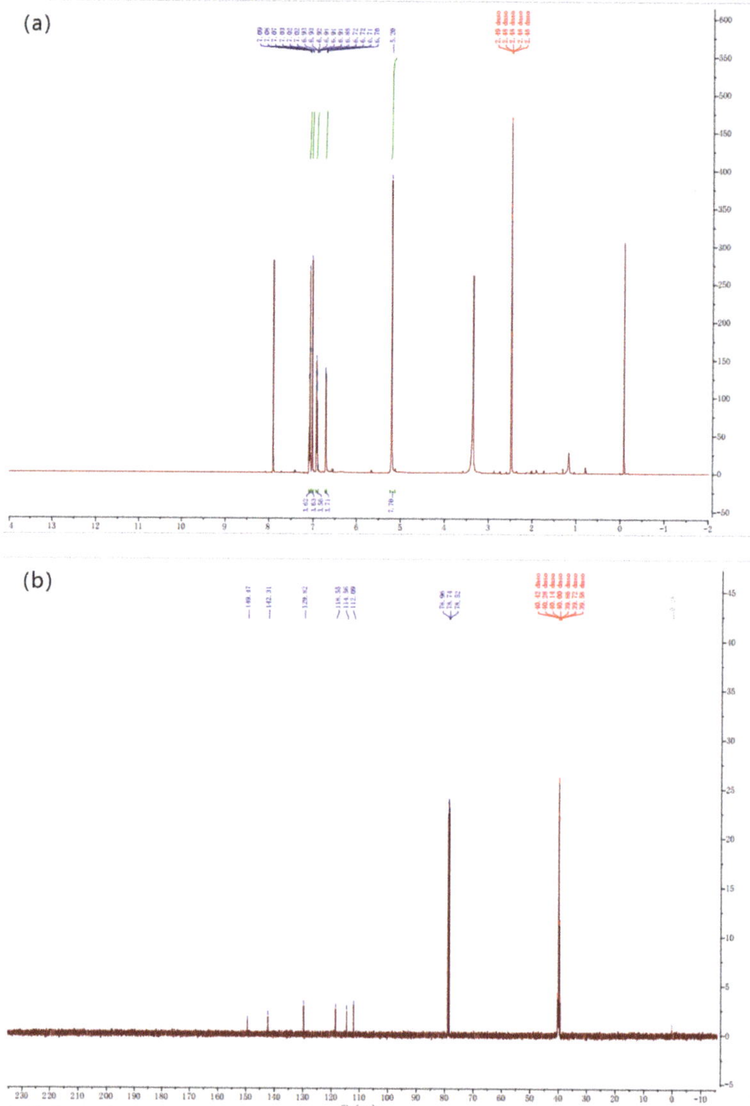

Figure 5. ^{1}H NMR (**a**) and ^{13}C NMR (**b**) spectra of complex (III).

2.5. Electrospray Mass Spectrum Analysis

The electrospray ionization mass spectrum peak of complexes (I)–(III) in ethanol solvent is shown in Figure 6, and the mass–charge (m/z) ratios are as follows: the molecular formula of complex (I) is [$C_4H_{18}CuO_{12}S_2$] (I), and the theoretical value of the mass–charge ratio is 384.84, while the actual value is 384.88 $[M+H]^+$. The molecular formula of complex (II) is [$C_{12}H_{18}N_4NiO_{11}S$] (II), and the theoretical value of the mass–charge ratio is 486.07, while the actual value is 485.92 $[M+H]^+$. The molecular formula of complex (III) is [$C_{24}H_{24}Cl_2N_4O_4S_2Zn$] (III), and the theoretical value of the mass–charge ratio is 631.86, while the actual value is 631.89 $[M-H]^+$.

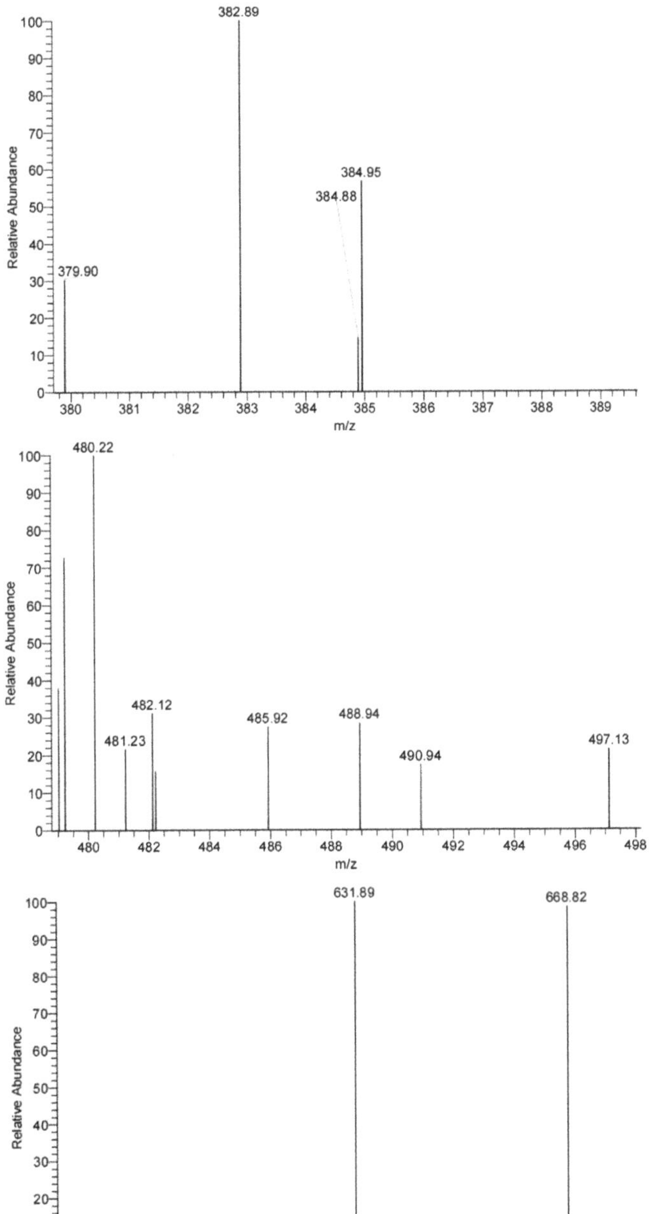

Figure 6. Electrospray ionization mass spectra of complexes (I)–(III).

2.6. Thermogravimetric Analysis for Complexes (I) to (III)

Thermogravimetric analysis (TGA) tests were conducted on complexes (I)–(III) within the temperature range of 30–900 °C, and the results are shown in Figure 7. The TGA curve of complex (I) shows four distinct stages of thermal decomposition. In the temperature range of 50–235 °C, complex (I) underwent continuous mass loss, and the first plateau appeared after this process, with mass losses of 2.68%, 40.91%, and 11.32%, respectively. These quality losses are mainly attributed to the decomposition of water molecules and ethyl sulfate ions in the complex. In the temperature range of 620–740 °C, the weight of complex (I) decreased again, and a second plateau period could be seen, with a mass loss of 17.98%. This stage of decomposition corresponds to the process of SO_4^{2-} in the complex decomposing into SO_3 gas. After these decomposition processes, the final remaining substance was CuO (calculated value: 20.48%, actual value: 19.23%). Complex (II) exhibited sustained weight loss in the temperature range of 35–535 °C, with a loss of 61.64% of the total mass, attributed to the decomposition of water molecules and ligands in the complex molecules. The final residual product was $Ni(NO_2)_2$ (calculated value: 30.92%, actual value: 30.63%). Complex (III) also underwent three distinct thermal decomposition processes, with the first stage occurring within the temperature range of 35–108 °C, resulting in a mass loss of 11.18%. This corresponds to the removal of two Cl^- ions from the complex molecules (calculated value: 11.22%). The temperature range of the second stage is 330–655 °C, with a total weight loss of 45.01%, which may correspond to the loss of two aniline molecules and two SO_2 molecules (calculated value: 50.33%).

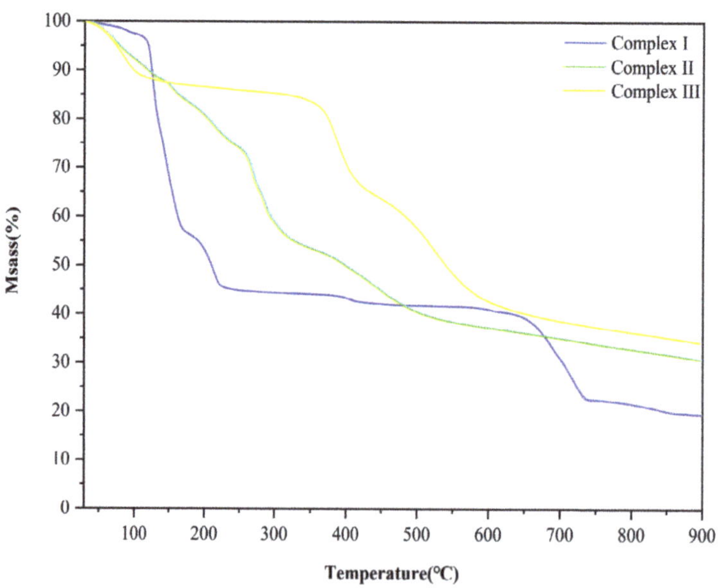

Figure 7. Thermogravimetric analysis (TGA) decomposition curves of complexes (I)–(III).

2.7. Cytotoxicity Assays of Complexes (I)–(III)

We conducted MTS cytotoxicity tests of complexes (I)–(III) against the human cancer cell lines lung cancer A549, liver cancer SMMC-7721, breast cancer MDA-MB-231, and colon cancer SW480 [35], and the results are shown in Table 2. The inhibition rates of complex (I) on two cancer cell lines, lung cancer A549 and liver cancer SMMC-7721, were tested. The table shows that the complex did not have a good inhibitory effect. In addition to complex (I), complexes (II) and (III) inhibited the cell growth of breast cancer cells MDA-MB-231 and colon cancer cells SW480. The inhibition rates of the two complexes on MDA-MB-231 cells were relatively good, at (22.87 ± 1.52)% and (19.86 ± 1.20)%, respectively. The inhibition rates on the colon cancer cell line SW480 were (18.68 ± 0.44)% and (18.68 ± 0.89)%. In 2022, Kundalkesha D Gaikwad et al. synthesized a related complex using 4,4′-diaminodiphenyl sulfone and Co (II) and found that the complex exhibited significant inhibitory effects on the cancer cell line A549. This may be attributed to the fact that cobalt complexes can inhibit the activity of some key enzymes and proteins, effectively hindering the growth of cancer cells [36]. In contrast, nickel and zinc complexes have weaker effects on enzyme and protein inhibition, and their anti-cancer mechanisms are not rich or diverse enough. In addition, in the experimental results, complexes (I)–(III) did not exhibit good biological activity, which may be related to the insufficient stability of the complexes. The stability of metal–organic complexes in biological environments is one of the important factors affecting their biological activity. From the thermogravimetric curve (Figure 7), it can be observed that the structures of complexes (I)–(III) begin to change within the temperature range of 30–35 °C. The instability of this structure may be one of the reasons for the poor biological activity of the complexes. In the future research work of our laboratory, we will focus on studying the above-mentioned issues and increasing the thermal stability of the complex by adjusting its structure.

Table 2. Cancer cell inhibition rates of complexes (I)–(III).

Complex	A549		SMMC-7721		MDA-MB-231		SW480	
	Cell Inhibition (%)							
	Average	SD	Average	SD	Average	SD	Average	SD
I	10.49	0.74	7.86	1.87	-	-	-	-
II	15.51	0.37	6.86	0.76	22.87	1.52	18.68	0.44
III	15.88	0.90	3.52	0.90	19.89	1.20	18.68	0.89

3. Experimental

3.1. Materials and Methods

Bis(phenylacetyl) disulfide and 3,3′-diaminodiphenyl sulfone were purchased from Beijing Bailingwei Technology Co., Ltd. (Beijing, China), and the metal salt was purchased from China National Pharmaceutical Group Chemical Reagent Co., Ltd. (Shanghai, China). All of these drugs were of analytical purity and had undergone no further purification or separation operations. Elemental analysis of the complex was performed via a VARIO ELIII elemental analyzer (German Elemental Analysis Systems, Hanau, Germany). The crystal data were obtained using a Bruker D8 Venture four-circle single-crystal X-ray diffractometer (Oxford Diffraction, Oxford, England). The UV–visible absorption spectrum was recorded using a CARY 5000 UV–Vis near-infrared spectrophotometer (Agilent Technologies, Santa Clara, CA, USA), and the infrared spectrum was obtained using a Magna–IR 750 Fourier-transform infrared spectrometer (Thermo Nicolet Corporation, Waltham, MA, USA). ESI–MS and NMR data were obtained using a Vanquish Q Exactive Plus liquid chromatography quadrupole electrostatic field orbital trap mass spectrometer (Thermo Fisher Scientific, Waltham, MA, USA) and a 500 MHz Bruker Advance (III) spectrometer (Agilent Technologies, Santa Clara, CA, USA), respectively. The chemical shifts of the ^{1}H NMR and ^{13}C NMR data are expressed in ppm, with DMSO-d6 as the solvent and $\delta = 2.5$ ppm. The thermogravimetric data were obtained with a NETZSCH TG209F1 thermal analyzer (PerkinElmer, Waltham, MA, USA) at a heating rate of 10 °C/min, with nitrogen gas used as the atmosphere during the experiment.

3.2. Synthesis of Complex (I)

Bis(phenylacetyl) disulfide (0.3020 g, 1 mmol) and $Cu(NO_3)_2{\cdot}3H_2O$ (0.4831 g, 2 mmol) were weighed separately and dissolved in 50 mL of anhydrous ethanol. The mixture was refluxed and stirred at 100 °C for 24 h and filtered while it was heated, and the obtained filtrate was slowly evaporated at room temperature. After approximately three days, the target complex (I) precipitated as blue columnar crystals. The samples were washed three times with petroleum ether and *n*-hexane, vacuum dried for 30 min, and weighed to 0.1526 g, and the yield was 39.6%, with an m.p. of 88.5–89.0 °C. Additional measurements were as follows: ESI-MS: 384.88 $[M-H]^+$; IR (KBr, υ, cm^{-1}): 3491(-O-H), 2990, 2899(-C-H), 1380 (-C-H), 1235 (-C-C), 1065, 1020 (-S=O), 626, 592 (-Cu-O). In $[C_4H_{18}CuO_{12}S_2]$(I) analysis, the calcd % values were as follows: S, 16.62; C, 12.44; H, 4.66. The found % values were as follows: S, 16.60; C, 12.83; H, 4.65.

We followed the above experimental method and reacted the ligand bis(phenylacetyl) disulfide with other metal salts ($Co(NO_3)_2{\cdot}6H_2O$, $Zn(NO_3)_2{\cdot}6H_2O$, $ZnCl_2$, and $Mn(CH_3COO)_2{\cdot}4H_2O$) at the same molar ratio, reaction temperature, and reaction time. The filtrate was filtered and allowed to evaporate naturally at room temperature. After approximately 1–2 days, white needle-shaped crystals precipitated from the filtrate, with a melting point of 112.0–114.3 °C.

3.3. Synthesis of Complex (II)

We weighed 3,3′-diaminodiphenyl sulfone (0.2482 g, 1 mmol) and $Ni(NO_3)_2{\cdot}6H_2O$ (0.5817 g, 2 mmol) at a molar ratio of 1:2 and dissolved them in 50 mL of anhydrous ethanol. The mixture was fixed at 100 °C for 24 h and then filtered. The filtrate was evaporated naturally at room temperature, and after a period of time, light green crystals of complex (II) were obtained. The mass was 0.3367 g, and the yield was 69.4%, with an m.p. of 103.6–104.5 °C. Additional measurements were as follows: ESI-MS: 485.92 $[M+H]^+$; IR (KBr, υ, cm^{-1}): 3512(-N-H), 3400(-O-H), 3180, 3095, 2925 (-C-H), 1600, 1435(-C=C), 872, 795, 715 (-C-H), 1297, 1142, 1045 (-S=O), 1380 (-N-O), 615(-Ni-O), 555, 525 (-Ni-N). In $[C_{12}H_{18}N_4NiO_{11}S]$(II) element analysis, the calcd % values were as follows: N, 11.54; C, 29.68; H, 3.710. The found % values were as follows: N, 11.84; C, 29.25; H, 3.971.

3.4. Synthesis of Complex (III)

The synthesis methods used for complexes (II) and (III) are essentially the same; that is, 3,3′-diaminodiphenyl sulfone (0.2482 g, 1 mmol) and $ZnCl_2$ (0.2726 g, 2 mmol) were precisely weighed and dissolved in 50 mL anhydrous ethanol. The mixture was refluxed at 100 °C for 24 h and then hot-filtered. The filtrate was evaporated naturally at room temperature. After a period of time, a light brown complex (III) was obtained with a mass of 0.2645 g and a yield of 83.6%, with an m.p. of 269.0–270.0 °C. Additional measurements were as follows: ESI-MS: 631.89 $[M-H]^+$; IR (KBr,υ, cm^{-1}): 3585, 3515(-N-H), 3130, 3065, 3010 (-C-H), 1605, 1480(-C=C), 855, 790, 712 (-C-H), 1300, 1150, 1092 (-S=O), 615(-Zn-O), 545, 522 (-Zn-N). In $[C_{24}H_{24}Cl_2N_4O_4S_2Zn]$(III) element analysis, the calcd % values were as follows: N, 8.85; C, 45.51; H, 3.79. The found % values were as follows: N, 8.97; C, 45.75; H, 4.07.

3.5. X-Ray Structure

The crystal data of complexes (I)–(III) were collected using a Bruker D8 Venture type four-circle single-crystal X-ray diffractometer at room temperature via GaK α (λ = 1.34139 Å) rays. The programs SHELXT [37] and SHELXL-2018/3 [38] were used to determine and refine the structure of the complex. To facilitate the analysis of the crystal structure, anisotropic refinement was performed on nonhydrogen atoms in the complex molecules, whereas constrained isotropic refinement was performed on hydrogen atoms, omitting hydrogen atoms in the complex. The crystal data, typical bond lengths, and bond angles of the complexes are shown in Tables 1 and 3.

Table 3. Crystal data and refinement parameters for compounds (I)–(III).

Complex	I		II	III
Empirical formula	$C_4H_{18}CuO_{12}S_2$	S_8	$C_{12}H_{18}N_4NiO_{11}S$	$C_{24}H_{24}Cl_2N_4O_4S_2Zn$
Formula weight	385.84	256.48	485.07	632.86
Temperature	200(2) K	200(1) K	102(2) K	200(2) K
Wavelength	1.34139	1.34139	1.34139	1.34139
Crystal system	Orthorhombic	Orthorhombic	Monoclinic	Monoclinic
Space group	Pbca	Fddd	$P2_1/n$	P2/n
a/Å	9.7657(8)	10.4397(12)	12.7468(6)	12.0055(7)
b/Å	7.3279(7)	12.8489(12)	7.6955(4)	5.0078(3)
c/Å	19.7608(17)	24.482(3)	19.1076(12)	22.2135(14)
α/°	90	90	90	90
β/°	90	90	104.886(2)	98.700(2)
γ/°	90	90	90	90
Volume	1414.6(2)	3284.0(6)	1811.42(17)	1320.13(14)
Z	4	16	4	2
D_{calcd} g/cm^3	1.812	2.075	1.779	1.592
μ (mm^{-1})	10.521	12.841	6.977	3.288
F(000)	796	2048	1000	648
2θ range (°)	11.08–144.54	10.004–103.808	3.277–61.972	3.442–56.749
Reflections collected	17,501	1176	25,756	16,456
Independent reflections	2104	529	25,756	2615
Final R indices [I >= 2σ(I)]	R1 = 0.0291, wR2 = 0.843	R1 = 0.1015, wR2 = 0.2766	R1 = 0.0663, wR2 = 0.1605	R1 = 0.0354,wR2 = 0.0993
Final R indices [all data]	R1 = 0.0321, wR2 = 0.0870	R1 = 0.1295, wR2 = 0.3736	R1 = 0.0913, wR2 = 0.1708	R1 = 0.0450, wR2 = 0.1030

3.6. Cytotoxicity Assay

The A549, SMMC-7721, MDA-MB-231, and SW480 cancer cell lines used in the experiment were all obtained from ATCC (Manassas, VA, USA). The cells were cultivated and inoculated with 10% fetal bovine serum culture medium at 37 °C. The MTS method was used to detect cell viability in the experiment [39]. The principle involves the use of succinate dehydrogenase in the mitochondria of cells to reduce the MTS and generate soluble formazan compounds. The number of formazan compounds produced was used to estimate the number of live cells.

4. Conclusions

In this work, two sulfur-containing ligands were directly reacted with $Cu(NO_3)_2{\cdot}3H_2O$, $Ni(NO_3)_2{\cdot}6H_2O$, and $ZnCl_2$ to synthesize three types of complex crystals: $[C_4H_{18}CuO_{12}S_2]$(I), $[C_{12}H_{18}N_4NiO_{11}S]$(II), and $[C_{24}H_{24}Cl_2N_4O_4S_2Zn]$(III). The structure of the complex was characterized and analyzed using modern analytical techniques, such as EA, XRD, FT-IR, UV, NMR, ESI-MS, and TGA, and the cytotoxicity of the complex crystals was preliminarily studied. Complexes (I)–(III) were initially screened for cytotoxicity using the MTS method to explore their potential as anticancer agents. The experimental data showed that complex (II) had a relatively good inhibitory effect on breast cancer MDA-MB-231. A simpler experimental method for preparing elemental sulfur was also discovered during the experiment. At present, the laboratory is still improving the experimental protocol by optimizing the ligand structure, selecting the appropriate metal center, and improving the synthesis method, with the expected results of improved cytotoxicity synthesis, better selectivity, and fewer toxic side effects in the future.

Author Contributions: Y.Y.: Writing—original draft, validation, investigation, data curation, conceptualization, visualization, and writing—original draft. D.L.: Assisted in completing the experiments. M.L.: Writing—review and editing, writing—original draft, visualization, validation, supervision, software, and resources. All authors have read and agreed to the published version of the manuscript.

Funding: This work was supported by Hefei University of Technology and the State Key Laboratory of Photochemistry and Plant Resources of West China.

Data Availability Statement: CCDC: 2385433(I), 2385441(II), and 2385435(III) supplementary crystallographic data were utilized for this study. These data are available free of charge from the Cambridge Crystallography Data Centre via www.ccdc.cam.ac.uk/data_request/cif (accessed on 20 September 2024).

Conflicts of Interest: The authors declare that they have no known competing financial interests or personal relationships that could have appeared to influence the work reported in this paper.

References

1. Aziz, K.N.; Ahmed, K.M.; Omer, R.A.; Qader, A.F.; Abdulkareem, E.I. Organometallic complexes and reaction methods for synthesis: A review. *Rev. Inorg. Chem.* **2024**, *44*, 685–698. [CrossRef]
2. Abdolmaleki, S.; Khaksar, S.; Aliabadi, A.; Panjehpour, A.; Motieiyan, E.; Marabello, D.; Faraji, M.H.; Beihaghi, M. Cytotoxicity and mechanism of action of metal complexes: An overview. *Toxicology* **2023**, *492*, 153516. [CrossRef]
3. Marakina, E.I.; Andrienko, O.S.; Sachkov, V.I.; Medvedev, R.; Amelichkin, I.; Shcherbakov, P.S. Transition metals in the process of complexation with organic ligands containing heteroatoms in the structure. *J. Phys. Conf. Ser.* **2021**, *1989*, 012027. [CrossRef]
4. Wu, Y.; Wu, T.; Huang, Y. A review: Biological activities of novel cyanopyridine derivatives. *Arch. Pharm.* **2023**, *356*, 2300067. [CrossRef] [PubMed]
5. Mendez-Arriaga, J.M. Recent advances in biological and catalytic applications of metal complexes. *Inorganics* **2024**, *12*, 249. [CrossRef]
6. Liu, X.; Manzur, C.; Novoa, N.; Celedón, S.; Carrillo, D.; Hamon, J.R. Multidentate unsymmetrically-substituted Schiff bases and their metal complexes: Synthesis, functional materials properties, and applications to catalysis. *Coord. Chem. Rev.* **2018**, *357*, 144–172. [CrossRef]
7. Ndagi, U.; Mhlongo, N.; Soliman, M.E. Metal complexes in cancer therapy—An update from drug design perspective. *Drug Des. Dev. Ther.* **2017**, *11*, 599–616. [CrossRef] [PubMed]
8. Lee, L.C.C.; Lo, K.K.W. Shining new light on biological systems: Luminescent transition metal complexes for bioimaging and biosensing applications. *Chem. Rev.* **2024**, *124*, 8825–9014. [CrossRef] [PubMed]
9. Li, D.; Shi, W.J.; Hou, L. Coordination polymers of copper(I) halides and neutral heterocyclic thiones with new coordination modes. *Inorg. Chem.* **2005**, *44*, 3907–3913. [CrossRef]
10. Pósa, V.; Hajdu, B.; Tóth, G.; Dömötör, O.; Kowol, C.R.; Keppler, B.K.; Spengler, G.; Gyurcsik, B.; Enyedy, É.A. The coordination modes of (thio)semicarbazone copper(II) complexes strongly modulate the solution chemical properties and mechanism of anticancer activity. *J. Inorg. Biochem.* **2022**, *231*, 111786. [CrossRef]
11. Cascajosa-Lira, A.; Andreo-Martínez, P.; Prieto, A.I.; Baños, A.; Guillamón, E.; Jos, A.; Cameán, A.M. In vitro toxicity studies of bioactive organosulfur compounds from Allium spp. with potential application in the agri-food industry: A review. *Foods* **2022**, *11*, 2620. [CrossRef] [PubMed]
12. Petropoulos, S.; Di Gioia, F.; Ntatsi, G. Vegetable organosulfur compounds and their health promoting effects. *Curr. Pharm. Des.* **2017**, *23*, 2850–2875. [CrossRef]
13. Bhattacherjee, D.; Raina, K.; Mandal, T.K.; Thummer, R.P.; Bhabak, K.P. Targeting Wnt/β-catenin signaling pathway in triple-negative breast cancer by benzylic organotrisulfides: Contribution of the released hydrogen sulfide towards potent anti-cancer activity. *Free Radic. Biol. Med.* **2022**, *191*, 82–96. [CrossRef] [PubMed]
14. Shoaib, S.; Ansari, M.A.; Ghazwani, M.; Hani, U.; Jamous, Y.F.; Alali, Z.; Wahab, S.; Ahmad, W.; Weir, S.A.; Alomary, M.N.; et al. Prospective epigenetic actions of organo-sulfur compounds against cancer: Perspectives and molecular mechanisms. *Cancers* **2023**, *15*, 697. [CrossRef]
15. Samaei, N.M.; Yazdani, Y.; Alizadeh-Navaei, R.; Azadeh, H.; Farazmandfar, T. Promoter methylation analysis of WNT/β-catenin pathway regulators and its association with expression of DNMT1 enzyme in colorectal cancer. *J. Biomed. Sci.* **2014**, *21*, 73. [CrossRef]
16. Debnath, T.; Deb Nath, N.C.; Kim, E.K.; Lee, K.G. Role of phytochemicals in the modulation of miRNA expression in cancer. *Food Funct.* **2017**, *8*, 3432–3442. [CrossRef]
17. Soltani, S.; Amin, G.R.; Salehi-Sourmaghi, M.H.; Schneider, B.; Lorenz, S.; Iranshahi, M. Sulfur-containing compounds from the roots of Ferula latisecta and their cytotoxic activities. *Fitoterapia* **2018**, *124*, 108–112. [CrossRef]
18. Balakrishnan, S.; Duraisamy, S.; Kasi, M.; Kandasamy, S.; Sarkar, R.; Kumarasamy, A. Syntheses, physicochemical characterization, antibacterial studies on potassium morpholine dithiocarbamate nickel (II), copper (II) metal complexes and their ligands. *Heliyon* **2019**, *5*, e01687. [CrossRef]

19. Yekke-Ghasemi, Z.; Ramezani, M.; Mague, J.T.; Takjoo, R. Synthesis, characterization and bioactivity studies of new dithiocarbazate complexes. *New J. Chem.* **2020**, *44*, 8878–8889. [CrossRef]
20. Singh, A.; Barman, P.; Gogoi, H.P. Thioether-based novel transition metal complexes: Synthesis, DNA interaction, in vitro biological assay, DFT calculations, and molecular docking studies. *Bioorg. Chem.* **2023**, *132*, 106343. [CrossRef] [PubMed]
21. Czylkowska, A.; Pitucha, M.; Lanka, S.; Raducka, A.; Rogalewicz, B.; Szczesio, M.; Świątkowski, M.; Żarczyński, A.; Klepacz-Smółka, A.; Szczytko, J.; et al. Triazole-based Mn(II), Fe(II), Ni(II), Cu(II) and Zn(II) complexes as potential anticancer agents—Physicochemical properties, in silico predictions and in vitro activity. *Polyhedron* **2024**, *261*, 117106. [CrossRef]
22. Luo, M.; Zhang, J.C.; Yin, H.; Wang, C.M.; Morris-Natschke, S.; Lee, K.H. One-step templated synthesis of chiral organometallic salicyloxazoline complexes. *BMC Chem.* **2019**, *13*, 51. [CrossRef]
23. Kim, G.; Na, C.W.; Myung, Y. Facile one-pot synthesis of nickel nanoparticles by hydrothermal method. *Materials* **2022**, *16*, 76. [CrossRef]
24. Cepus, V.; Borth, M.; Seitz, M. IR spectroscopic characterization of lignite as a tool to predict the product range of catalytic decomposition. *Int. J. Clean Coal Energy* **2016**, *5*, 13–22. [CrossRef]
25. İlkimen, H.; Salün, S.G.; Gülbandılar, A.; Sarı, M. The new salt of 2-amino-3-methylpyridine with dipicolinic acid and its metal complexes: Synthesis, characterization and antimicrobial activity studies. *J. Mol. Struct.* **2022**, *1270*, 133961. [CrossRef]
26. Wang, Y.; Mao, X.; Chen, C.; Wang, W.; Dang, W. Effect of sulfuric acid concentration on morphology of calcium sulfate hemihydrate crystals. *Mater. Res. Express* **2020**, *7*, 105501. [CrossRef]
27. Yadav, V.; Niluroutu, N.; Bhat, S.D.; Kulshrestha, V. Sulfonated poly(ether sulfone) based sulfonated molybdenum sulfide composite membranes: Proton transport properties and direct methanol fuel cell performance. *Mater. Adv.* **2020**, *1*, 820–829. [CrossRef]
28. Mabrouk, W.; Charradi, K.; Maghraoui-Meherzi, H.; Alhussein, A.; Keshk, S.M.A.S. Proton conductivity amelioration of sulfonated poly ether sulfone octyl sulfonamide via the incorporation of montmorillonite. *J. Electron. Mater.* **2022**, *51*, 6369–6378. [CrossRef]
29. Kumar, D.; Kapoor, I.P.S.; Singh, G.; Fröhlich, R. Preparation, characterization, and kinetics of thermolysis of nickel and copper nitrate complexes with 2,2′-bipyridine ligand. *Thermochim. Acta* **2012**, *545*, 67–74. [CrossRef]
30. Sakthivel, R.V.; Sankudevan, P.; Vennila, P.; Venkatesh, G.; Kaya, S.; Serdaroğlu, G. Experimental and theoretical analysis of molecular structure, vibrational spectra and biological properties of the new Co(II), Ni(II) and Cu(II) Schiff base metal complexes. *J. Mol. Struct.* **2021**, *1233*, 130097. [CrossRef]
31. Aravindan, P.; Sivaraj, K.; Kamal, C.; Vennila, P.; Venkatesh, G. Synthesis, molecular structure, spectral characterization, molecular docking and biological activities of (E)-N-(2-methoxy benzylidene) anthracene-2-amine and Co(II), Cu(II) and Zn(II) complexes. *J. Mol. Struct.* **2021**, *1229*, 129488. [CrossRef]
32. Yu, T.; Zhu, Z.; Bao, Y.; Zhao, Y.; Liu, X.; Zhang, H. Investigation of novel carbazole-functionalized coumarin derivatives as organic luminescent materials. *Dyes Pigments* **2017**, *147*, 260–269. [CrossRef]
33. Gaikwad, K.D.; Ubale, P.; Khobragade, R.; Deodware, S.; Dhale, P.; Asabe, M.R.; Ovhal, R.M.; Singh, P.; Vishwanath, P.; Shivamallu, C.; et al. Preparation, characterization and in vitro biological activities of new diphenylsulphone derived schiff base ligands and their Co(II) complexes. *Molecules* **2022**, *27*, 8576. [CrossRef] [PubMed]
34. Chen, L.; Tan, J.; Lan, T.; Huo, H.; Sun, L.; Jiang, Y.; Zhang, N.; Li, C.; Wang, J. Effect of ligand structure on ethylene oligomerization over salicylaldehyde imine nickel complexes: Experiments and calculations. *Appl. Organomet. Chem.* **2023**, *37*, e7276. [CrossRef]
35. Mosbah, H.K.; Ibrahim, A.B.M.; Zidan, A.S.A.; Aly, A.A.M.; Saber, S.H. La(III), Ce(III), Pr(III) and Eu(III) complexes with fenamic acid based ligands: Preparation, spectral and thermal analysis and evaluation of their cytotoxicity in MDA-MB-231 breast cancer cells. *J. Iran. Chem. Soc.* **2024**, *21*, 1681–1689. [CrossRef]
36. Ambika, S.; Manojkumar, Y.; Arunachalam, S.; Gowdhami, B.; Sundaram, K.K.M.; Solomon, R.V.; Venuvanalingam, P.; Akbarsha, M.A.; Sundararaman, M. Biomolecular Interaction, Anti-Cancer and Anti-Angiogenic Properties of Cobalt(III) Schiff Base Complexes. *Sci. Rep.* **2019**, *9*, 2721. [CrossRef] [PubMed]
37. Amer, A.A. Synthesis of some new polyfunctionalized pyridines. *J. Heterocycl. Chem.* **2017**, *55*, 297–301. [CrossRef]
38. Dyachenko, I.V.; Dyachenko, V.D.; Dorovatovsky, P.V.; Khrustalev, V.N.; Nenajdenko, V.G. One-pot synthesis of thieno[2,3-b]pyridine and pyrido[3′,2′:4,5]thieno[3,2-d]pyrimidine derivatives. *Russ. J. Org. Chem.* **2020**, *56*, 974–982. [CrossRef]
39. Zhang, L.; Yin, H.; Zhang, J.C.; Luo, M.; Meng, X.G. Synthesis, crystal structure and anticancer activity of 4-chloro-2-methoxybenzoic acid transition metal complexes. *J. Mol. Struct.* **2024**, *1316*, 139080. [CrossRef]

MDPI AG
Grosspeteranlage 5
4052 Basel
Switzerland
Tel.: +41 61 683 77 34

Inorganics Editorial Office
E-mail: inorganics@mdpi.com
www.mdpi.com/journal/inorganics

www.ingramcontent.com/pod-product-compliance
Lightning Source LLC
Chambersburg PA
CBHW061512210926
51850CB00008B/1890